BUILDING SMALL BARNS, SHEDS, & SHELTERS

Monte Burch

Storey Books
Schoolhouse Road
Pownal, Vermont 05261

The mission of Storey Communications is to serve our customers by publishing practical information that encourages personal independence in harmony with the environment.

Acknowledgements

Like most books, this one is the result of the efforts of many individuals and organizations. Assistance has come from the United States Department of Agriculture Cooperative Extension Service, American Plywood Association, Portland Cement Association, T. A. Haigh Lumber Co., Ashline Construction Co., Kaiser Aluminum Corporation, S. T. Griswold Co., Inc., F. W. Webb Co. and Nordic Holsteins. Material and information has been provided generously by many individuals, including Robert Bennett, Jim Dickerson, Richard Pratt, Grant Wells, Phyllis Hobson, Mary Twitchell, Stu Campbell and Leigh Seddon. Special thanks are due to the property owners whose small barns and outbuildings are featured in this book. Photographers whose work appears here include Eric Borg, Middlebury, Vermont, Grant Heilman, Lititz, Pennsylvania and Nancy and Mike Bubel, Wellsville, Pennsylvania. The illustrations are by Bob Vogel, Hinesburg, Vermont.

Printed in the United States by Vicks Lithograph
45 44 43 42 41 40 39

Library of Congress Cataloging-in-Publication Data

Burch, Monte
Building small barns, sheds & shelters.

Includes index.
1. Building — Amateurs' manuals. 2. Farm buildings — Design and Construction — Amateurs' manuals. I. Title.
II. Title: Building small barns, sheds, and shelters.
TH4955.B87 1982 690'.89 82-15439
ISBN 0-88266-245-7

Contents

Introduction

Grant Heilman Photography

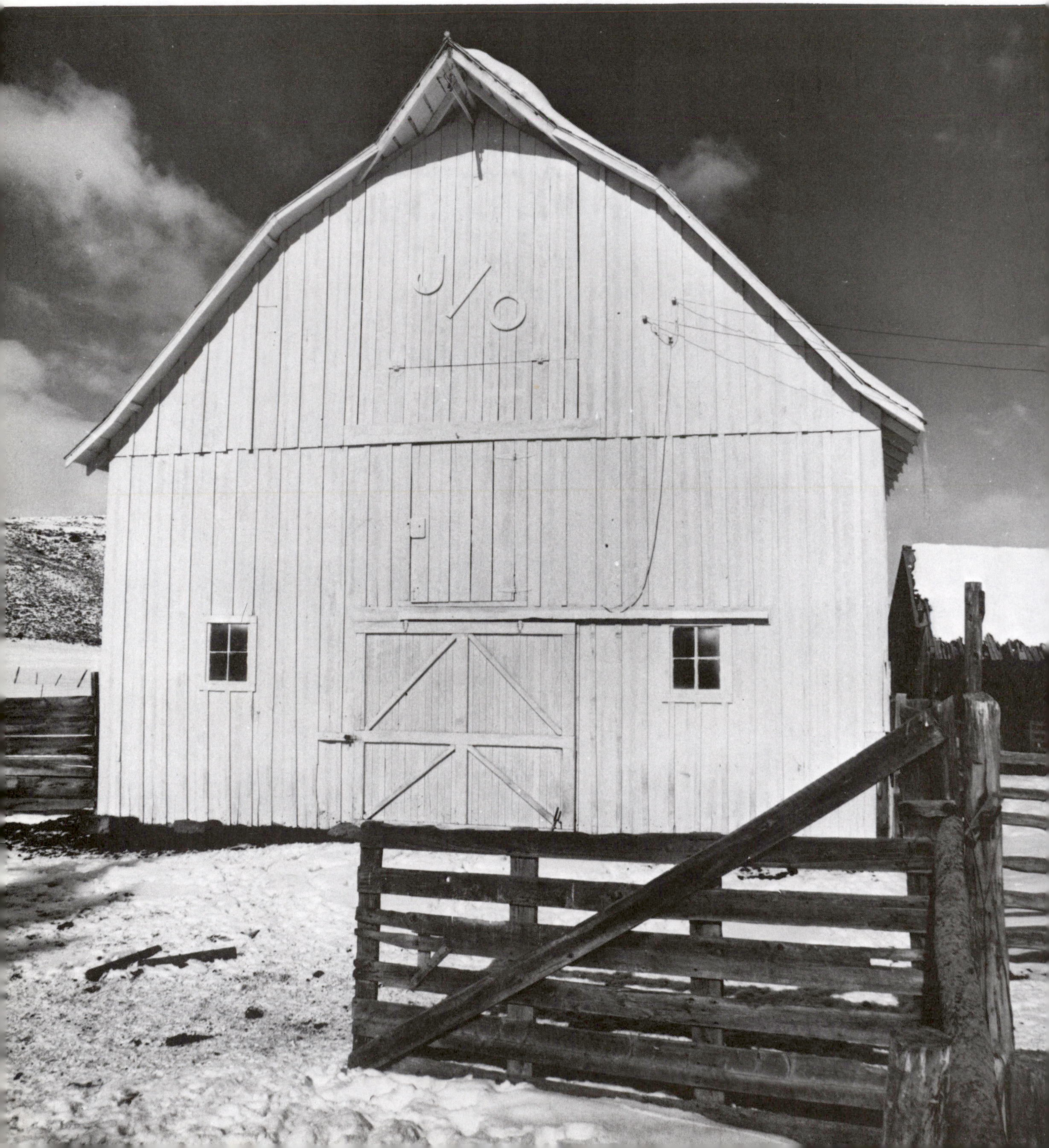

As a youngster growing up on a small farm in central Missouri, I learned an important rule: to be a jack-of-all-trades was essential. Without versatile skills, success in a rural farming community was almost impossible. I watched my Dad, grandfather and lots of uncles and cousins as they used skills that city folks would find unfamiliar and unnecessary. Although their efforts were sometimes clumsy, building a barn or chicken house, patching an old sagging fence, building a log corral and even repairing plumbing and wiring were not activities they could delegate to others. In those days, almost no one would journey all the way out to the country for small repair jobs and, even if they did, the service was much too expensive. And what was true then remains true today: in a rural setting, being able to use your own hands to build or repair barns, outbuildings and fences can spell the difference between costly frustration and pleasant success.

My first building experiences occurred more than 25 years ago. For the older men in the family, I was the "go-for," the one sent for another handful of nails, for a dropped hammer. As I grew older and improved my skills, I was allowed to do some of the nailing, a job I enjoyed thoroughly. For some reason, nailing down flooring was especially gratifying. Eventually, I joined my father in the construction business. Together, we built homes, schools and motels during the day, then worked on our own barns and outbuildings in the evening.

Drawing from these experiences, I have written this book for the many homeowners who might want to build a new barn, shed or animal shelter. Those who want to repair a rundown building might find parts of this book helpful. This book is not for big-time farmers who can hire agricultural contractors to build a $30,000 barn; it's for those who simply want to raise a few cows, horses, pigs, rabbits, chickens, or some other poultry or small livestock. The structures are built to suit these animals, but the building methods described are basically standard construction techniques.

My family lives on a working farm with about 50 head of cattle, a registered Duroc hog operation, lots of chickens and a few horses—some for breeding and some for pleasure. Many of the farm-building plans in this book are adapted from, or identical to, designs used by my family, relatives, friends or neighbors. The designs and construction methods are time-proven. A few of the designs were provided by the Extension Service, an excellent source of plans and information for the rural homeowner.

In most cases, buildings presented in this book may be built and used almost anywhere. Our Missouri winters can be tough:

it drops to well below zero with a couple of feet of snow on the ground and often stays that way for a couple of months; about six months later, we get 110-degree weather with no rain. The structures I describe can withstand this weather, so they're probably suitable for most parts of the country without modification.

There are skills involved in building anything, even a small dog house, not to mention a barn. The person who has had no more carpentry experience than trying to tighten the hinges on a door with a screwdriver will probably find the task difficult at first. However, barns are not as demanding as, say, small pieces of intricate furniture. Even if you can't drive a nail without bending it over or denting the wood, for the most part, these errors won't show. And you'll gain more and more experience as construction progresses.

My aim is not to teach carpentry or how to use and select tools, but to illustrate the methods of constructing various outbuildings. The first eight chapters provide general carpentry and construction guidelines; chapters nine through fifteen include plans of various barns, sheds, shelters and outbuildings. Sixteen covers fencing. Almost anyone, even someone with minimal carpentry skills, should be able to build a suitable small building such as a chicken house or hog shed. Larger buildings could follow. To build even the largest barn in this book, requires no special materials or costly equipment, just good common sense and a slow, methodical approach to the job.

The plans are for buildings found on small farms or rural properties. Many can be altered or adapted for multiple uses. For example, the small horse barns can be adapted for use as dairy barns for cows. On the other hand, some of the buildings are specialized and intended for one purpose only. The hog-farrowing house and woodshed are examples.

Even if you don't wish to use the specific plans in this book, you can adapt them to suit your purposes by altering their designs and dimensions. And that, in sum, is what successful small farming is all about—adapting to suit your environment and problems. In any case, I can only hope that this book is a means for you to gain more independence as well as the satisfying feeling of creating and building with your own hands.

M. B.

CHAPTER 1

Planning

Grant Heilman Photography

As you travel about your state, take a look at the homes and farms, their barns, sheds and shelters. What are the different shapes of these outbuildings? What are the roof angles and materials? How are the buildings spaced? These are some questions to think about as you begin contemplating a barn or outbuilding of your own. When searching for an attractive, functional design, the countryside is a good resource.

The first and most important step in any building endeavor, whether it be a woodshed or 20-stall horse barn, is planning. First, develop a site plan that presents the building layout, property lines, other buildings, soil conditions and the like. Next complete working drawings of the building, showing exactly how it is to be constructed. With these plans, you can then apply for building permits, zoning variances or other legal permits you might need. Finally, prepare a construction budget in order to plan for financing.

The Site Plan

Perhaps you purchased an existing farm with well-planned outbuildings. Most likely, though, you're stuck with an inefficient, haphazard or unsightly arrangement. We've lived on our farm for about ten years and when we purchased it there was nothing but an old house, a run-down barn and several heaps of stone where outbuildings once stood. We made the mistake of rebuilding without a site plan. Only later—after we planted fruit trees in a good horse pasture and built a barn over the only good site for an access road to the valley—did we discover our mistake. But, you needn't repeat our errors. Once you have determined the rough size and shape of the barn or outbuilding you are going to build, locate it so that it makes the best use of your land and the natural elements and so that it is convenient and aesthetically pleasing. Do this on a site plan that considers the following:

Building position. While it's best to locate outbuildings conveniently, they should also be far enough apart, and far enough from your house, to reduce the chance of fire spreading from one to another. A minimum separation of 75 feet is advisable; this distance provides plenty of room for fire-fighting equipment to pass between buildings.

For other, perhaps obvious reasons, it's a good idea to place livestock buildings at a good distance from—and downwind of—the main house. There's nothing quite like the odor of a pig pen on a hot, muggy summer day.

Sun and shade. The orientation of your outbuildings with respect to the sun and prevailing winds will in part determine how warm they are in winter and how cool in summer. In a well-designed and insulated barn, passive solar heating can often provide all the necessary winter heat for livestock.

Buildings for pigs, poultry, and other livestock should provide maximum winter protection and summer ventilation. For best natural heating, face the long dimension of the building toward the south and provide adequate window area to capture the sun's heat. Ideally, this long, open side of the building should also face away from the prevailing winter winds to minimize heat loss. Often winter winds are from the north, but to obtain specific information for your area, contact the nearest National Weather Service office.

Summer cooling and ventilation are important, especially in southern climates. The coolest buildings are generally long buildings with the long dimension facing east and west. Shade trees also help, but locate structures to avoid continuous shadows (and damp cold) in feed lots and livestock areas. Again, plan windows and doors to take advantage of the prevailing summer winds to ventilate and cool the building.

Topography and soils. When choosing a building site, look at the lay of the land and the soils carefully. Ideally, the building site should be fairly level and well drained year round.

The type of soil present will determine the size and type of foundation you should use. Among the best soils for supporting a building is a course sand and gravel mixture; the worst is poorly-drained clay. (For more on soil bearing capacities, see p. 36.) Usually, the soil type is fairly uniform over a large area, but sometimes it varies from foot to foot. If you encounter a bad clay soil, check other possible building sites in hopes of finding a sandy, well-drained soil.

Animal shelters such as hog or dairy barns should *not* be uphill of your water supply. Nor should they be located so runoff from the lots or feeding areas will contaminate creeks, streams or ponds. Even a small number of animals can generate a lot of manure and a serious sanitation problem unless drainage from the area is controlled and adequate disposal methods are provided for. Check with your local Extension Service office for advice and laws governing feedlot runoff and waste disposal.

When planning a small barn, shed or shelter, consider the building's position, the sun and shade, topography and soils, access and utilities.

Access and utilities. As you begin to think about the details of your outbuilding, give some thought to access. Where will the loading and feed-storage areas be located? Plan these so they're easy to reach with a truck at all times of the year. If a cattle chute is located behind a barn, it may be impossible to get a truck to it in mud season.

One thing I've learned from experience is that a circular driveway makes access easier. After backing over tricycles, gas cans and many other items with a large farm truck, I finally put in a circular driveway. Now I can drive through without wasting a great deal of time and energy turning a long stock trailer around. (See illustration, p. 4.)

If you're thinking of building a livestock barn, calculate the water requirements for the animals to determine the adequacy of existing or proposed water supplies. If there's a chance your well will be inadequate, drill a new one *before* beginning construction of your new building. Approximate daily water requirements of various farm animals are shown in Table 1-1.

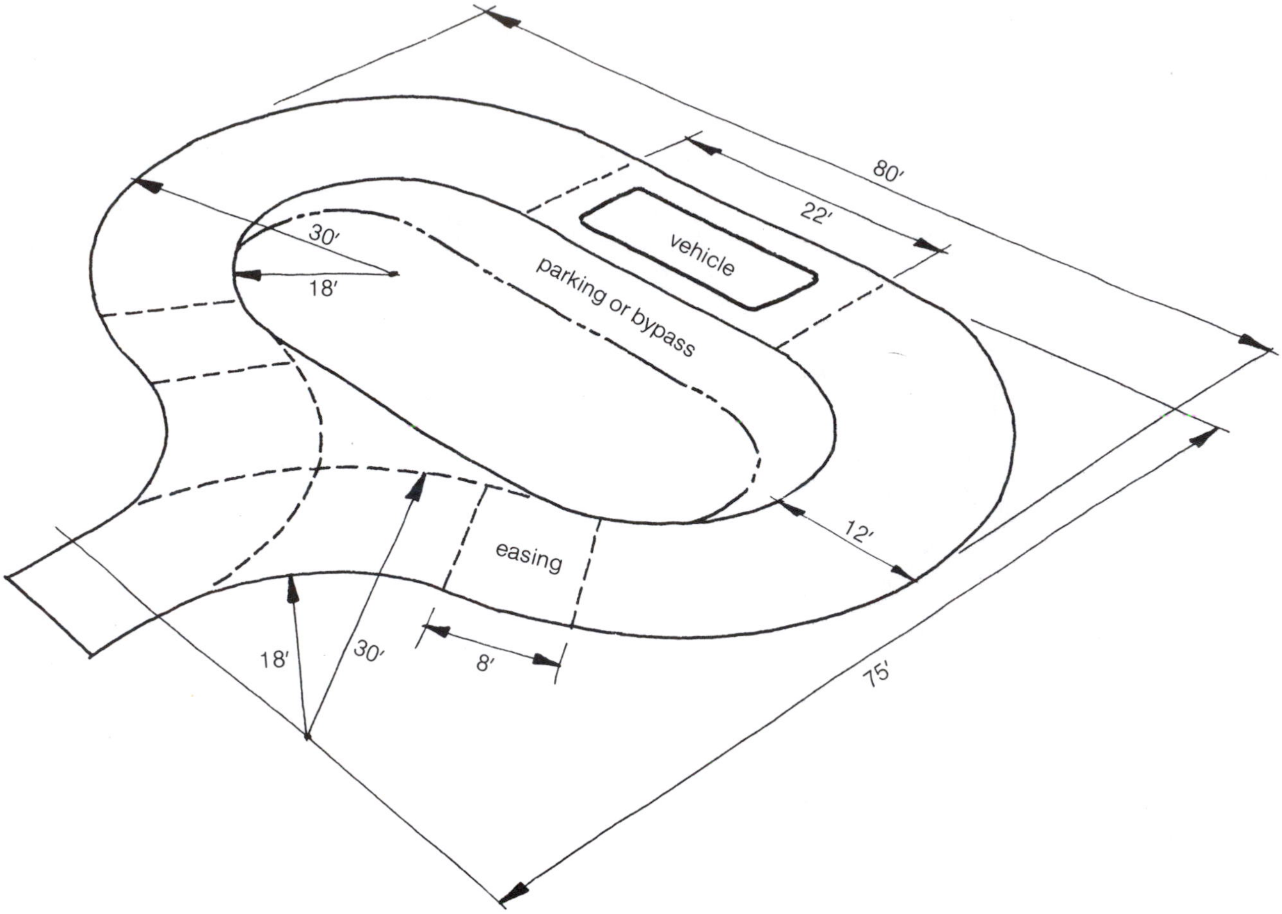

A circular drive is a convenient way to approach a house and outbuildings. Adjust dimensions to suit your property and vehicles.

Table 1-1 Approximate Daily Water Requirements.*

Water Use Per Animal	Gal/Day
Milking cow	20–25
Dry cow	10–15
Calves (1–1½ gals./100 lbs. body weight)	6–10
Swine, finishing	3– 5
nursery	1
sow and litter	8
gestating sow	6
Beef animal	8–12
Sheep	2
Horse	12
100 chicken layers	9
100 turkeys	15
Rabbit (10 lbs.)	1 pt.
Rabbit doe (10 lbs.) with litter	2 qts.

**Midwest Plan Service Structures and Environment Handbook* (Aimes, Iowa: Midwest Plan Service, 1980).

For information about electrical service, contact your Extension Service engineer and the local utility company. The Extension Service can give you advice on your specific needs, and the utility can give you cost estimates on bringing in additional power. It's also not a bad idea to think about emergency power supplies; a portable, gas-operated generator is one solution.

Building Plans

Before beginning construction, make rough sketches, then detailed plans of your proposed barn or outbuilding. The sketches should show the general shape and floor plan of the building. To develop a floor plan, first consider the building's functions and the flow of inside and outside activities. An activity list is a useful way to outline the prerequisites for a building. Some typical activities include housing livestock and poultry, storing feed and equipment and sheltering a car, truck or tractor.

A second consideration is the style and appearance of the building you wish to erect. To help you select a style, you might want to scan through Chapters 4 and 5 on framing and roofing respectively, and Chapters 9 through 15, which include plans for various barns, sheds and shelters.

Once you have determined your preferred style and present and future activities, it is necessary to size the spaces according to their function. For example, each animal requires a mini-

Small barns, sheds and shelters may be built with many different styles, shapes and materials. One possibility is this steel barn, suitable for horses.

An unusual style was used for this octagonal barn. More widely-used styles are the saltbox (opposite page, top) and the small gable-roof outbuilding (opposite page, bottom).

mum amount of space; with insufficient room, problems develop, serious problems such as diseases and, with poultry and swine, even cannibalism. Grain bins and equipment storage must also be sized to handle anticipated needs efficiently. Appendix A lists space requirements for various animals, feeds and beddings.

When the space requirements and building size have been decided, sit down with some graph paper and an architectural scale ruler and begin your working drawings or plans. The usual scale of such drawings is ¼ inch equals 1 foot, though they can be made smaller or larger depending on the size of the building.

In most cases, drawings needed to build a barn or outbuilding include a foundation plan, a floor plan showing location of walls and partitions and an elevation or section showing the wall and roof details. If you have doors and windows on every side of the building, four separate elevations might be in order to help build them and to give you a sense of what the building will look like before you frame the walls.

If there are tricky details such as sliding doors, special beams or complicated roof angles, it is best to make separate, detailed drawings of these before you start. A detail that might take half an hour for you to figure out at a desk can easily kill half a day when it interrupts construction. A well thought-out design and an accurate set of drawings will eliminate expensive, frustrating mistakes.

Permits and Financing

Before constructing a barn or large outbuilding, it may be necessary to obtain a building permit. Usually, town or municipal officials can give you advice on this question. In the best instances, the local building codes are written to insure that structures are safe and that the building won't offend or infringe upon the rights of neighbors. Sometimes local ordinances specify that a building must be a certain distance from property lines or that certain uses are not permitted. To be sure of your exact building codes and zoning requirements, contact your local building inspector.

Your detailed site plan and working drawings will be of great help to you if you need to get local approval for your building. If you must get a zoning variance, the zoning administrator will want to see your site plan with distances to property lines, roads and bodies of water clearly indicated. The building inspector will want to see your working drawings so he can determine whether the foundation is adequate and whether framing, wiring and plumbing meet code requirements.

Your plans will also come in handy when you try to estimate the cost of constructing the building or apply for a bank loan. A quick way to estimate the cost of building materials is to take the plans to a local building supplier. He will give you a free estimate and often some free design help. A concrete supplier can give you an estimate for the foundation materials needed or a contract price for the complete foundation.

For any building project costing more than a few hundred dollars, prepare a budget that itemizes all costs in as much detail as possible. With a budget, you can plan to pay your monthly construction bills on time and keep track of whether expenses are running over or under estimates.

The least-costly way to finance a building is through personal savings. A large project can be built in several stages over a period of a few months to several years to avoid taking out an expensive loan. If you must apply for a bank loan, however, make sure you have your plans and budget ready to present to the bank officer. Banks are hesitant to loan to an owner-builder putting up an outbuilding that may not have great resale value. Unless you have tangible assets that the bank can secure the loan with, the bank's decision will likely hinge on the thoroughness of your plans and budget.

PLANNING CHECKLIST

Siting
- ☐ locate building on site to complement other structures
- ☐ consider soil composition, topography and drainage
- ☐ consider sun, shade and prevailing winds
- ☐ plan for access and utilities
- ☐ consider relationship of building to outdoor functions

Building Planning
- ☐ organize activity list for building's functions
- ☐ determine style: shed, lean-to, gable, gambrel roof, etc.
- ☐ determine position of doors and windows considering lighting
- ☐ determine width of aisles, interior doors, stalls, etc.
- ☐ draw scale foundation and floor plans and side elevations
- ☐ choose siding, roofing and finish materials
- ☐ draw plans for electrical and other services as needed
- ☐ draw details of special or hard-to-build items
- ☐ prepare materials list

Permits and Financing
- ☐ apply for building permit or zoning variance as necessary
- ☐ get estimate on materials from building supplier
- ☐ prepare itemized budget for entire project

CONSTRUCTION BUDGET

Item (labor & mat.)	Estimated Cost	Actual Cost
Excavation		
Footings		
Foundation		
Concrete floor		
Framing		
Roofing		
Siding and trim		
Windows and doors		
Plumbing		
Wiring		
Heating/air conditioning		
Utilities		
Septic system		
Building permits		
Landscape and grading		
Equipment rental		
Allowance for overruns		
Total		

CHAPTER 2

Materials, Tools and Power Equipment

Borg

TRADITIONALLY, AMERICAN SETTLERS and homesteaders have been adaptable people. Often they built barns, sheds and outbuildings with whatever tools and materials were readily available. And this long-standing tradition continues. Stone, sod and logs are some of the natural building materials used today; others are sometimes salvaged and recycled. For example, scraps of wood recovered from an old shed may be fine for a chicken house. The materials, tools and equipment you select will depend, in large part, on the amount you wish to spend on the building project.

Materials

Native Lumber

Native lumber, sometimes called green lumber, is dimensional wood that has been rough sawn at a local sawmill. It is neither planed nor dried and is therefore quite wet and heavy. Its primary advantage is that it costs one-third to one-half less than kiln-dried wood purchased from a building supplier.

Native lumber is perfect for barns and outbuildings where its slightly uneven dimensions and rough texture do not matter. It is also ideal for traditional board and batten siding. Because green lumber shrinks a bit when it dries, cracks will develop between siding boards. But battens cover these cracks and seal out weather. Always use cement-coated or galvanized nails when building with green lumber because regular nails may pull out when the lumber shrinks.

Most sawmills cut all types of wood, from hardwoods such as oak and maple to softwoods such as pine, spruce and hemlock. In the northeast and northwest, spruce and hemlock are the common building materials. In the south, pine and even oak are popular choices. For a more complete description of various woods and their principal characteristics, see Appendix B.

Native lumber is cut at the mill into 1- and 2-inch stock that is usually available in widths up to 10 to 12 inches. Large beams such as 6 x 6s are also available at most mills, while building suppliers seldom carry them.

Store green lumber carefully if it is not used as soon as it comes from the mill. Lay boards in "stickered" piles so they dry properly without warping, twisting and cracking. A *sticker* is a small piece of 1-inch wood that separates each layer of boards so air can circulate through the pile.

Using a small, portable sawmill is one way to cut timber. You might be able to rent or borrow this equipment.

If you're lucky enough to own some land with marketable timber, you can have this custom sawn for your own buildings. Our farm has about 60 acres of timber with a good number of marketable saw logs. We cut and haul these logs to a local mill where they are sawn to the exact dimension we need. We could purchase a small portable sawmill and saw our own. If you try this, use caution; for the unskilled, this is a hard and sometimes dangerous job.

There are at least two ways to process your logs. The first is to sell the logs to a lumber buyer, then use the money to purchase cut lumber from a mill or building supplier. As a second choice, you could have a mill harvest, haul and cut the logs for a percentage of the lumber or its value.

County foresters can give you advice on whether you have marketable timber, how much it is worth and who to contact for custom sawing. I must add, however, that there are some unscrupulous people in the logging business and my advice is to be cautious. Make sure you get a fair price for your timber and that your forest is not damaged by careless logging. Professional foresters can help you avoid the "bad guys."

Kiln-dried Lumber

Standard softwood dimensional lumber, the most common material for small barn and outbuilding construction, is usually readily available and easy to work with. It's kiln-dried and milled to exact dimensions (features adding to its cost), and not nearly as heavy as green native lumber.

Before selecting your softwood dimensional lumber, think about how you will use it. For instance, you might not be particular about materials for a portable chicken house. If you build a hay barn or horse barn, though, you would want attractive, durable materials so the building will last a lifetime and be nearly maintenance free.

Softwood lumber sold by building suppliers or lumberyards has been sorted into various grades. Understanding these grades is both important and difficult. There are more than a dozen associations or trade organizations with different grading systems. But, I'll briefly describe a grading system I'm familiar with. The one used by your supplier may differ.

Yard lumber, most often sold by local building suppliers, includes 2 x 4s, 2 x 6s, 2 x 8s, 2 x 10s, 2 x 12s and 4 x 4s. Common lengths are from 8 to 20 feet. My supplier sorts lumber into grades such as No. 1, 2, 3 and 4 common, depending upon

Kiln-dried lumber is milled to exact dimensions. When planning your building and selecting lumber, be sure to note the difference between nominal and actual sizes.

WHAT'S A BOARD FOOT?

Lumber is usually measured by the board foot (bdf), which is a volume measurement of 144 cubic inches.

Thus a piece of 1-inch board that is 12 inches long and 12 inches wide is exactly 1 board foot.

Similarly, an 8-foot board that is 2 x 6 measures 2 inches times 6 inches times 8 feet divided by 12, or 8 board feet.

Kiln-dried Lumber

Nominal Size	Actual Size
1 x 2	¾ x 1½
1 x 3	¾ x 2½
1 x 4	¾ x 3½
1 x 6	¾ x 5½
1 x 8	¾ x 7¼
1 x 10	¾ x 9¼
1 x 12	¾ x 11¼
2 x 4	1½ x 3½
2 x 6	1½ x 5½
2 x 8	1½ x 7¼
2 x 10	1½ x 9¼
2 x 12	1½ x 11¼
4 x 4	3½ x 3½
6 x 6	5½ x 5½
8 x 8	7¼ x 7¼

the number of knots, blemishes and defects found in the board. No. 4 common is the cheapest grade and usually has a lot of open knots and other structural weak spots.

Structural lumber is harder and much more dense than the yard lumber and is used primarily for heavier framing members. It may not even be available in some local yards, although you can probably order it. Structural lumber is graded according to its density. There are five general classifications:

1. Dense select . . . . Southern pine and Douglas fir
2. Select . . . . Douglas fir
3. Select . . . . other softwood species, except Southern pine
4. Dense common . . . . Douglas fir and Southern pine
5. Common . . . . all softwoods

In almost all light framing, I use lumber 1 to 1½ inches thick or smaller and No. 1 common yard materials. For wall framing, spruce, pine or fir No. 2 common are most often used. For heavier beams, posts and girders, I use common structural or dense common.

All lumber is cut to the full dimension at the mill, but after drying and planing, dimensions are smaller. For example, a 2 x 4 usually measures 1½ x 3½ inches; 2 x 6s measure 1½ x 5½; 2 x 8s are actually 1½ x 7¼; and a 2 x 10 is really only 1½ x 9¼. In short, everything is half an inch or more smaller than its nominal measurements.

Treated Lumber

Use lumber pressure treated with a wood preservative for all framing members in contact with water, the earth or a concrete foundation. Pressure treatment protects the wood from rot, decay and insects. Although the initial expense of treated lumber is more (typically, 50 to 100 percent more than untreated), it lasts a great deal longer.

With a little effort, you can treat wood yourself and, in some cases, add 10 years to the life of a pole or framing member in contact with the ground. However, home-treated wood is not as durable as commercial pressure-treated lumber. For a description of home treatment of posts, see pp. 214–216.

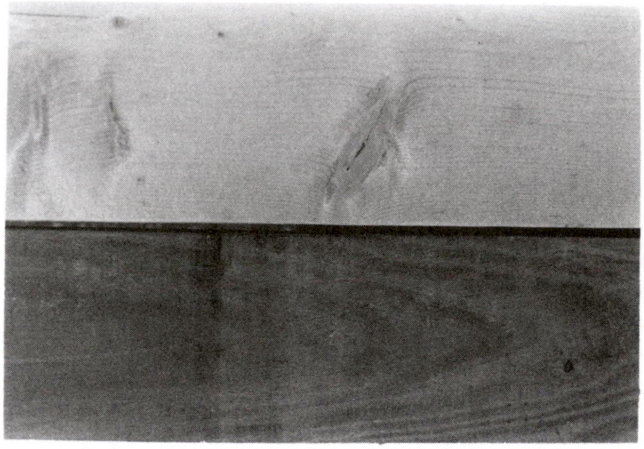

Pressure treated lumber appears a shade darker than nontreated lumber.

Plywood

Plywood, which comes in 4- x 8-foot sheets and thicknesses ranging from ¼ to 1 inch, is an excellent material for roof and wall sheathing and flooring. It is cheaper per square foot than boarding in some parts of the country; in others, it is more costly. But its strength and ease of installation make it a preferred material by many builders. Because the grains of the different laminations go in different directions, plywood is much stronger than solid wood, and it won't split or crack.

Plywood is available with softwood or hardwood faces. Hardwood-faced plywood is most commonly used for furniture and cabinets, while softwood-faced plywood is used for structural purposes. Plywood is available in several different grades and with either interior or exterior glue. Table 2-1 shows the different grades of plywood and their uses. By using the table, you can pick the best yet most economical plywood for a particular job.

Table 2-1 Plywood Grades for Exterior Uses*

Grade (Exterior)	Face	Back	Inner Plies	Uses
A-A	A	A	C	Outdoor, where appearance of both sides is important.
A-B	A	B	C	Alternate for A-A, where appearance of one side is less important. Face is finish grade.
A-C	A	C	C	Soffits, fences, base for coatings.
B-C	B	C	C	For utility uses such as farm buildings, some kinds of fences, etc., base for coatings.
303® Siding	C (or better)	C	C	Panels with variety of surface texture and grooving patterns. For siding, fences, paneling, screens, etc.
T 1-11®	C	C	C	Special 303 panel with grooves 1/4″ deep, 3/8″ wide. Available unsanded, textured or MDO surface.
C-C (Plugged)	C Plugged	C	C	Excellent base for tile and linoleum, backing for wall coverings, high-performance coatings.
C-C	C	C	C	Unsanded, for backing and rough construction exposed to weather.
B-B Plyform	B	B	C	Concrete forms. Re-use until wood literally wears out.
MDO	B	B or C	C	Medium Density Overlay. Ideal base for paint; for siding, built-ins, signs, displays.
HDO	A or B	A or B	C Plugged or C	High Density Overlay. Hard surface; no paint needed. For concrete forms, cabinets, counter tops, tanks.

* American Plywood Association

For all barn and outbuilding construction, use exterior plywood. It is bonded with a waterproof glue that is not affected by moisture, urine or silage. Plywood can be cut with standard circular saw blades, though special fine-toothed plywood blades are made that do not splinter the face surfaces.

In addition to plywood, there are several other sheathing panels that are made from recycled wood fibers. The most common is flake board or "chip" board made from glued chips of wood and sold in 4 x 8 sheets. Flake board, most often used for exterior sheathing for both walls and roofs, is considerably cheaper than plywood. While it is a rigid, structural material, it is not as strong as plywood.

Siding

There are many types of wood, metal and synthetic siding materials. They differ greatly in cost, durability and appearance.

Board and batten is a traditional barn siding, using 1-inch-thick cedar, spruce, pine or hemlock boards nailed in a vertical position. Cedar is undoubtedly the best choice because of its resistance to water and decay. Unfortunately, it is also the most expensive and may not be available from local sawmills. Spruce and pine, the next best choices for siding, are fairly resistant to weathering. Hemlock is a strong, attractive wood, but it is susceptible to grain separation, causing large layers to split and separate from a board.

Plywood siding makes an attractive exterior, as suggested in this small horse barn.

Two other types of vertical siding are shiplapped and tongue-and-groove boards. Shiplapped boards have ½- to ¾-inch half-laps on their edges, allowing the boards to overlap and form a seal. Tongue-and-groove boards have a tongue on one side and a groove on the other, allowing them to lock together.

Other wood siding materials include clapboards, drop siding and shingles. Clapboards and shingles tend to be expensive and labor intensive to install. They also must be applied over a full wall sheathing, whereas vertical boards can be attached to nailers spaced every two feet, eliminating the need for expensive plywood underlayment. Drop siding is similar to tongue-and-groove boards, but goes on horizontally, imitating the look of clapboards.

Finally, there are several plywood sidings available in 4 x 8 or longer sheets. T 1-11 is the most common plywood siding, but it also comes in different textures and without grooves. While ply-

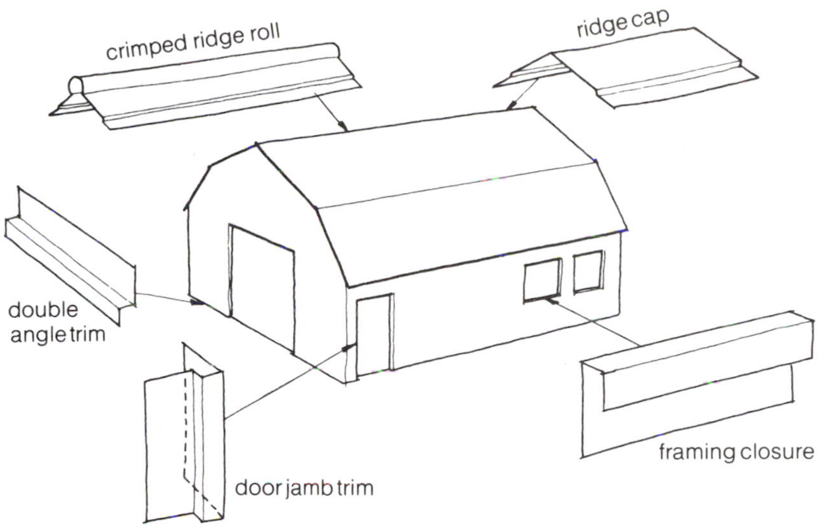

Metal barns are available with an array of accessories and parts. Some common accessories are identified here.

wood siding is expensive, it eliminates the need for wall sheathing and cuts labor time dramatically.

Galvanized-metal siding is a practical, easy-to-use and long-lasting material. Metal siding is sold in sheets, usually 34 or 38 inches wide and 6 to 12 feet long, with corrugations every 2½ inches. Twenty-eight-gauge steel is standard for roofing and siding applications. While once sold only with a metallic-grey surface, galvanized metal now is available with baked enamels and vinyl coatings in different colors.

Corrugated aluminum is another choice. But, I prefer not to use it because it tends to be more expensive and dents more easily than galvanized metal.

Panels are available cut to your specified length so that on-site cutting can be kept to a minimum. A full range of new accessories has advanced the ease of construction markedly. For instance, you can purchase all the pieces needed to trim doors and posts. The result is an extremely neat, weatherproof and nearly maintenance-free building.

Roofing Materials

Galvanized metal is also an excellent roofing material; it goes on quickly and will outlast asphalt shingles. While the cost per square foot of galvanized metal is slightly more than asphalt shingles, its overall cost is lower. This is because metal roofing can be applied directly to 1 x 3 nailing strips set on the rafters, whereas asphalt shingles require a solid deck (usually plywood).

Asphalt shingles, while not as popular today for barns and outbuildings because of their labor and material cost, are still an excellent and economical roofing material. Shingles come in a wide range of colors and are sold by the bundle, enough to cover 33 square feet or one-third of a *square* (10-x-10-foot area).

When properly applied, an asphalt-shingle roof is more wa-

Asphalt Roofing Materials

Product	Approx. shipping weight per square	Packages per square	Length	Width	Units per square	Side or end lap	Top lap	Head lap	Exposure
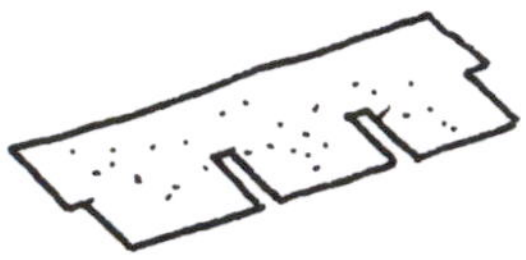 3 Tab square butt strip shingle	235 lb. 300 lb.	3 or 4	36″ 36″	12″ 12″	80 80		7″ 7″	2″ 2″	5″ 5″
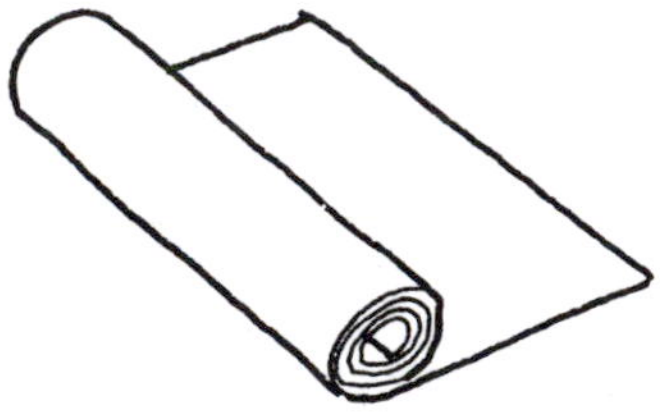 Saturated felt	15 lb. 30 lb.	1/4 1/2	144′ 72′	36″ 36″		4″ to 6″ 4″ to 6″	2″ 2″		34″ 34″
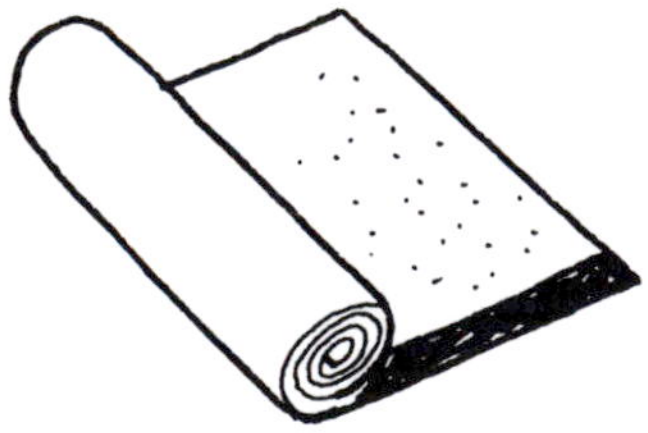 Mineral surfaced roll	90 lb. 90 lb. 90 lb.	1.0	36′	36″	1.0 1.075 1.15	6″ 6″ 6″	2″ 3″ 4″		34″ 33″ 32″
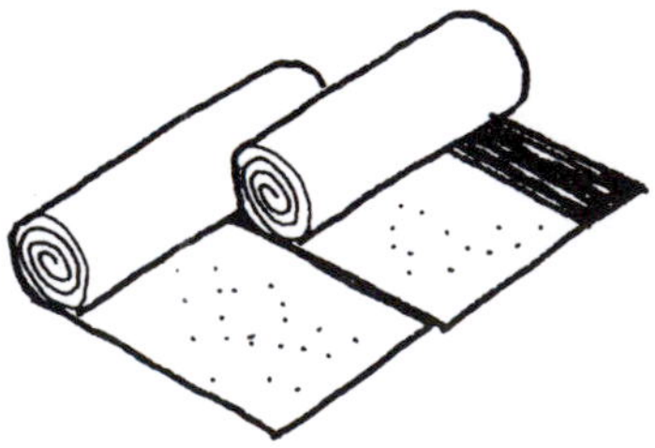 19-inch selvage double coverage	110 lb. to 120 lb.	2	36′	36″			19″	2″	17″

terproof and less likely to produce "sweating" conditions than a metal roof. It is also quieter in rain and hailstorms. Another reason to use asphalt shingles on a barn roof is to match the appearance of a nearby building such as the main house.

Asphalt roll roofing or *half-lap,* as it is sometimes called, is another economical roofing material. It has a composition similar to asphalt shingles, but it comes in 3-foot-wide continuous rolls. Roll roofing is slightly less expensive than shingles and easier to apply. Roll roofing is also available in colors from white to black.

Concrete blocks and cement make for a durable foundation that can be built in stages, in your spare time.

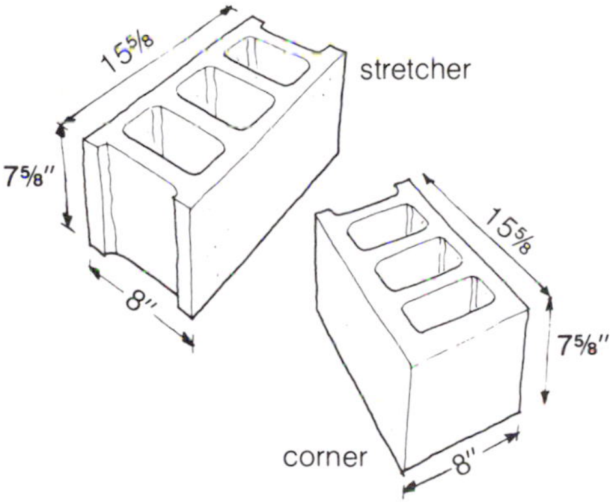

So-called stretcher *and* corner *blocks are the two most commonly used for concrete block walls.*

Concrete and Cement

Concrete is commonly used for building foundations in the form of solid walls or piers. Occasionally, concrete or concrete blocks are used for above-ground walls as well. While poured concrete is quite expensive and concrete blocks require a good deal of labor, they are the sturdiest and most durable of building materials. Milk barns are often built of concrete blocks because they are easy to clean and stand up to regular washings with water and disinfectant.

Concrete is made by mixing cement, sand and crushed stone along with water in the proper proportions. Cement is the binding agent that holds together either concrete or mortar. Portland cement is most commonly used today and is manufactured from limestone mixed with other pulverized rock. A 1:2½:3½ concrete mix is a standard mix for footings and foundation walls (1 part cement, 2½ parts sand and 3½ parts crushed stone or gravel). Concrete comes ready-mixed in 80-pound bags that yield ⅔ of a cubic foot of concrete. You can use bags for small jobs, but when you need more than ½ a cubic yard (27 cubic feet equal 1 cubic yard) mix your own with a power mixer or hire a concrete truck to deliver it ready to pour. (See "Estimating Concrete," p. 40.)

When mixing mortar for laying a block or brick wall, the proper proportions are 1 part Portland cement, ½ part hydrated lime and 4½ parts clean, screened sand. Mortar can also be bought ready-mixed in 80-pound bags.

Concrete blocks are available in a variety of sizes and shapes. However, the standard blocks for exterior wall construction are 8 inches wide, 7⅝ inches high and 15⅝ inches long. This allows room for a ⅜-inch mortar joint, giving a finished 8 x 8 x 16 block space. Blocks used for interior walls are normally 4 inches thick instead of 8.

Insulations suitable for small barns and outbuildings include fiberglass batts, loose fill and rigid foam.

Insulation

Livestock barns and outbuildings such as tool shops can be insulated to provide comfortable winter working conditions. Three of the most basic insulations are fiberglass batts, loose fill and rigid foam.

Fiberglass is the most common and least-expensive insulating material. It comes in batts, 16 or 24 inches wide, to fit between studs and rafters and in thicknesses from 3½ inches to 12 inches. Its insulating quality is measured as an R-value, such as R-11 for 3½-inch fiberglass or R-38 for 12 inch. It is now customary in cold northern climates to put at least 6 inches of fiberglass in the walls and 12 inches in the roof of a well-insulated building. For exact insulation specifications, consult an Extension Service office or university agricultural engineer.

Cellulose and vermiculite are loose-fill insulators for block walls or other spaces into which they can be poured or blown. Cellulose may be blown through holes drilled into wall cavities.

Rigid foams are the best insulators per inch of thickness. For example, polyisocyanurate foam has an R-value of 7.4 per inch. Such foams are quite expensive, however, and should only be used where space is a limiting factor such as in roofs or around foundations where moisture might damage other insulating materials. Most foams are available in 4 x 8 rigid sheets. Do not use spray foams. (For more on insulations, see Appendix C.)

Fasteners

Using the correct fasteners when building is as important as using the correct materials. Nails are the primary fastener, although screws, bolts and connecting plates are used in certain instances.

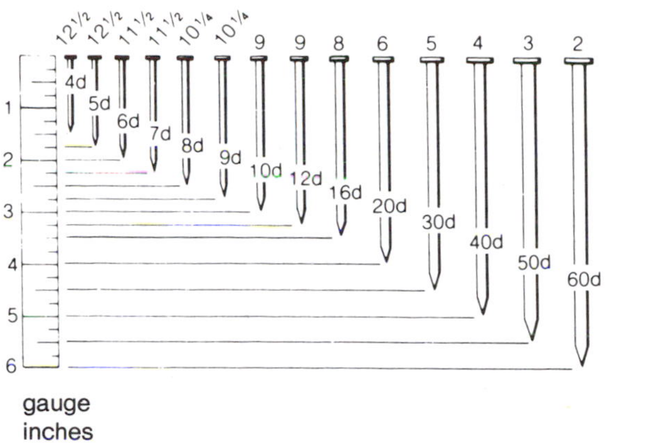

A nail is sized according to its length and gauge of its wire. The nails you'll use most often will be 6 ds, 8 ds and 16 ds.

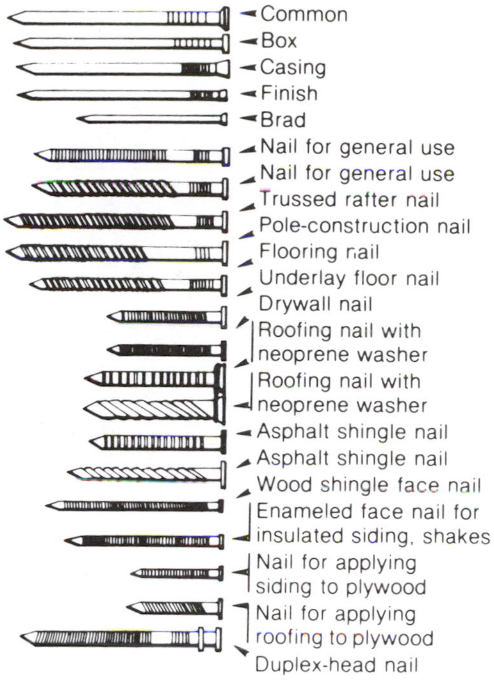

Sometimes it's hard to know what a particular nail looks like, or what it's for. This illustration should help.

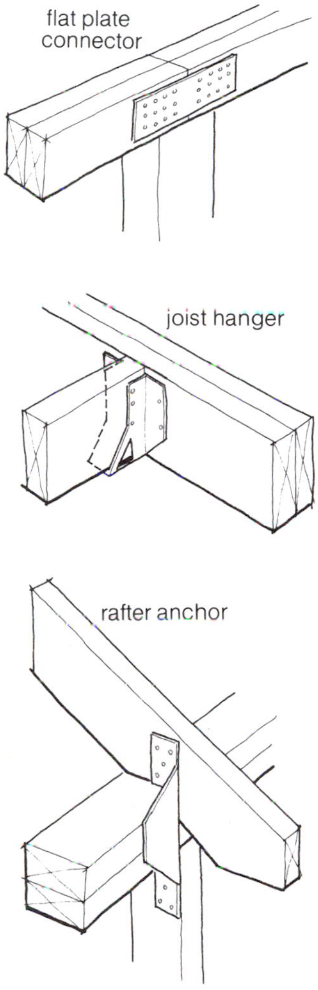

Common steel fastening plates include flat-plate connectors, joist hangers and rafter anchors.

Nails, available in many different sizes and styles, are measured by the "penny" (abbreviated "d") and sold by the pound. An eight penny (8d) nail, for example, is 2½ inches long, whereas a 10d nail is 3 inches long.

Nails with a special zinc coating are called galvanized nails. Use these for exterior nailing of siding and trim where rust and corrosion will cause staining or weaken the nail. Spiral-grooved nails, called pole-barn nails, should be used for pole framing. They are harder to drive and easier to bend than standard nails.

For securing posts, beams and joists, there is a wide variety of special steel fastening plates. The most common are called *joist hangers* used to secure floor joists. Not only do these plates make a secure joint, they cut labor time considerably.

Paints and Stains

Applying a finish to a barn or outbuilding makes sense from both aesthetic and economic points of view. Always cover siding with either paint, stain or a clear oil to protect it from the weather.

Paint, either latex or oil-based, protects wood very well. However, it is fairly expensive, it requires a lot of work to apply and, in severe northern climates, it must be applied frequently. For those who demand some color on their barn, a more economical choice is an oil stain that is easier to apply and lasts longer. While the initial expense of paint or oil stain might be the same, oil stains cost less in the long run. Oil stains are available in a wide range of colors.

For those who like natural wood colors, clear linseed oil is an excellent preservative and protection. It is probably the least-expensive finish available. In addition, because linseed oil is non-toxic, it can be used freely on wood that livestock come in contact with. You can either brush or spray it on.

Hand Tools

If you decide to build one of the barns or outbuildings described in this book, there are several essential or helpful tools. Some, such as post-hole diggers, are for specific jobs and therefore not always required. Simply select the tools that suit your building needs and construction methods. Purchasing cheap dime store tools is not worthwhile. It is important to obtain quality tools for convenience, craftsmanship and safety.

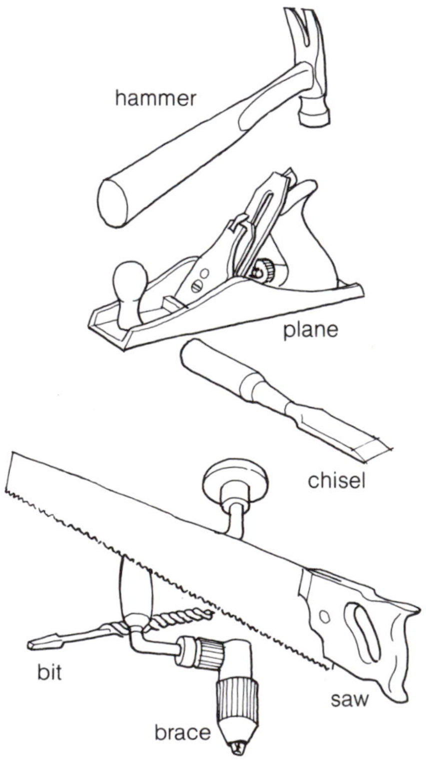

Some of the most widely-used hand tools are the hammer, plane, chisel, handsaw and brace and bit.

Hammers. The best all-around choice is a standard 16-ounce hammer. However, you may also want a heavier, 20-ounce ripping hammer as well; this is ideal for driving large spikes into native lumber. Although it may be tiring until you get accustomed to the extra weight, driving large nails with a 20-ounce hammer is easier than with a lighter hammer. The hammer should have a straight ripping claw that provides leverage for prying and removing boards. A heavy, short-handled sledge is helpful for driving stakes for form boards or for other chores requiring heavy pounding.

Hand saws. A standard, crosscut hand saw in coarse cut with 8 points to the inch is suitable for most barn construction. This should be a good, sturdy saw, preferably hollow or taper ground. Although these saws are expensive, they provide good clearance in native green materials and reduce the likelihood of pinching and binding.

A couple of specialty saws can really help, especially in areas without electricity. The first, a timber saw, sometimes called a *docking saw,* is used primarily for fast cutting of heavy timbers. It has 3½ points to the inch and is usually made of heavy-gauge steel. A second type is the *one-man saw.* Primarily for felling trees, this saw can also be used for fast crosscutting of heavy timbers and beams. (A chain saw is also suitable. And for those with electricity, a good circular saw will suffice for most construction-cutting jobs.)

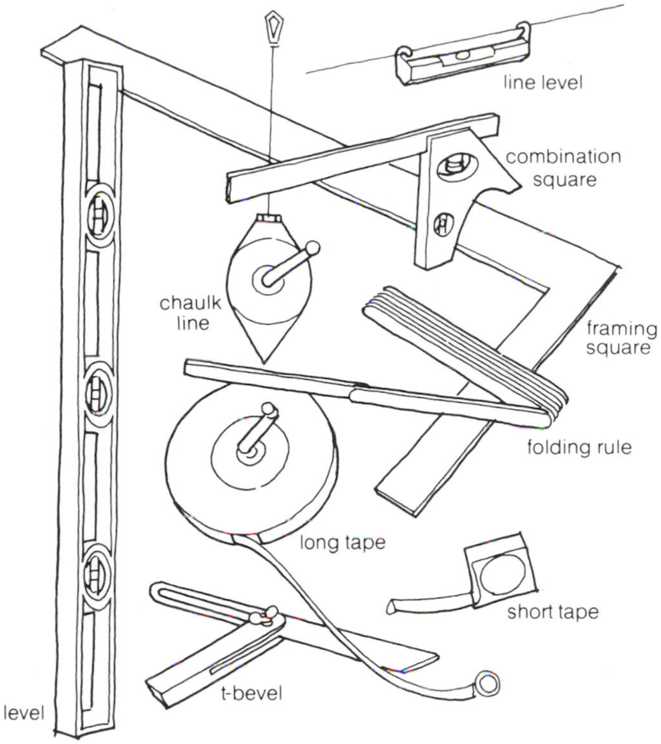

Measuring tools and levels, such as these, are essential to achieve construction accuracy.

Chisels. For notching studs and cutting mortises and tenons, obtain several different-width chisels. These should be heavy-duty, sturdy tools. If working with large beams, you may want a large, socket-firmer chisel. These are up to 19 inches long. A utility knife with a retractable blade is an all-purpose tool useful for cutting off splinters, sharpening pencils and completing many other cutting chores.

Planes. You may need a plane to smooth down wood for a joint, especially when making a tight, rodent-proof feed room. For this, a No. 5 jack plane is the best choice. In addition, a good draw knife can also be used to smooth down timbers for jointing: for instance, fitting a tenon into a mortise.

Boring tools. Boring tools are needed to install bolts, hinges, etc., and sometimes for cutting mortises. A good brace and bits are invaluable. But, make sure the brace is double acting (works when rotated in either direction) so you can use it in tight corners. The bits should be large, high-quality auger bits. In addition to the brace and bits, use a special auger fitted with a wooden handle for boring deep holes in heavy beams; this auger provides more leverage than a brace.

Measuring tools. The first and probably most important measuring tool is a good steel tape. Preferably, this should be at least 25 feet long and should be marked with feet and inches. In most cases, you need a good 50- or 100-foot tape to lay out a barn, and to make other long measurements. You may want an additional, 12-foot tape for small jobs; it's light and convenient, but for barn building it's a bit short.

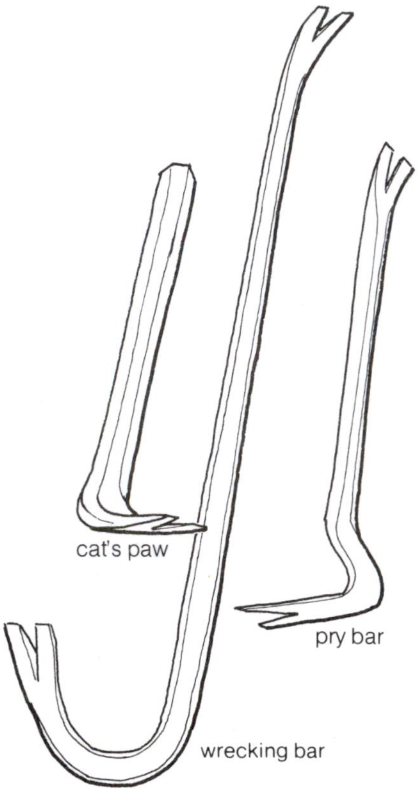

Three prying bars are, from left, the cat's paw, double-claw wrecking bar and pry bar.

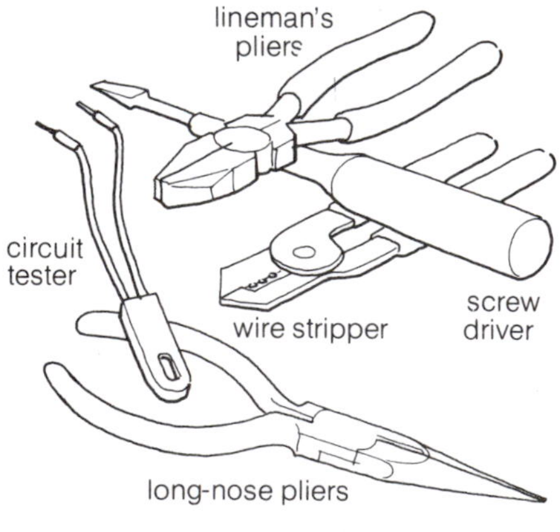

Helpful electrical tools are pliers, strippers, insulated screwdrivers and circuit testers.

Another essential tool is a large square, commonly called a *framing square,* with a 24-inch long body and a 16-inch tongue. The square is used for marking boards to be cut, for laying out rafters and for determining the squareness of a joint. A small combination square is convenient for marking boards, and a T bevel for marking angle cuts.

Levels. Several levels are usually needed. The first is a standard, 2-foot aluminum model. This may be used for most construction work, however, a 3- or 4-foot level is essential for many chores, particularly laying out sill plates or plumbing barn walls. Naturally, the longer levels are more expensive.

Use a string level to set a level line between corners of a building, or as a guide for installing concrete blocks. A combination plumb bob/chalk line is useful for marking cut lines and determining the plumbness of corners or walls. Keep several lengths of string on hand; mason's line is preferred because of its strength and durability.

Pry bars. A nail puller called a *cat's paw* is essential for removing nails from wood. Another good tool for this task is a pry bar. You will also want a 3-foot crow bar with a double-angle claw on one end for prying nailed boards loose.

Masonry tools. Tools for working with concrete include a float for smoothing off foundation tops, a trowel for applying mortar and an edger for rounding corners to prevent chipping. A mason's hammer or regular hammer and cold chisel are needed for cutting concrete blocks.

Electrical tools. For installing wiring, get a good pair of lineman's pliers. These should have insulated handles and be large enough so you can cut through plastic-sheathed cable easily. A pair of needle-nosed pliers will also come in handy for bending wire leads.

A good combination stripper-crimper can reduce a great deal of time when stripping wire ends for connections. In my opinion, the plier type is much better than the knife style. My brother still has a nasty scar from a careless moment with a knife stripper.

Make sure all your screwdrivers have insulated handles, and that you have at least one large one with a long handle. A small power-testing light is a reassuring tool to have so you can determine when a wire is "hot."

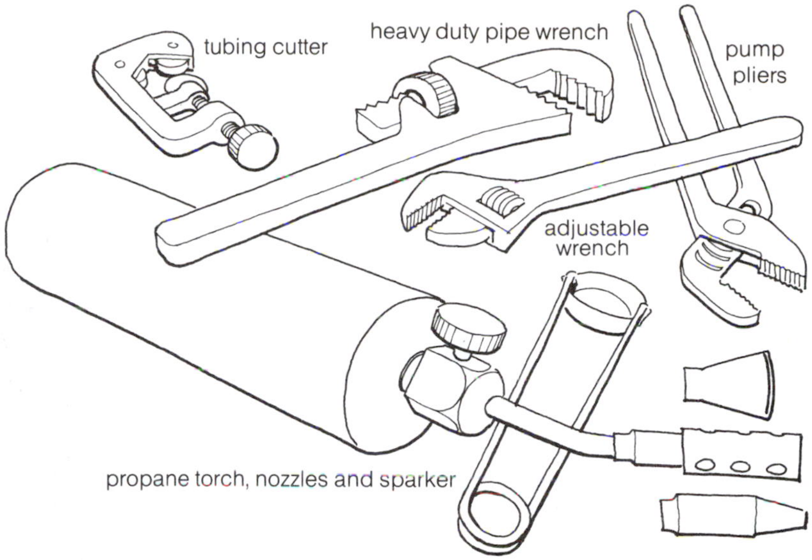

If you plan to do plumbing work, a basic tool set would include those shown here.

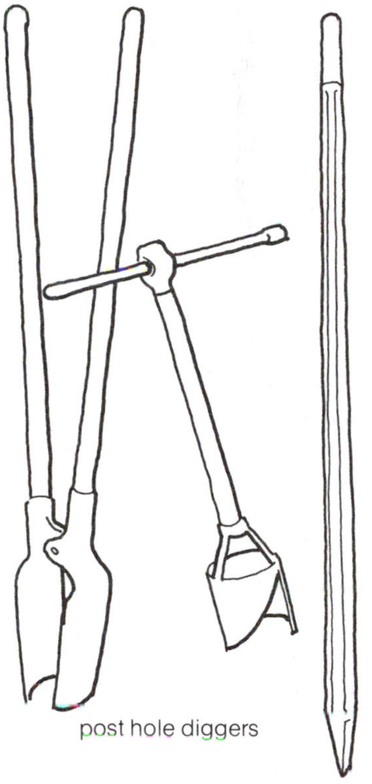

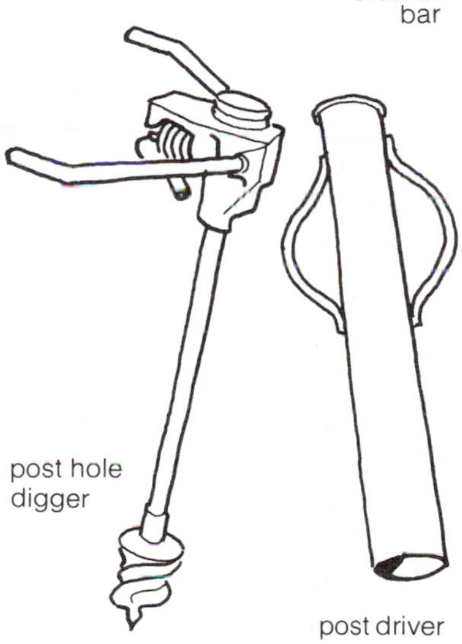

Often, these digging tools may be obtained from rental businesses or supply centers.

Plumbing tools. For installing copper pipe, you will need a pipe cutter, flux, solder and a propane torch. A pair of large pump pliers and leather gloves are useful for holding hot pipes. In addition, a 10-inch adjustable crescent wrench and a 10-inch pipe wrench are necessary for working with threaded joints or old galvanized pipe.

Fencing and digging tools. I seem to always need extra digging tools such as axes, picks and shovels. For digging fence-post holes or holes for foundation piers, use a post-hole digger that has two handles and a pair of clam-type blades. Another type of hole digger has an auger which works well, but I prefer the clam-type digger for tough soils.

Use a homemade post driver for driving steel posts in the ground and a heavy-duty sledge or "post maul" with a long handle for driving sharpened wooden posts. A post driver is made by welding handles on a piece of 30-inch steel pipe 3 inches in diameter. Weld a cap on one end and fill it with about 6 inches of lead. This tool will drive steel posts in even the most stubborn spots.

A good selection of shovels includes a short- and long-handled shovel, a round-pointed shovel and a short-handled spade. An axe or hatchet and a large sledge hammer are useful for cutting and driving small posts. A heavy steel bar, often known as a *breaker bar,* is a digging tool that will help you remove large rocks.

Fence stretcher. There are several kinds of fence stretchers. For light work, the kind with the rope pulleys will do; however, for long runs of heavy fencing, you need a heavy-duty style. For woven-wire fencing, use homemade fence clamps. Fence clamps are two 2 x 4 boards held together with bolts. Woven-wire fenc-

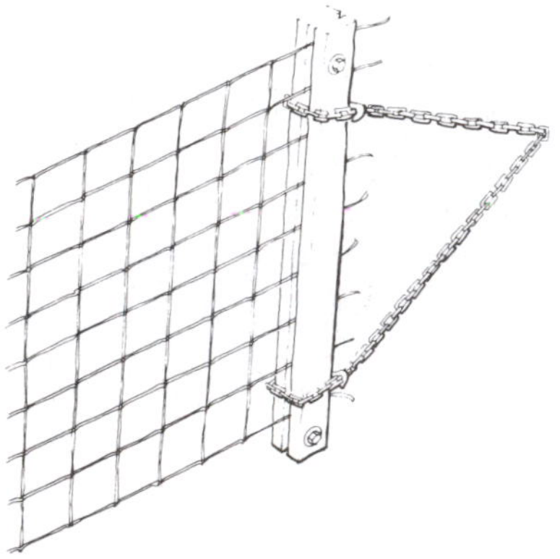

The chain end of a fence stretcher may be secured to a vehicle, to pull the fence taut.

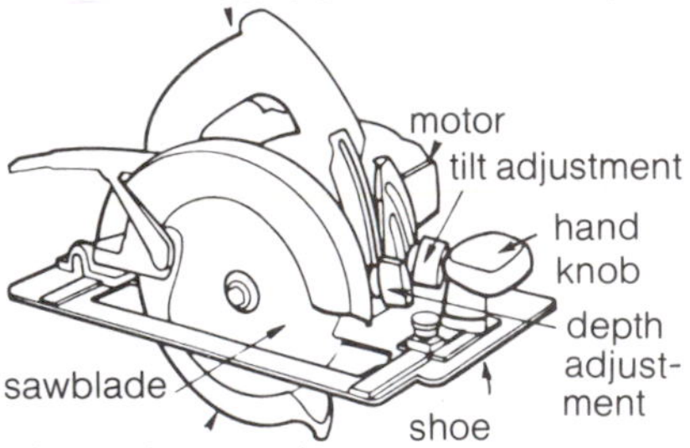

The shoe or base of a portable electric saw can be tilted to one side to permit angle cuts, and it be adjusted to let the blade cut at different depths.

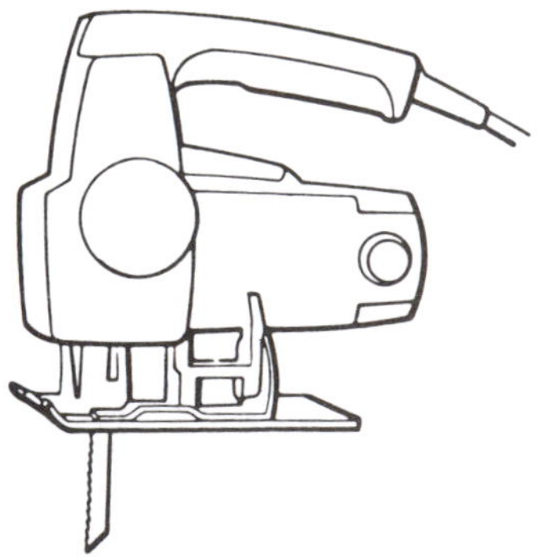

Hole cutting is best done with a saber saw.

ing is placed between the two boards and the boards are bolted tightly. Then a chain is placed around both ends of the clamp and a fence stretcher, tractor or vehicle is used to pull the fencing taut.

A pair of fencing pliers can quicken a great many jobs including twisting and cutting heavy wires and removing staples.

Saw horses. Two good saw horses are helpful for supporting material to be cut and for laying out rafters. You can make saw horses with scrap wood or lumber and saw-horse clamps.

Power Equipment

You can build almost any small structure without power equipment, but items such as power saws and post-hole diggers quicken the job and make it much easier. Often, good electric-powered equipment may be rented.

Electric saw. Probably the most important power tool for building barns, sheds, shelters and other outbuildings is a portable electric saw. Choose a heavy-duty, *double-insulated* saw to protect against shock. A good choice is a 7½-inch, 2⅛-horsepower saw. This saw has plenty of power and cuts deep enough for most dimensional lumber used for barns. When equipped with a good carbide blade, it handles enough material to wear you out by the end of a day. You can also equip it with blades that will cut metal roofing and barn coverings.

Heavy-duty saber saws and portable electric drills are sometimes useful. I use a drill with a ½-inch *chuck,* the mechanism that grips the shank of a bit. Some carpenters prefer a smaller, ⅜- or ¼-inch chuck, saying this size is a bit more versatile and suitable for most commonly-used bits. However, the larger ½-inch chuck has a slower speed and may offer a bit more stability for heavy drilling jobs.

Chain saw. One essential power tool, especially for building pole barns and fences, is a chain saw. You can use it to trim poles or posts after they have been installed. Even if dimensions are correct, the holes may not all be at a perfect depth or the ground may slope slightly. A simple solution is to merely trim tops of poles level with a chain saw.

Post-hole digger. There are two basic types of powered post-hole diggers. The first, powered by a small gasoline motor, does

KEEP IT SHARP

Sharp woodworking tools are a necessity if you are going to work efficiently and skillfully. A dull saw will wander through the wood, making it impossible to get a square cut, and a dull chisel will gouge wood rather than shave it off in thin strips.

Circular-saw blades and hand saws are best taken to a professional sharpener. With the proper tools and jigs, it is a fairly easy job to do yourself, but a professional can usually do it quicker, more accurately and for only a few dollars a blade.

However, you can sharpen chisels and plane blades in your shop or right on the job very easily. All that is needed is a grinding stone and some lightweight oil. When sharpening a chisel, first grind the bevel, then hone the cutting edge to make it razor sharp.

To grind the bevel, place the chisel or plane blade on an oiled stone with the beveled edge down and at an angle of 20 to 30 degrees. Keeping the angle constant, move the blade back and forth with short easy strokes until the nicks are gone from the cutting edge.

Now the blade can be honed by raising the grinding angle to 30 to 35 degrees and moving the blade in a figure-eight pattern across the stone. This will produce a second bevel which gives a razor sharp cutting edge. Now, the blade should be able to slice a piece of paper.

A chain saw may be used for many purposes, including trimming posts or poles used to frame an outbuilding. With an adapter or portable chain-saw mill, you can use a chain saw to cut logs into boards.

a fast job of boring in ground that doesn't have too many rocks. Usually, it may be rented. A better type fits on the back of a tractor with a three-point hitch. Numerous small tractors such as the *Ford 8N* are ideal for this and many other farm chores. This tractor can also be equipped with fence winders, post drivers and even cement mixers.

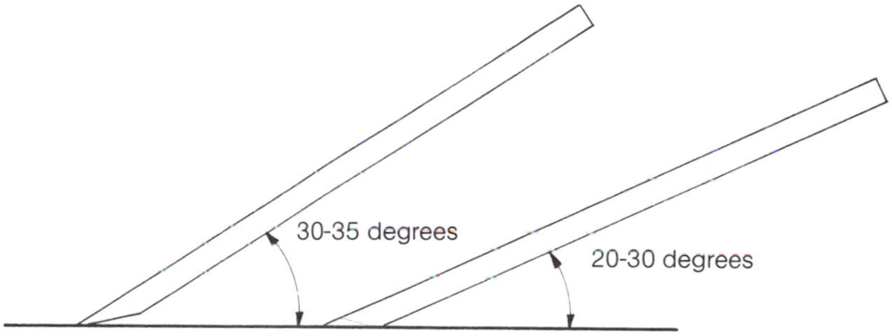

Hold the blade at a 20- to 30-degree angle when grinding (right), then use a 30- to 35-degree angle for honing.

CHAPTER 3

Footings and Foundations

Borg

THE FOUNDATION IS the most important part of any building. It should be designed to permit the most economical use of materials, while strong enough to support the weight of the building and to withstand the rigors of the climate. If you are building a small chicken coop, elaborate concrete foundations are obviously unnecessary. Sills or runners made of pressure-treated wood may be all that are needed. Posts set on concrete piers, are fine for supporting single-story sheds or small barns. But they won't do for a two-story barn. Larger buildings, especially those with load-carrying floors, need a continuous wall foundation.

When selecting a foundation, bearing capacity of the soil is an important factor to consider. Concrete can withstand pressures of up to 2,500 pounds per square inch, but soft clay soil can only support about 14 pounds per square inch, or 1 ton per square foot. Thus the bearing weight of soil is often the limiting factor, and concrete footings that support the foundation piers or walls must be large enough so they don't sink into the ground.

Also, you must compensate for cold climates. Freezing ground expands, which can cause a foundation to heave and crack if its footings are not below the frost line, the depth to which the ground freezes. The frost line varies from zero in the deep south to more than 5 feet in the north. Foundation footings must be placed below this line. Accurate information for your area is available from the National Weather Service or any knowledgeable builder.

Other considerations are your budget and how much you want to do yourself. A concrete-pier foundation is one of the cheapest and easiest to build. If you need full foundation walls, then you can choose between a poured concrete foundation or a block wall. Poured concrete is expensive, but it goes in quickly with a minimum of labor. Concrete blocks, on the other hand, are cheaper but require much more work. Ultimately, your choice will depend on your budget, how much time you have and how strong your back is.

Because of the engineering calculations involved, you may want to seek professional advice on your foundation. An Extension Service engineer can take a look at your proposed building, the site and the soils, and advise you on the proper footing depth and wall sizes. You may also want to hire a professional to put the foundation in. This is especially true if you are pouring concrete walls or floors that require forms, reinforcing steel and skilled finishing work. Floors with drains or plumbing (Chapter 8) are also tricky and require skilled workmanship.

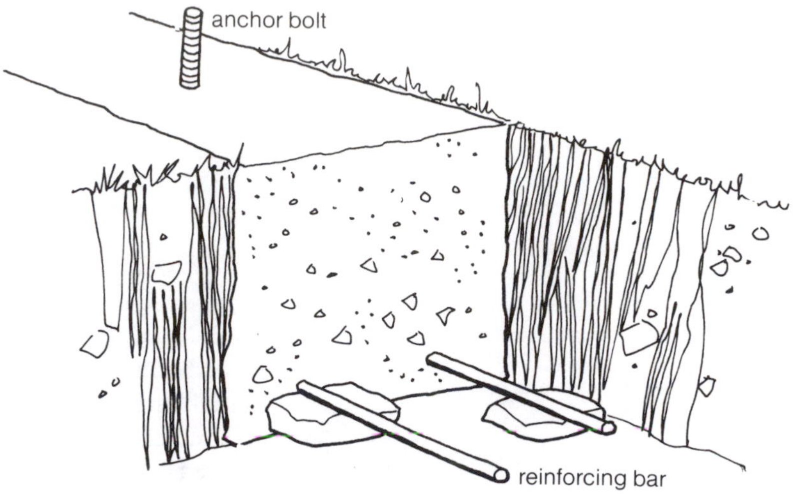

Trench footings are strengthened with reinforcing bar (rebar). Anchor bolts are used to secure the sills to the foundation.

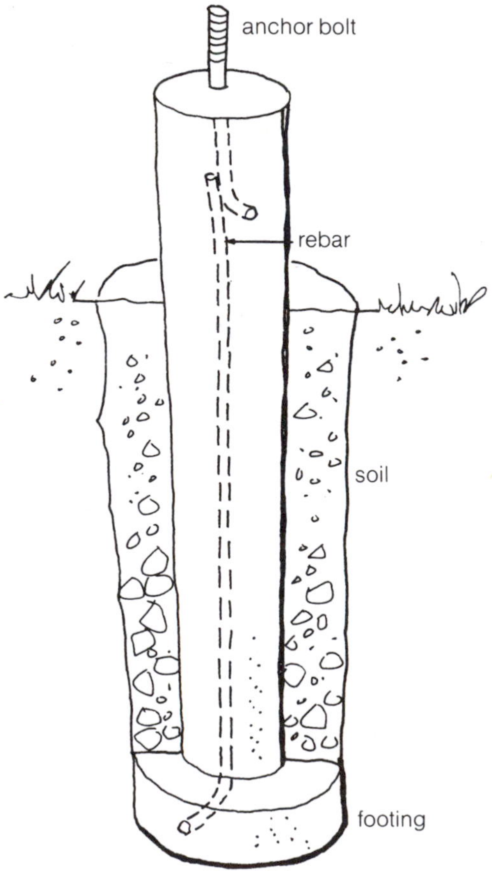

Piers, with anchor bolts or brackets, are often used with small barns.

Five Different Foundations

There are five different types of concrete foundations to choose from: each has a particular advantage and usefulness.

Trench footings. Simple continuous concrete footings that prevent the structure from sinking into the ground may be poured into trenches. This low-cost foundation is suitable for small buildings in southern climates where there is little or no frost. The footings are usually 8 to 16 inches wide, depending on soil conditions and the weight of the building, and they stick up several inches above the ground to keep the sill plate and framing members dry.

Pier foundation. For small or large buildings that don't have a concrete floor, concrete and even wooden piers can be used to support posts or large sill beams. The piers must be placed on footings that are usually twice the diameter of the piers and below the frost line. If wooden posts are used, they must be pressure treated.

Pier foundations have the advantage of minimizing excavation and the need for concrete. A standard cylindrical concrete pier is 8 inches in diameter and has a 16-inch diameter footing. Holes for the piers can be dug with either a post-hole digger or backhoe, and the concrete can be poured into cardboard forms available from any building supply company. These forms are often called *Sonotubes,* their original brand name.

Pole buildings are most often built on pier foundations, with metal anchor brackets attaching the poles or posts to the concrete piers. Stud-framed buildings can also be built on sill beams spanning the piers (see p. 61). For small sheds, you can use 6-inch piers spaced every 8 feet and 6 x 6 sills. For larger barns, use 8-inch piers spaced from 4 to 8 feet apart, depending on the building size, and 8 x 8 sill beams.

Poured concrete wall. For large, permanent structures, full 8-inch concrete walls with 16-inch wide footings are desirable.

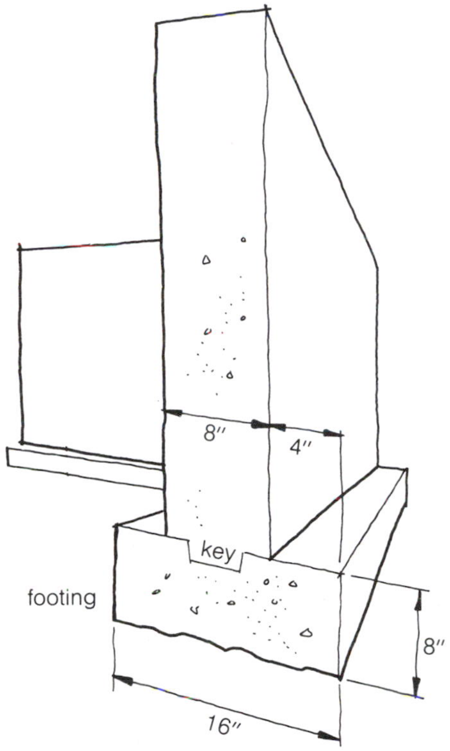

Typically, concrete-wall foundations have footings twice as wide as the wall.

The footings are always placed below the frost line; steel reinforcing bars (known as *rebar*) are placed in the concrete wall and anchor bolts are imbedded in the top to secure the wooden sill plates. (For more on reinforcing, see pp. 46–48.)

While this type of foundation is the most expensive, it is the sturdiest and gives you the option of having a basement or underground space for storing roots and other crops that like a cool, damp climate. Often, just a corner of a barn is excavated for storage space rather than the entire building area.

Concrete block wall. If you need a full foundation, another option is to build concrete block walls. This foundation is less expensive than poured concrete and may be laid block by block as time permits. Like other foundation walls, the blocks are laid upon 16-inch footings of poured concrete set below the frost line. Concrete block walls do not have as much lateral strength as a poured concrete wall. This drawback can cause problems when the foundation is set in wet clay soils that expand as they freeze, putting lateral stress on the wall. To reduce this stress, make sure block foundations are well drained, back filled with gravel and reinforced with rebar.

Floating slab. Concrete floors are often recommended for milk rooms, feed and tack rooms and other rooms requiring sanitary conditions and ease of cleaning. In southern climates, you can pour a concrete slab in conjunction with trench footings to form a solid, yet economical, foundation and floor.

In northern climates, contractors often pour similar foundations known as *floating slabs*. These move slightly when frost

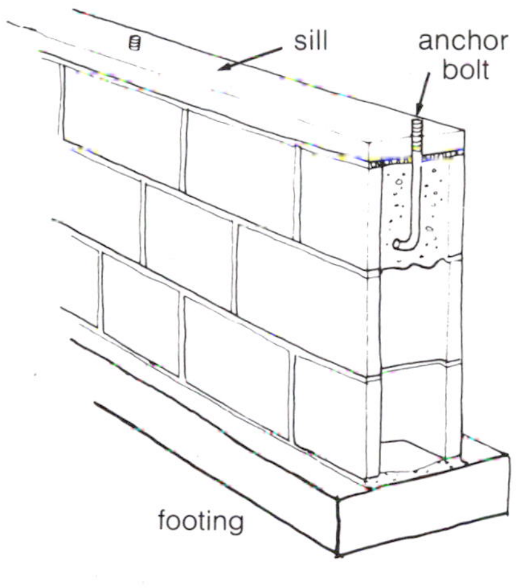

A concrete-block foundation. Rebar is placed inside the block cores.

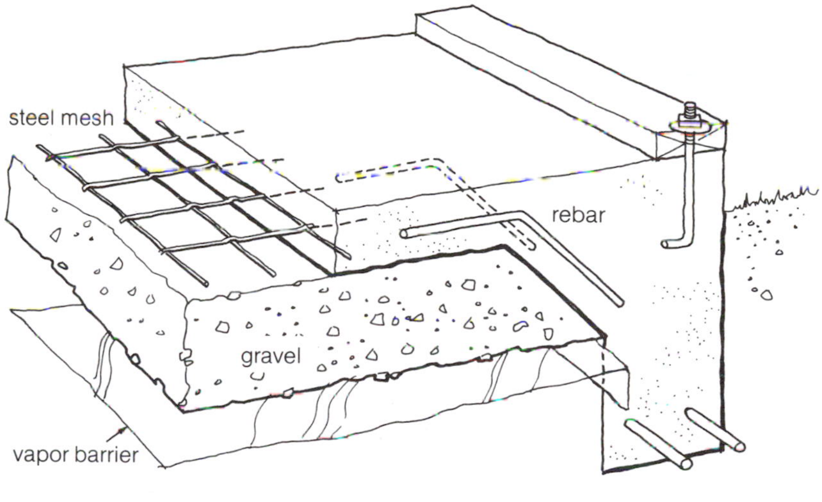

A floating slab rests on a bed of gravel; the thick "turndowns" at its edges keep the slab from shifting.

heaves occur. They are primarily for small, light outbuildings such as garages, sheds and greenhouses. Because such slabs do not have footings below the frost line, they must be reinforced heavily with steel mesh and bars so they do not crack as they move.

Foundation Layout

Regardless of the foundation you select, the first step is to prepare the site and lay out the building lines. Site preparation consists of simply clearing the site of all woody vegetation, making sure it is fairly level and removing large rocks. If there is any doubt about the water table and drainage at the site, dig a test pit to see what kind of soils you'll be building on and how well drained they are. An Extension Service agent can help you analyze your soils and their suitability for building.

Once you have figured out roughly where your building will stand and which way it will face, you can lay out the exact building lines for the excavation work. Keep in mind that careful layout work will prevent headaches later on. Don't hurry. Laying out a large building is relatively simple but it takes time and attention to detail.

To get started, collect four wooden stakes, perhaps 2 x 4s about 2 feet long, and obtain a 50- or 100-foot steel measuring tape. One stake with a nail in its top will be located at each corner of the building, to indicate the outside line of the foundation walls.

First, measure one long wall of the building and drive two stakes into the ground, one at each corner (or end) of the wall. (Align the wall according to your site plan.) Drive a nail into the center of the top of each stake, at the exact corners of the wall. Extend a mason's string tautly between the two stakes and tie it to the two nails.

Some builders continue the layout by measuring the remaining walls and placing stakes at the exact corners. Then, to assure that the corners are square, they measure the long diagonals to see if they are the same length. However, there's another way to proceed:

To lay out an adjoining wall and square it with the first, you can use the triangle rule. The rule is that any triangle that has sides with 3 feet, 4 feet and 5 feet (or multiples thereof) will make up a right-angle triangle. The square corner will be where the 3- and 4-foot sides meet. When using this method, it's important to use the appropriate maximum expansion of the 3-4-5

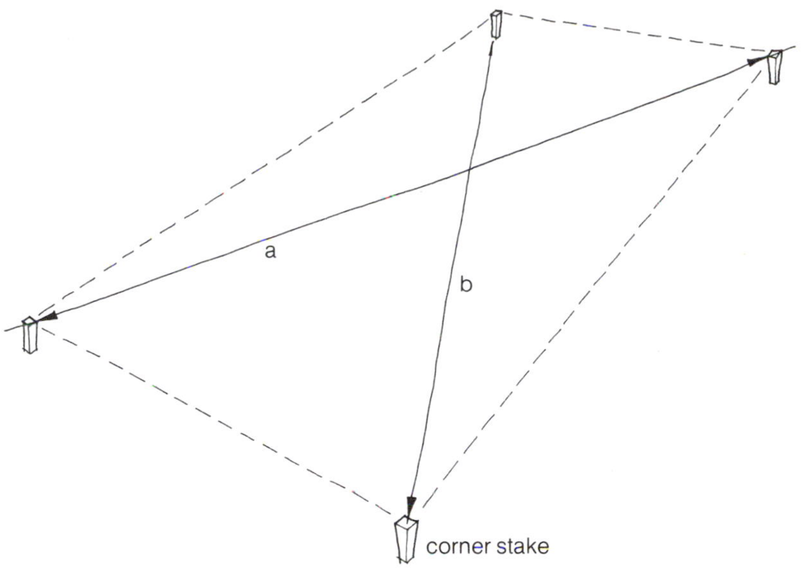

One way to check the squareness of a foundation layout is to assure that the diagonals (a and b) are equal.

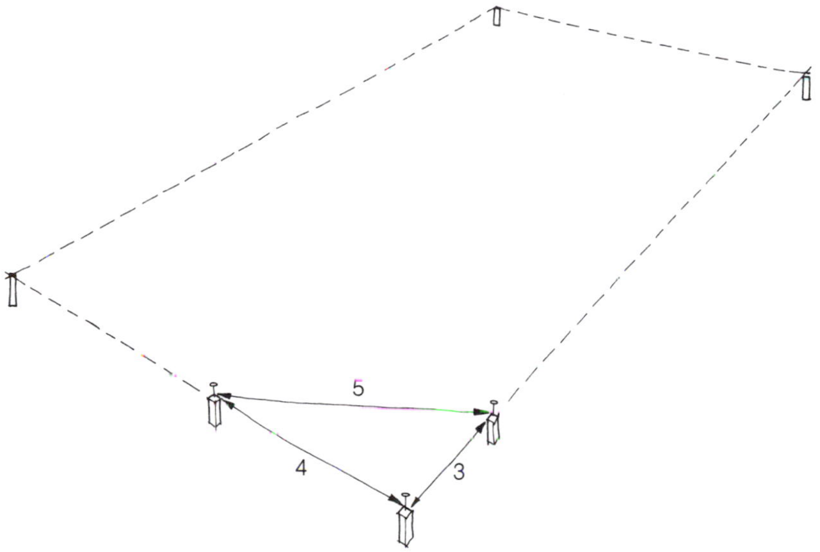

Or, you can use the 3-4-5 triangle method. For larger foundations, use greater multiples (9-12-15, for example) to enhance accuracy.

The 3- 4- 5 Triangle

1. Go to one corner, measure down the string 3 feet (or multiples thereof), drive a stake into the ground and, at exactly 3 feet, pound a nail into the top of the stake.
2. Ask a friend to take another string (or measuring tape if you have one), attach it to the corner stake and measure 4 feet. Mark at that length. Extend the string or tape in the direction of the adjoining wall.
3. Hook a measuring tape over the nail on the stake set at 3 feet. Extend the tape 5 feet toward the adjoining wall.
4. When the 5-foot and 4-foot points intersect, a right-angle triangle is formed and the corner is square. At the point of intersection, drive another stake. This establishes a line for the building's adjoining short wall and makes it square with the long wall.
5. Once the right-angle triangle is formed, extend the string or a tape beyond the 4-foot point, as necessary, to reach the full dimension of the wall. If you're planning on a side wall of 20 feet, for example, extend the string to that distance. Again, drive a stake into the ground at the new corner, and pound a nail into the stake at the exact wall dimension.
6. Repeat steps 1 through 5 until foundation outline is complete. If you anticipate laying out several foundations, you might make a permanent 3-4-5 triangle from scrapwood to facilitate layout of the corners.

rule to fit the building. For a 30- x 20-foot building, for example, use an 9-12-15-foot triangle; otherwise, errors can be large. To establish a 3-4-5 triangle and thereby ensure that a second wall is perpendicular to the first, follow the steps outlined in the accompanying box.

Now that the outside corners of the building are marked exactly with stakes, it is time to set the batter boards. Batter boards are simply short boards nailed to 2- to 3-foot long stakes to form a right-angle corner. The purpose of batter boards is to hold the building lines and to keep their location marked while the foundation trenches or holes are excavated. Locate the batter boards at least 4 feet outside the building corners so that a backhoe can dig a trench without disturbing them.

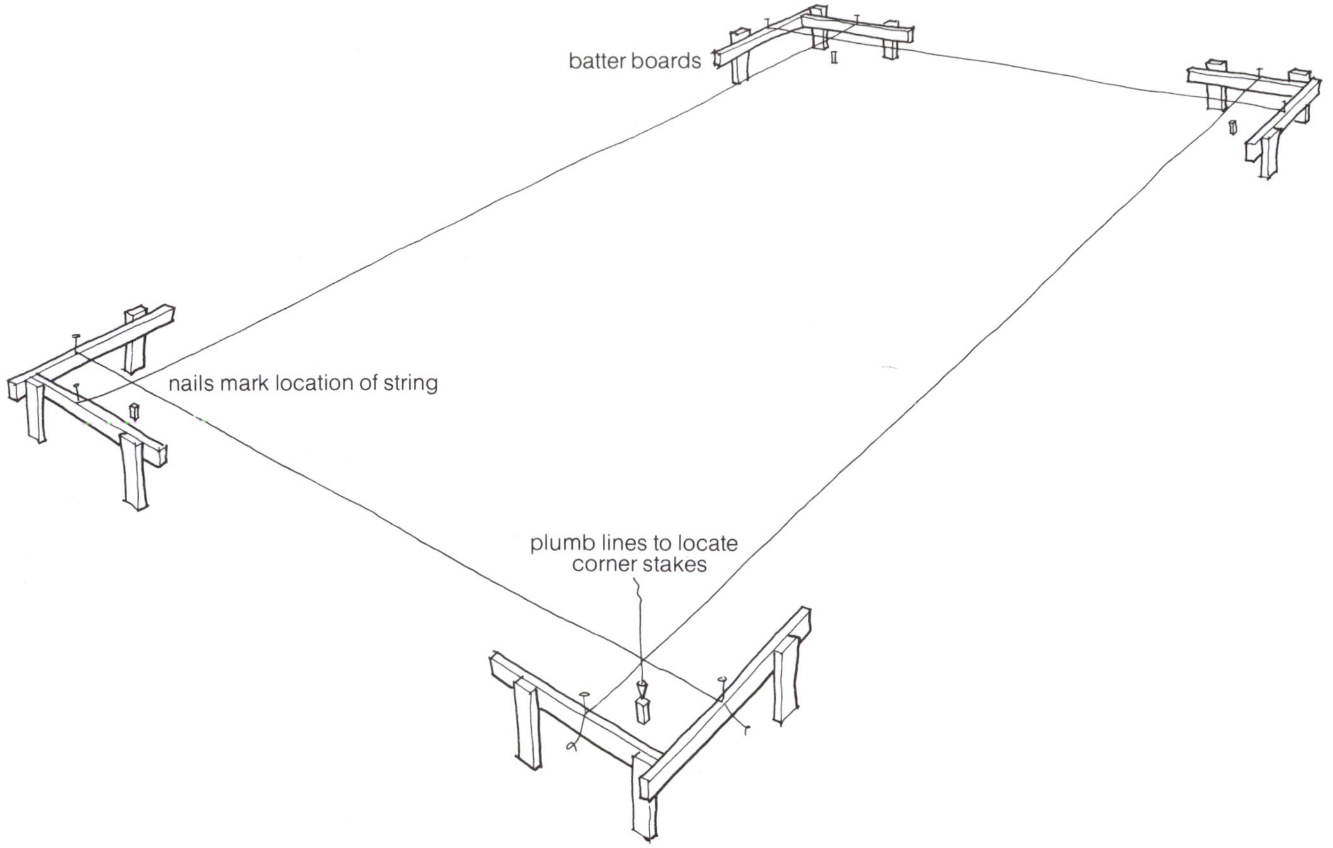

Batter boards form right triangles at the foundation corners. To provide room for excavation equipment, they should be set back at least 4 feet from the foundation.

Attach a new set of strings to the batter boards so that they cross exactly on top of the corner stakes. The position of the strings on the batter boards can be marked with nails securely driven into the tops of the boards. The strings can now be removed and the corner stakes taken up in order to excavate the foundation. If you need to find the exact building line again during excavation, you can quickly reattach the strings to the batter boards.

As you lay out your building, keep in mind what the string line really marks. It should mark the *outside* edge of the foundation wall or piers, not their center or inner edge. This is important so that a building wall designed to be exactly 24 feet (and therefore take exactly six sheets of 4-foot-wide plywood siding) does not turn out to be 24 feet, 8 inches and require a full new sheet of plywood that is mostly wasted. When digging for concrete piers, it is especially important to remember that the string marks the outside of the pier.

Excavation

Excavating can either be done by hand or by a backhoe. For small jobs such as digging a shallow trench foundation or digging holes for concrete piers, it is often sensible to use a shovel and post-hole digger and do the work by hand. If you have a deep wall foundation or rocky soil, however, you'll probably want to hire a backhoe. A backhoe may cost $25 an hour, but in four hours it can do more digging than you can do in a week. And if you run into large boulders, you must have power equipment.

To locate foundation holes for piers, first mark the center of each pier with a small stake. To do this, set up your building lines on the batter boards and measure in the diameter of the pier and drive a stake to mark its center. Do this for all the piers, then remove the building lines. Now dig holes centered on the stakes and at least as wide as the footing will be. Thus for an 8-inch pier with a 16-inch footing, dig a hole 16 inches in diameter.

Trenches for poured concrete walls should be about 4 feet wide so you can move around in them and place the concrete forms. Since the normal concrete wall is 8 inches thick, the center of this trench is 4 inches inside the building lines. Small stakes driven on the outside edge of the trench lines will help the backhoe operator stay in line.

If you are digging a large foundation, ask the backhoe operator to keep the topsoil separate from the lower horizons as much as possible. This topsoil can then be set aside for the final grading and landscaping, and you won't have to try to grow grass in gravel or clay. Always be present when the backhoe is digging to make sure the trench is deep enough and to handle surprise problems such as coming across an old water line.

To determine the depth of your excavation, measure to the frost line from the *highest* corner of the building. After the foundation is in, it can be backfilled and graded to create an even soil depth around the perimeter. If you're working with a sloped site that can't be evened out with a little backfill, seek professional advice on the installation of a stepped foundation. With this foundation, footings may be laid below the frost line, without extensive excavation.

Footings

Footings are the fundamental building supports carrying the foundation wall or piers. For farm and residential construction, they are normally twice the width of the foundation wall and one half as thick. Thus for a normal 8-inch foundation, the footings should be 16 inches wide and 8 inches thick. For an 8-inch pier foundation, the footings should be 16 inches in diameter and 8 inches thick.

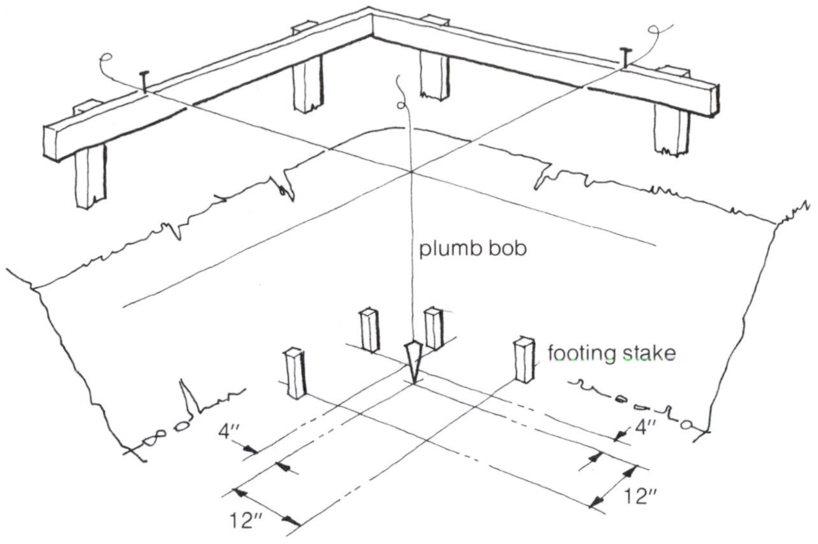

Drop a plumb bob to relocate the foundation, then measure 4 inches to one side and 12 to the other. This establishes the outline for a 16-inch wide footing.

If you are building a large barn, it is wise to rough check the bearing strength of the soils and footings to make sure they can support the entire load. Accompanying tables give the customary loads of a typical building and the bearing strength of different soils. These tables can be used to get a rough estimate of the total building weight and therefore its bearing weight per square foot of foundation area. Check this weight against the bearing strength of your particular soil to make sure the foundation footings are adequate.

Table 3-1 Bearing Capacities of Soils*

HARD ROCK	UP TO 40 TONS PER SQUARE FOOT		
Soft rock	"	8	"
Coarse sand	"	4	"
Hard, dry clay	"	3	"
Fine clay sand	"	2	"
Soft clay	"	1	"

* Eccli, Eugene *Low-Cost, Energy-Efficient Shelter* (Emmaus, Pennsylvania: Rodale Press, 1976).

Once you have excavated your foundation trenches or holes to the proper depth, re-attach your layout strings that mark the edge of the building. Use a plumb bob to drop a vertical line from the string to the bottom of the trench. This is the outside of your foundation wall or 4 inches from the center of the footings if you are using an 8-inch wall. Using the plumb bob, mark 4 inches to the outside and 12 inches to the inside. The footing will go in this 16-inch space.

Mark the inside and outside of the footings at each of the building corners with stakes and attach two sets of strings to

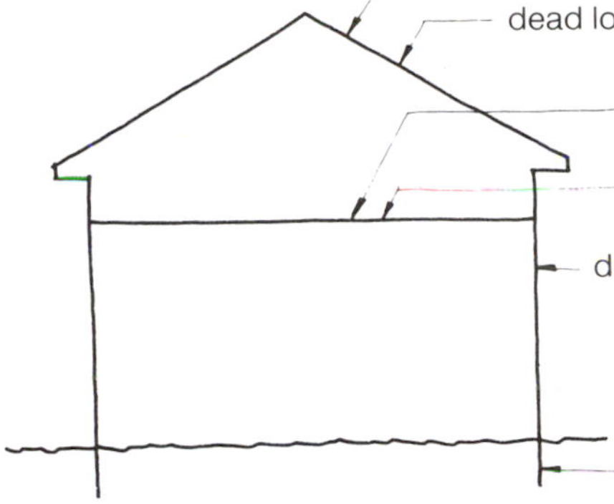

Loads on 20- x 24-foot barn.

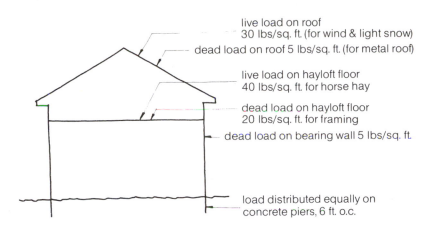

Calculating Building Loads and Footing Sizes

Here is a simple example of how to approximate building loads to determine the necessary size of foundation footings. In the example, we will use a 20 x 24 barn with a dirt floor and hayloft above. The foundation supports are concrete piers set 6 feet on center with footings 16 inches in diameter.

The accompanying illustration shows standard building loads. "Live" loads are usually defined as those not associated with the building or its framing; for example, hay, wind, snow and people. "Dead" loads include the building framing and inherent material weight. Using standard figures for building loads, as shown in the illustration, we can calculate the following loads for a 20- x 24-foot barn:

1. Roof load =
   live load, dead load, area
   (30 lbs./sq. ft. + 5 lbs./sq. ft.) x 670 sq. ft. = 23,450 lbs.

2. Hayloft floor load =
   live load, dead load, area
   (40 lbs./sq. ft. + 20 lbs./sq. ft.) x 480 sq. ft. = 28,800 lbs.

3. Exterior wall load (side bearing walls only) =
   live load, dead load, area
   (0 lbs./sq. ft. + 5 lbs./sq. ft.) x (8′ x 24′ x 2) = 1,920 lbs.

4. Total foundation load on both bearing walls = 54,170 lbs.
   ÷ 2 = 27,085 lbs. (weight on one wall)
   ÷ 5 = 5,417 lbs. (weight on one pier)

5. Load per sq. ft. of footing soil =
   5,417 lbs. ÷ 1.4 sq. ft. (footing area for 16-inch diameter pad)
   = 3,869 lbs. per sq. ft.

Since sand/clay or hard clay soils can support 2 to 3 tons per square foot (4,000–6,000 pounds) as shown in Table 3-1, the 16-inch diameter footings here would be adequate for these soils. If you were building in soft clay soils, however, it would be wise to increase the footing size until you were under 2,000 pounds per square foot bearing weight. Footings 24 inches in diameter would effectively do this, having a bearing surface of 3.14 square feet.

outline the perimeter of the footing. You can now build footing forms to hold the concrete in place using 2 x 8 lumber secured with stakes in the ground. Make sure the stakes are on the outside of the boards, that the boards are 16 inches apart and that their tops are perfectly level. If you need to raise the boards off the ground a bit to level them, you can simply backfill with a little dirt to seal cracks on the bottom where concrete might seep out.

If you are pouring footings for concrete piers or wooden posts, you can simply use the circumference of the holes as the forms for the concrete. Before you pour the concrete, however, make sure the hole is where you want it, and that you can get a pier or post plumb and centered on the footing so its outside edge lines up with the outside edge of the building.

When you pour footings, *make absolutely sure they are on solid, undisturbed ground that has been scraped free of all loose dirt.* This is especially important to remember when pouring pier footings since dirt and loose debris can easily fall into a hole and go undetected. Footings poured on loose dirt will later settle causing the foundation and the building to settle and sometimes crack. If for some reason, the soil beneath the footings has been disturbed, be sure to tamp it down thoroughly.

The footing and foundation wall or pier must be "tied" together for strength, and this is done in two ways. For wall footings, a "key" is molded into the footing by depressing a 2 x 4 on edge into the wet concrete. Push the 2 x 4 down about 1½ inches so part of the board remains above the surface. This will make removal of the board easier. Remove the board when the concrete has set. The wall will be poured into the 1½-inch indentation. The indentation "locks" the wall in place and helps prevent water from flowing between the wall and the footing joint (see p. 31). For a pier foundation, a short piece of ⅜-inch reinforcing bar can be stuck in the wet footing to tie the pier and footing together (see p. 30). Also see pages 46 and 47 on reinforcement and anchors.

Concrete

When putting in a foundation, you must first decide whether to mix the concrete by hand or have ready-mix concrete delivered by truck. Your decision will be based on the cost of ready-mix concrete, the size of the job and whether you can get a concrete truck to the building site.

Excluding labor, ready-mix concrete is quite a bit more expensive than concrete you can mix by hand in a wheelbarrow or machine mix in a power mixer. In general, however, for jobs

In most cases, using ready-mix concrete from a truck is the simplest way to pour a foundation.

that require a continuous pour of 1 cubic yard or more, buying ready-mix concrete is the right choice. A concrete truck with chutes can back right up to your foundation, deliver the concrete exactly where you want it and pour your walls in an hour or less.

Usually, ½ to 1 cubic yard is the minimum load a concrete supplier will deliver, and you should establish beforehand what truck charges or minimum load charges you will be billed for. (The maximum load one truck can usually carry is 5 yards.) Everything should be in place and set to go before the truck arrives. Normally, you have one hour to unload the truck and after that a substantial hourly truck fee is charged. Also, the concrete may set up quickly so make sure the forms are securely in place, that the truck can reach all sections of the foundation with a 10-foot chute, and that you have shovels and a wheelbarrow handy in case you have to transport concrete by hand to a far corner.

Concrete for individual footings or small foundations can be mixed by hand in a wheelbarrow or power mixer. Never use a wheelbarrow to mix more than 1 cubic yard. A good-sized wheelbarrow will hold only about 1½ cubic feet of concrete, meaning you'll need to mix about 20 wheelbarrows full before you get 1 cubic yard. For small jobs such as pouring pier footings, however, a wheelbarrow and pre-mixed bags of concrete are fine. Pre-mixed concrete is available in 80-pound bags; each bag makes about ⅔ of a cubic foot of concrete. One bag of concrete mix is just right for a 16-inch diameter footing, 6 inches thick.

If you're mixing more than a yard of concrete, get either an electric power mixer or one that can be driven from a tractor power take off (PTO), and buy the concrete ingredients in bulk. Concrete is a combination of Portland cement, sand and gravel

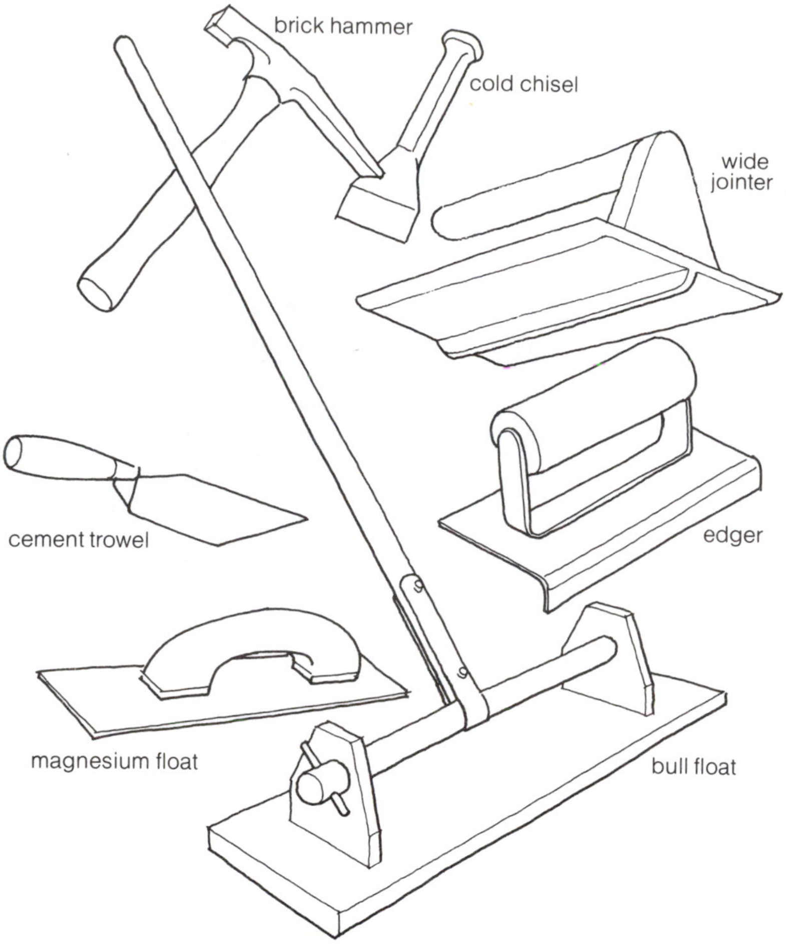

When working with concrete or concrete blocks, these are some of the masonry tools you'll be likely to use.

ESTIMATING CONCRETE

When you order concrete by the truck, it is measured in yards. A yard is 27 cubic feet or a volume measuring 3 x 3 x 3 feet. When figuring the amount of concrete needed to fill any square or rectangular area, the following formula can be used if all the measurements are in feet:

$$\text{Cubic yards} = \frac{\text{width x length x thickness}}{27}$$

For example, the concrete needed to pour a slab 9 x 18 feet and 4 inches thick would be:

$$\text{Cubic yards} = \frac{9 \text{ x } 18 \text{ x } \frac{1}{3}}{27} = \frac{9 \text{ x } 18}{27 \text{ x}3} = \frac{6}{3} = 2 \text{ cubic yards}$$

If you are trying to figure the volume of a Sonotube foundation, you need to substitute the cross-sectional area of the Sonotube for the length x width measurements. The area of a circle is pi x radius^2. Thus the concrete needed for one 6-inch Sonotube that is 4 feet deep would be:

$$\text{Cubic yards} = \frac{3.14 \text{ x } 0.25^2 \text{ x } 4}{27} = \frac{0.785}{27} = 0.029 \text{ cubic yards}$$

You would need 34 Sonotubes just to use up one yard of concrete!

mixed with water. The standard mixture usually specified for foundations is 1:2½:3½, or 1 part Portland cement, 2½ parts sand and 3½ parts gravel. If the sand is damp (and bulked up), use a 1:2:4 mixture.

The sand and gravel should be clean and free of trash, leaves and other debris. Also, it should be screened to remove small pebbles and to assure uniform size. Gravel or crushed stone can be obtained from concrete suppliers, sand and gravel dealers or building supply companies.

Because the size of gravel or crushed stone varies in different locations, it may be necessary to change the amount of cement in your mix. Generally speaking, when gravel is smaller than the normal 1½-inch size, it is good practice to use more cement. When gravel size is a maximum of 1 inch, add ¼ bag of cement to a five-bag mix; when gravel is a maximum of ¾-inch, add ½ bag.

When mixing concrete, first place the cement, sand and gravel in the wheelbarrow or mixer and mix thoroughly. An old hoe with a couple of holes cut in it and a shortened handle makes an excellent tool for mixing small amounts.

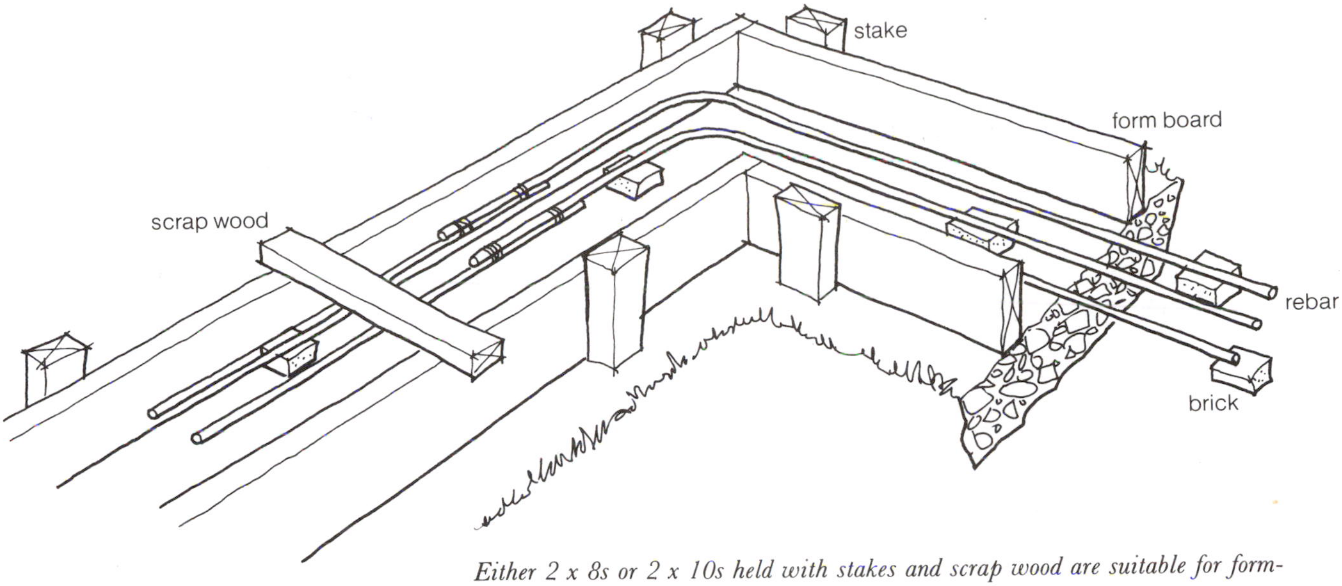

Either 2 x 8s or 2 x 10s held with stakes and scrap wood are suitable for forming concrete footings.

When the aggregates and cement are mixed thoroughly, and no dark or light streaks remain, add water. The amount of water is *very* important and ultimately determines the strength of the concrete. Too much water makes the finish concrete weak and flakey. Too little and the solution may not mix properly, or it may set up too quickly. Usually, you should add five gallons of water for each 94-pound bag of cement. If the sand is wet, use less.

Don't add the water all at once. Add just a little at a time and allow the concrete to mix thoroughly, then add more as needed. To test the mix to see if it has the right amount of water, pull the hoe through the concrete in a series of jabbing motions. If the mix is correct, the little ridges pulled up will stay. If it is too wet, the concrete will slump back quickly. Add more gravel and cement. If the mix is too dry, the ridges won't be smooth and even.

Working With Concrete

When working with concrete, always wear rubber gloves and rubber boots. Rubber gloves will protect your hands from the abrasive and caustic action of cement which can easily wear away skin after a few hours of contact. Rubber boots will allow you to step in the concrete, water and mud without getting wet, and your good leather work boots won't be ruined.

When pouring footings, piers, slabs or walls, forms must be used to keep the concrete in place. As I mentioned earlier, 2 x 8 or 2 x 10 boards held in place with stakes are fine for footings. For piers, use *Sonotubes.* For concrete slabs, use the foundation wall as a form or 2 x 8 boards secured with stakes. For concrete walls, more elaborate forms are necessary.

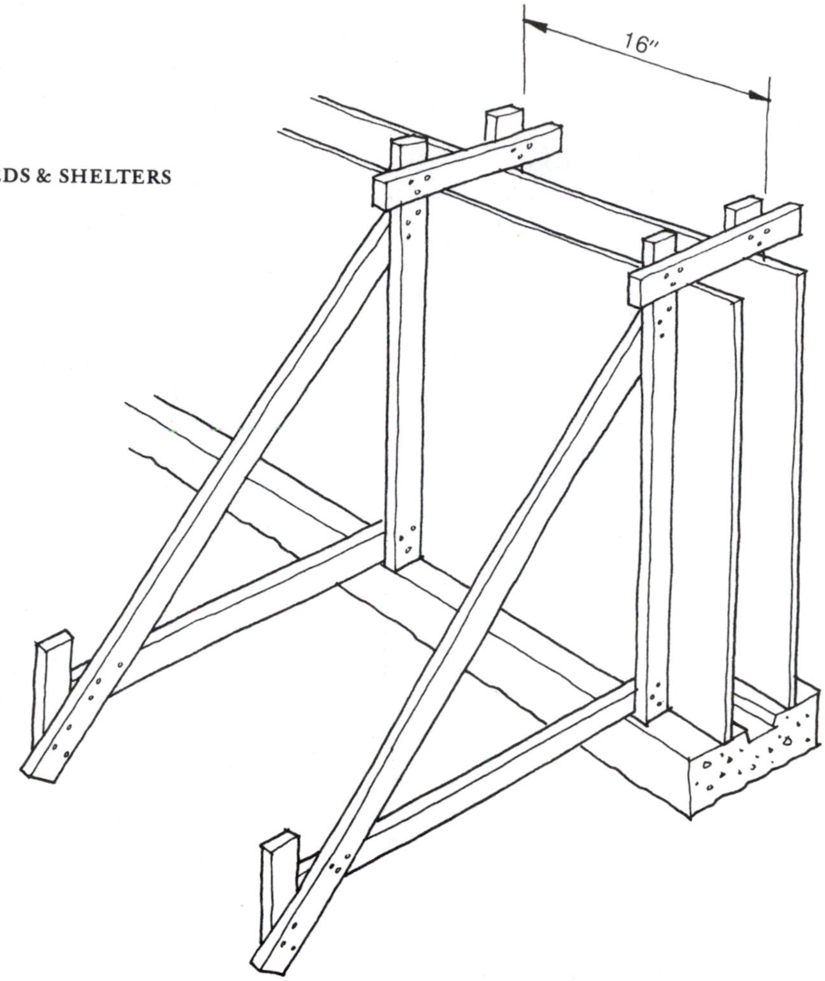

Wall forms should be well braced. If a wall is more than 4 feet high, it's usually best to rent forms from a rental company or supplier.

You can make wall forms from ¾-inch plywood, or 1-inch boards, supported by 2 x 4 studs every 16 inches and braced at every stud. This type of form can support concrete up to 4 feet high. The plywood panels should be 8-inches apart and tied together on top with 1 x 4 boards. Oil the inside faces of the plywood with linseed oil or old crankcase oil before pouring the concrete. That keeps the concrete from sticking.

For forms more than 4 feet high, you'll need double 2 x 4 horizontal bracing tied together with steel rods. Because these forms are complicated and expensive to build, it is often wise to rent them from a tool-rental company or concrete supplier. Standard panels are available in 4- and 8-foot heights that go together quickly with steel pins.

When the concrete is ready, pour it into the forms continuously to avoid cracks, voids and weak spots. These blemishes can best be avoided by using ready-mix concrete for large building foundations. Try to pour the concrete exactly where you want it so you won't have to move it again with shovels and rakes. This will make it easier on your back and improve the concrete's strength.

Once the footing or wall forms have been filled, poke a shovel around the outside edges of the concrete to knock out air bub-

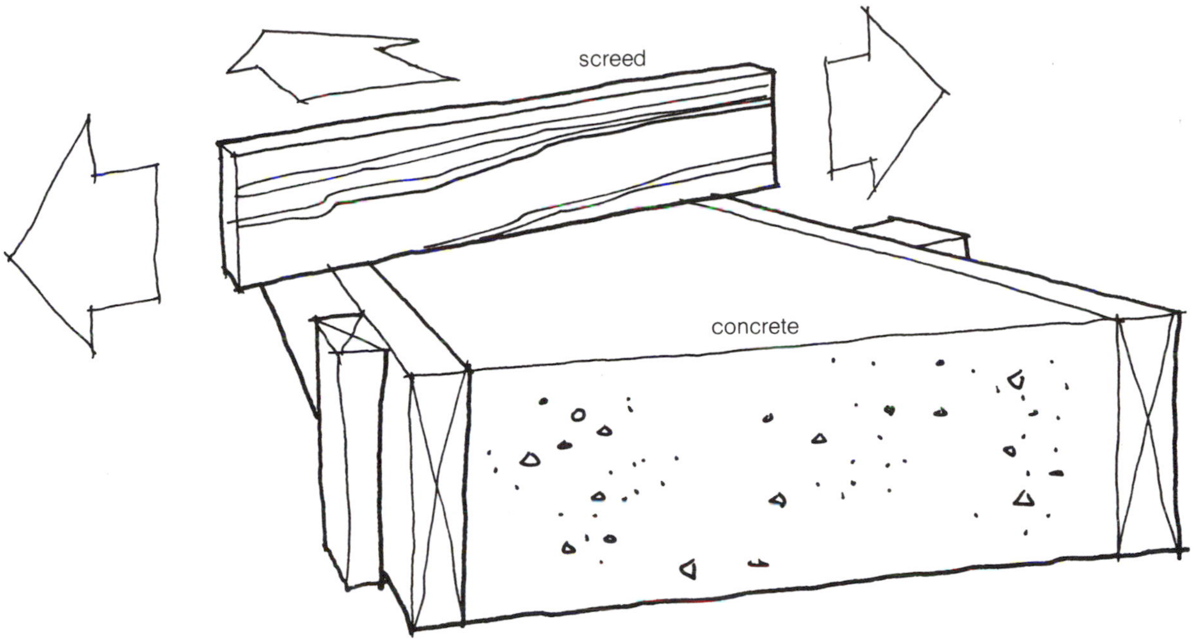

Smooth off concrete with a screed. As the board is pushed along the tops of the form boards, it is also moved laterally in a sawing motion.

bles and help the concrete settle evenly. Do not overdo this and cause settling of the aggregate to the bottom.

Using a short piece of 2 x 4 or a magnesium float, strike or screed the excess concrete off the top of the forms and smooth until the concrete is level. Now move the 2 x 4 or float in a sawing motion along the top of the forms. This operation is called *floating* concrete; it raises the water and cement paste to the top to give a smooth finish. Do this until all the large aggregate is submerged, but don't attempt to make the top as smooth as glass.

Slabs. A concrete slab is finished in the same way, but requires a good deal of skill and patience to get a good finish. After pouring the slab, usually to a depth of 4 inches, use a rake to roughly level out the surface. With a helper, take a long, straight 2 x 4 and use this to screed off excess concrete and to level the surface further. Push the 2 x 4 back and forth with a sawing motion and slowly move down the slab about 1 inch per stroke.

Next, float the slab using a bull float with a long handle. Your skill in handling the float will determine the appearance of the floor. After all the surface water has disappeared and the concrete has begun to harden, drag a push broom with stiff bristles lightly over the surface to give the concrete a slightly textured surface. This is especially advisable for barn floors that will be wet and slippery.

If you are pouring a large concrete slab more than 10 or 12 feet wide, pour it as three separate sections so you can screed it and level it properly. First, divide the floor area into thirds using 2 x 6 form boards to hold the concrete. Secure these to

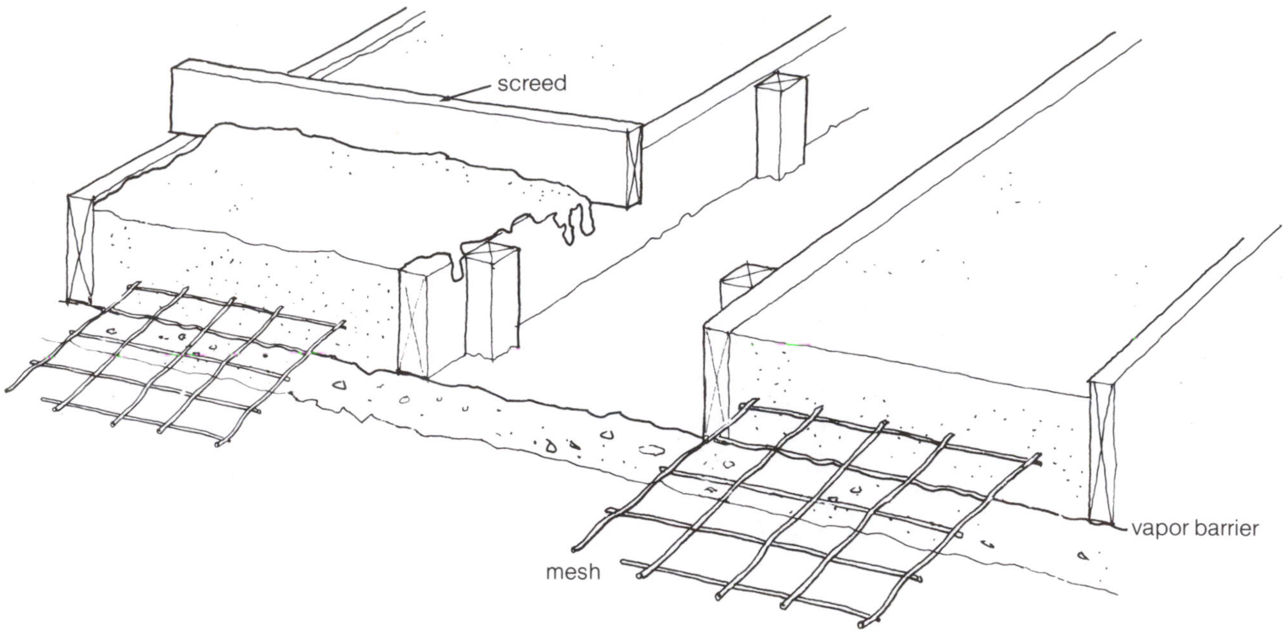

When pouring a large slab, it helps to divide the job into three sections with form boards. Pour the two outside sections first, then the middle.

stakes driven in the ground and make sure the top edges of the boards are level at exactly the height of the finished floor.

Pour the two outside sections first. Use a rake to tamp the concrete and force out any air bubbles, and then screed the concrete using the 2 x 6 forms as a guide. Float the two concrete sections and then let them harden for 24 hours before removing the forms and pouring the center section.

Concrete slabs must have expansion joints around their perimeter to keep them from cracking the outside foundation wall when they heat up and expand. A thin, fibrous asphalt material is available for just this purpose, or you can use 1-inch rigid foam that helps insulate as well. Large slabs should also have expansion joints every 10 to 12 feet. These can be made using a long-handled groover that forms 1½-inch deep grooves in the wet concrete. After the concrete has hardened, expansion joints can also be cut with a special masonry saw and blade.

Piers. The first step, when pouring piers, is to position the *Sonotubes.* Using the layout lines as a guide, place the cardboard tubes on top of the footings and carefully backfill them with earth until they are braced, making sure they are plumb with the outside lines of the building.

Mark the height the concrete will be poured to in each tube. There are several ways to do this. One is to use a string and line level that rides on the string. Starting in the corner where the ground is highest, make a mark 6 inches off the ground on the tube. This will be the top of the foundation and the bottom of

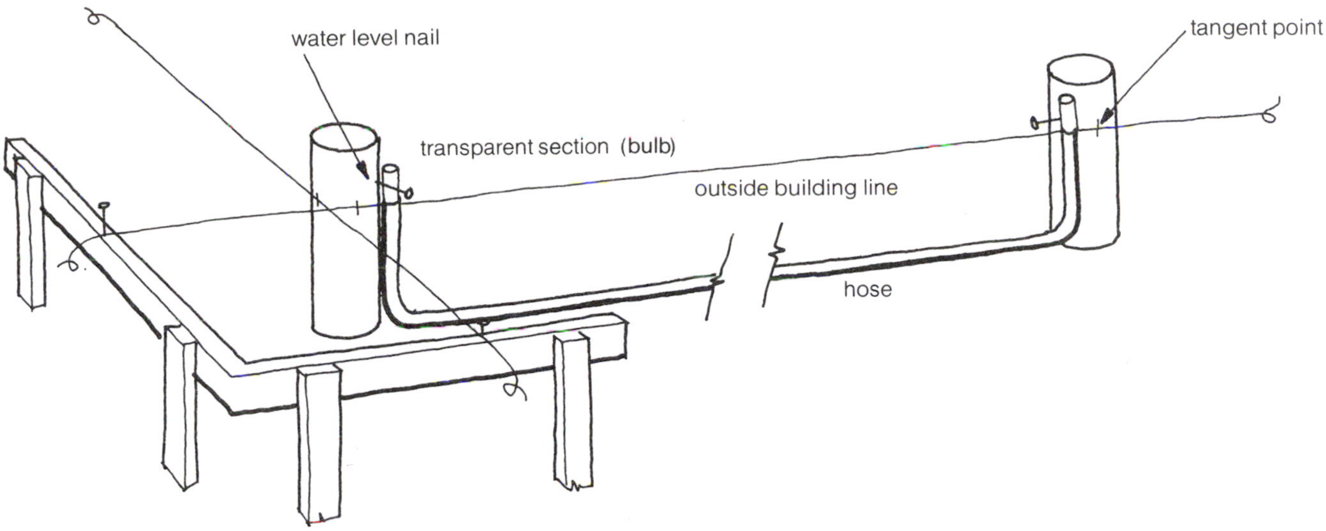

You can use a water hose to establish level points, then mark each tube with a nail. Cut off the tubes a few inches above nail markers before pouring concrete.

the sill. With the line level attached, take the string from corner to corner, marking the foundation tubes where the level string crosses them. Because the line levels are sometimes inaccurate and hard to read, a better way to check level is with a water hose and a bulb attachment at each end. Using a water level as illustrated here, you can mark the tops of all the foundation tubes exactly and quickly. When marking tubes, I usually drive a nail through at the correct height. Cut off the tubes a few inches above the nail marks and you are all set to pour. The concrete truck can simply back up to each tube, and using a chute, fill it up.

"Curing" Concrete

Concrete must harden or "cure" before forms are removed or weight is put upon it. Within 24 hours of pouring, concrete will be hard enough to take off the form boards in order to start waterproofing and backfilling the foundation. However, the concrete is still "green" and extremely susceptible to chipping or cracking.

If possible, leave the form boards on three or four days and cover any exposed concrete with burlap or old bags that can be wetted down periodically. Concrete must cure and dry slowly, otherwise it will be weak and crumbly. Try to avoid pouring concrete slabs in the hot, noon sun, and always have a water hose on hand so you can dampen the surface if it starts drying too fast.

Concrete must also not be allowed to freeze. If you are pouring in the fall or winter and there is a chance of frost, insulate the concrete with a covering of old tarps and hay. In cold weather, use heated water and aggregate to make sure the con-

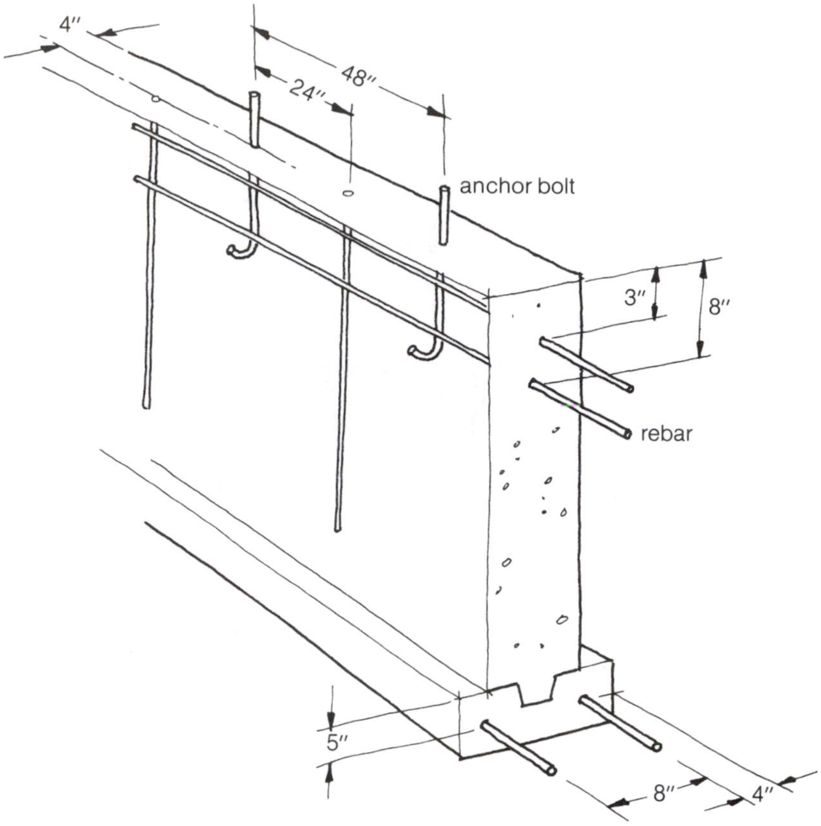

Concrete walls and footings are strengthened with rebar placed as shown here. Note the spacing of the rebar and anchor bolts.

crete is warm enough to set up. Add calcium chloride to the concrete mixture if you want to lower its freezing point. Winter concrete work is best left to experienced contractors, since it is easy to ruin an entire foundation if it does not cure properly.

Concrete Reinforcement and Anchors

Concrete footings, piers, walls and slabs should be reinforced with steel. Also, anchor bolts should be inserted into the concrete when it is wet. These secure the structure's sill to the top of the foundation.

Reinforcing helps prevent cracks due to the freezing action of the ground or large point loads such as a tractor. Two standard reinforcing materials are No. 4 reinforcing bar (rebar), and steel wire mesh, usually 6- by 6-inch No. 10. Rebar, used in footings walls and piers, is placed in the proper positions in the concrete when it is poured. The accompanying illustration shows proper reinforcement of a concrete wall. For reinforcement of slabs, see p. 31; for piers, see p. 30.

Use No. 10 wire mesh with concrete slabs. To prepare for pouring a slab, first place a 6-mil polyethelene vapor barrier on well tamped or undisturbed soil. Spread several inches of crushed stone carefully and evenly on top of the barrier, then lay wire mesh on top of the gravel. When pouring the concrete, lift the wire mesh up and position it in the middle of the slab supported by rocks, about 2-inches off the gravel.

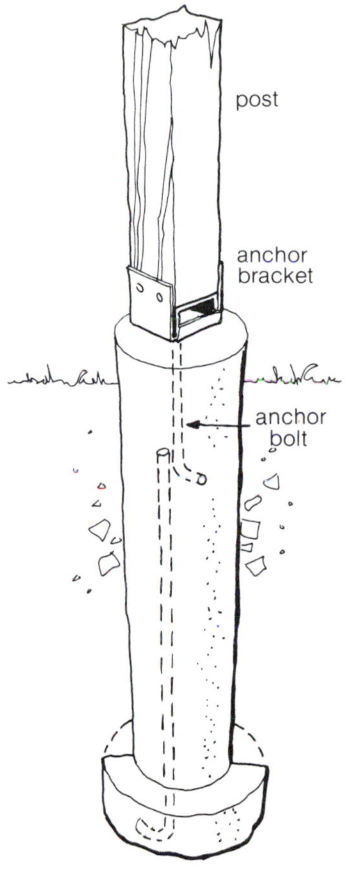

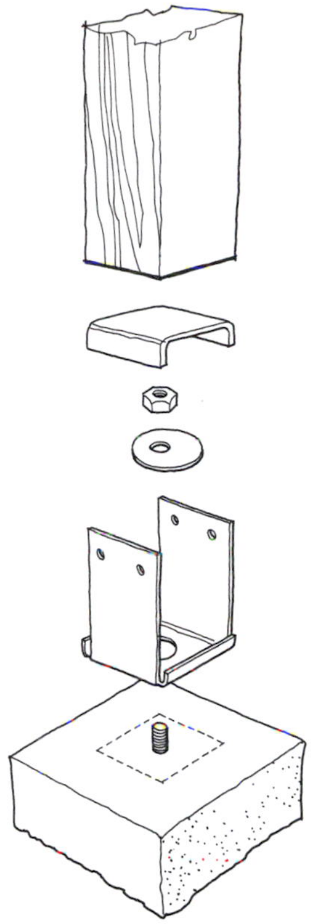

Anchor brackets like these are suitable for securing posts, used with pole buildings. The illustration at right shows an exploded view of the bracket, which permits adjustment of the post's position.

Often construction diagrams indicate that the vapor barrier should go on top of the gravel. I prefer to put it under for two reasons. When the vapor barrier is on top, it is often punctured by the gravel and wire mesh as you prepare to pour the concrete. If it has been punctured badly, the concrete can get under the plastic and form huge air pockets that force the vapor barrier to the surface. When this happens, the only thing to do is to pop all the pockets with a shovel, thereby destroying the vapor barrier.

When you pour a floating slab, use extra reinforcing ties to hold the slab and footings together.

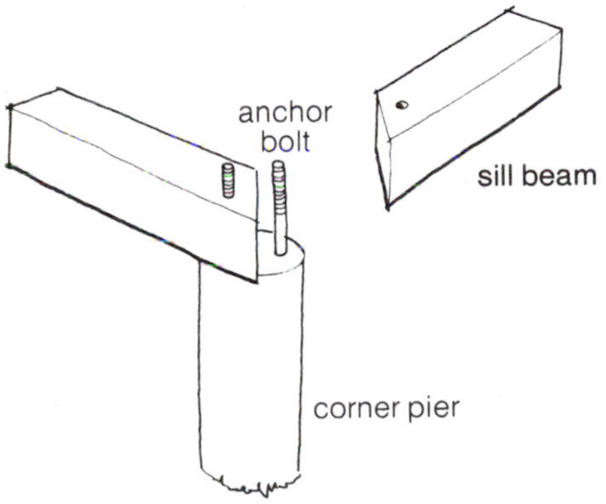

For large sill beams, embed two anchor bolts in corner piers.

Anchors. On the top of concrete slabs, walls and piers, anchor bolts are needed to attach the building's sill plates. These bolts are 8 inches and longer. They have a threaded top for attaching a washer and nut and a J-hook at the bottom which is set in the concrete.

For standard stud framing on a concrete wall, set the anchor bolts at 4- to 6-foot intervals and let them stick up out of the concrete about 3 inches. Remember to set two anchor bolts near, but not on, the corners of the building, to catch the ends of the sill plates as they meet. (If they're right at the corners, the bolts will interfere with the corner studs.) Anchor bolts must

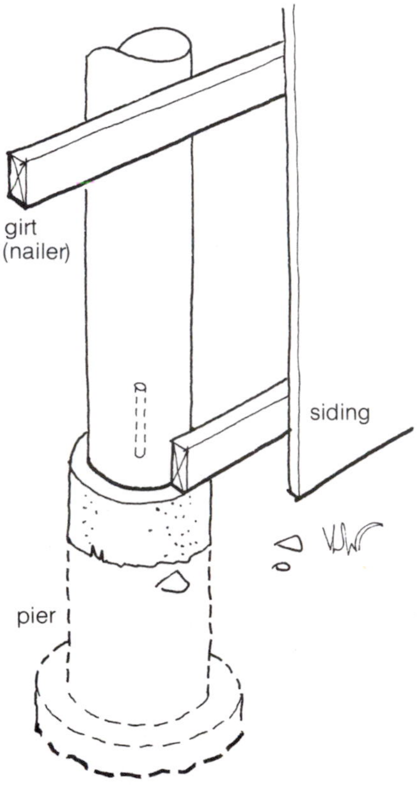

When positioning an anchors, compensate for girt width, so siding can drop down past the top edge of the pier.

be set as soon as the concrete has been poured, so make sure they are on hand before you start pouring.

Anchor bolts are also needed for pier foundations, and should be large enough to stick into the concrete at least 4 inches and to go through the sill plate. If you are using 8 x 8 beams for sills, you will need 16-inch long anchor bolts. If you cannot find bolts this long, you can make them by adding J-bends on one end of ½-inch threaded rods.

Place a bolt in the center of each tube, high enough so it will stick up an inch above the sill so a washer and nut can be put on. The only exception is the corner piers where the two sills will come together with a 45-degree bevel cut. There, two anchor bolts should be imbedded in the pier to hold both pieces of wood. In this case, the corner anchor bolts will unavoidably interfere with the corner studs. To remedy this problem, you can drill holes in the beams to "sink" the tops of the anchor bolts and nuts. Or, you can notch the corner studs, to make them fit around the anchors.

For pole buildings, there are special anchor brackets to hold the posts securely on top of concrete piers. The brackets take either square or round posts, and are slightly adjustable to help align the posts correctly. Anchor bolts can also be used as pins to secure posts, but they are not as secure and much harder to align. Usually, poles set directly in the ground on top of concrete footings require no mechanical fasteners since the poles are held in position by the backfilled earth.

Positioning anchors or anchor brackets accurately is a critical step. If the brackets are not placed correctly, the outside of the pole-building framing will not line up with the outside of the foundation piers. Always compensate for the thickness of the girts that attach horizontally to the poles and hold the siding in place. For example, if 2 x 4 girts are used, the poles must be set 1½ inches in from the outside of the piers so that the siding can drop down past the top of the pier and keep rain from collecting on its top.

Moisture and Thermal Protection

Standing water is the downfall of many foundations. Water expands clay soils, putting tremendous pressure on foundation walls, and when it freezes, the stress is even greater. To avoid heaving and cracking, always locate foundation walls in well-drained soils or install perimeter drains.

In poorly-drained soils, lay plastic perforated drainage pipe

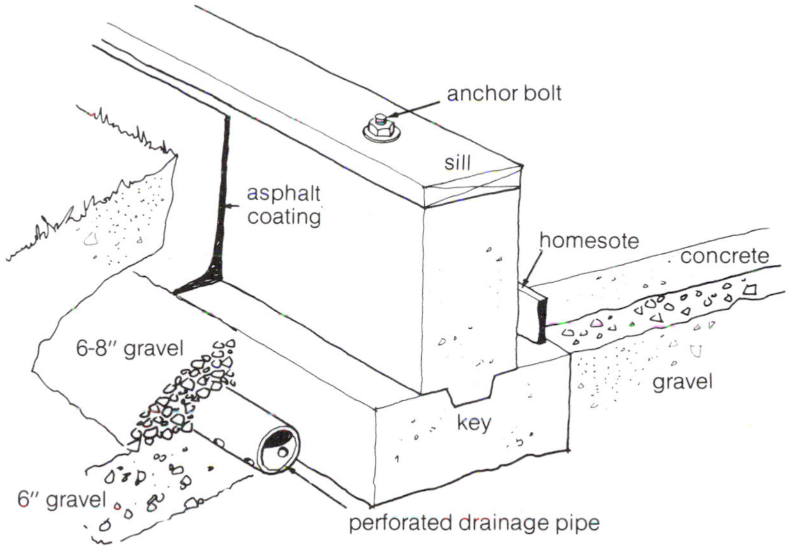

Place perforated drainage pipe (holes facing down) next to footings.

around the footings after the foundation walls have been poured and tarred. Tarring with asphalt coating helps seal and protect the walls, a good way to keep water from seeping through and wetting the interior or cellar space.

Lay the perforated drainage pipe next to the footings on at least 6 inches of clean crushed stone or pea gravel. Connect all sections of the pipe with couplings and elbows to make a continuous loop around the building. The perforated holes should face down and the loop should drain toward one corner at a minimum slope of 1 inch in 20 feet. From this corner, a discharge pipe should lead the water away from the foundation to a dry well or to a grade discharge at a lower elevation. Cover the pipes with another 6 to 8 inches of crushed stone to prevent them from clogging with mud and silt.

Insulation. Always insulate the foundation of a heated building. Even if the building only has a dirt floor foundation, insulation reduces heat loss through it and keeps the building much more comfortable. Rigid foam, known as *blueboard,* is the standard foundation insulating material. Usually it is 2 inches thick and available in 2- x 8-foot sheets with a tongue and groove for a tight fit. It is applied with a concrete adhesive.

After the foundation wall has been tarred and the perimeter drains set, use adhesive to secure the blueboard to the walls, starting at the top of the footings. Install it right up to the building's sill plate or to where the wall insulation begins. Where the foam is above ground, cover it with plywood or plaster to protect it from the sun and weather.

In northern climates, concrete floor slabs are also insulated. Standard practice is to lay down 2-inch rigid foam under the concrete at least 8 to 10 feet in from the outer walls. If a warm

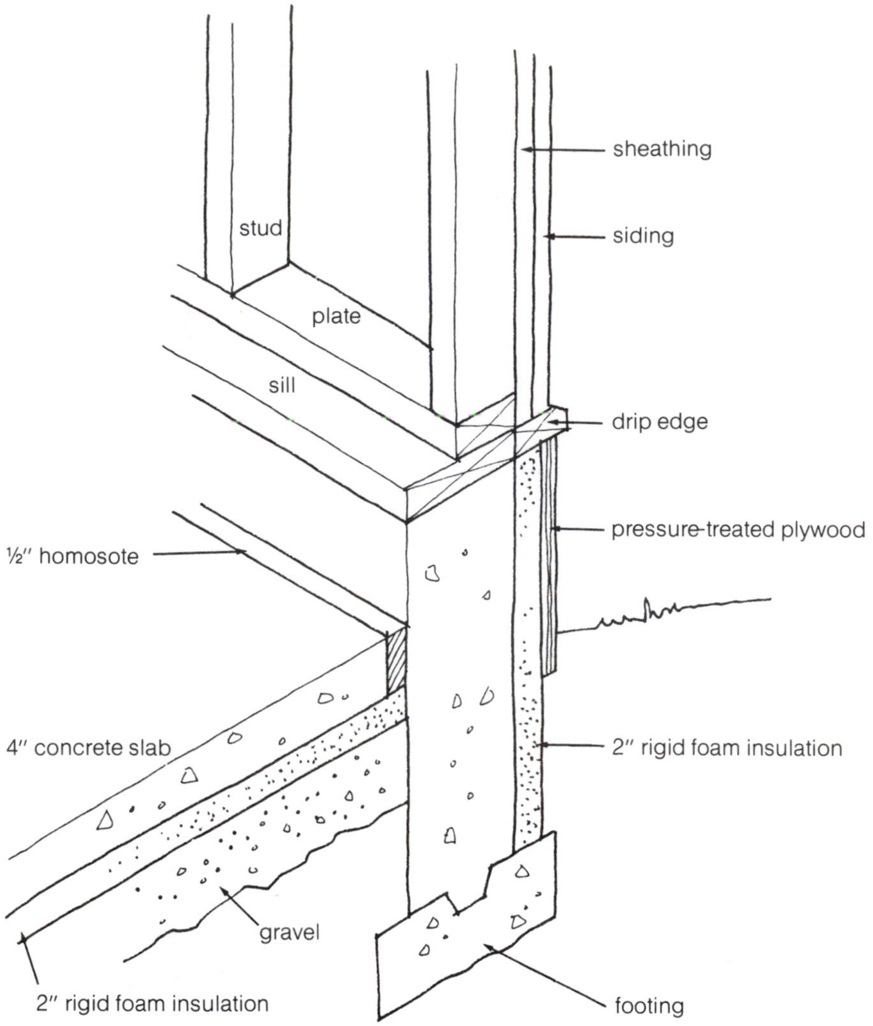

Here's a typical insulated wall foundation with insulated slab. Note that the rigid foam for the wall is protected by a drip edge and pressure-treated plywood.

slab is desired for livestock, insulate the slab completely from the ground. Foundation insulation is relatively expensive per square foot, but, in northern climates, it will add comfort and pay for itself in a few years in reduced heating bills.

Insulating pier foundations. When constructing a stud-framed barn or pole building on piers, you face a problem: how to insulate and seal the area between the bottom of the siding and the ground? To prevent rot, the sill plate and the siding are at least 6 inches off the ground, leaving a large airspace for wind and snow to enter the barn.

The solution is to nail 8-inch pieces of pressure-treated plywood onto the sides of the sill plate so the pieces are buried 2 inches in the ground. Use aluminum drip cap where the plywood meets the siding to keep water from flowing into the joint.

This plywood skirt gives you a surface to attach the interior insulation. Rigid foam boards can be glued onto the inside of the plywood between the concrete piers. Unlike fiberglass, this 2-inch thick foam will not absorb moisture and will not photodegrade since it is protected from the sun by plywood.

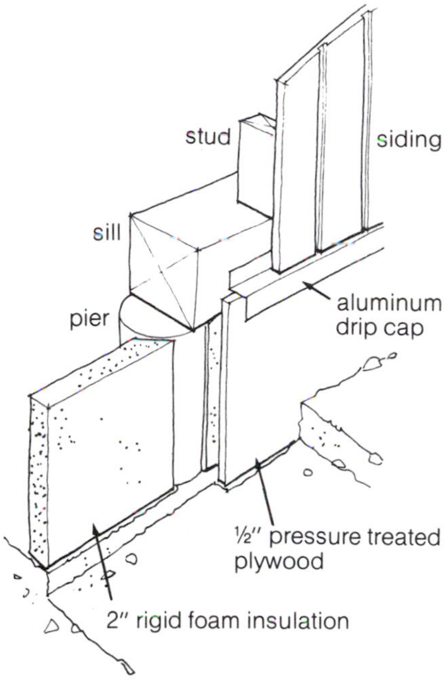

Rigid insulation for pier foundations can be installed between piers, then protected with a ridge cap and pressure-treated plywood.

Either oil or paint the wood skirt or cover it with a cement coating. The last alternative gives the impression of a normal foundation wall.

Concrete Block Foundations

Many owner-builders prefer concrete block foundations to poured walls. They're easier to build, in some ways, and they are usually cheaper. This depends on the current price of concrete relative to concrete blocks and, if you seek help from a professional mason, the cost of this service. The greatest advantage of a block foundation is that you can build your walls as time permits.

If you use a block foundation, make the wall lengths multiples of full-sized blocks if possible. This eliminates the necessity of block cutting. The standard nominal block size is 8 x 8 x 16 inches. The actual size is 8 x 7⅝ x 15⅝ inches. This allows for a ⅜-inch mortar joint between blocks and between rows.

You can strengthen block walls by cementing reinforcing rods vertically inside the block cores or by adding galvanized hardware cloth or metal tie bars to the top of a course (horizontal layer). Always add a layer of reinforcing mesh on top of the third-to-last course to strengthen the top of the wall and to suspend the mortar for the anchor bolts. Use 18-inch anchor bolts, placed every 4 feet, for block walls. Insert them in mortar, 15 inches into the wall, so that 3 inches are left to hold the sill.

As with concrete, you can mix mortar in either a wheelbarrow or a power mixer. Type M mortar, used for foundation walls, is a mix of 1 part Portland cement, ¼ part hydrated lime and 3 parts clean sand (80-pound bags of mortar mix are also available from building suppliers). When mixing mortar, always remember to dry mix it first, add just enough water to give it a pliable consistency, and then test it for proper stiffness with a hoe. If it is pliable, yet pulls up easily with a hoe and stays without slumping, it is mixed properly.

Laying blocks. After completing the footings, put your building strings back on the batter boards to re-establish the outside corners of the foundation. At the point where the strings intersect, drop a plumb bob to determine the exact corners of the foundation wall. Place one block at each corner, then position strings from the top outside corner of one block to another. This will give you a straight and level line for laying the first row of blocks.

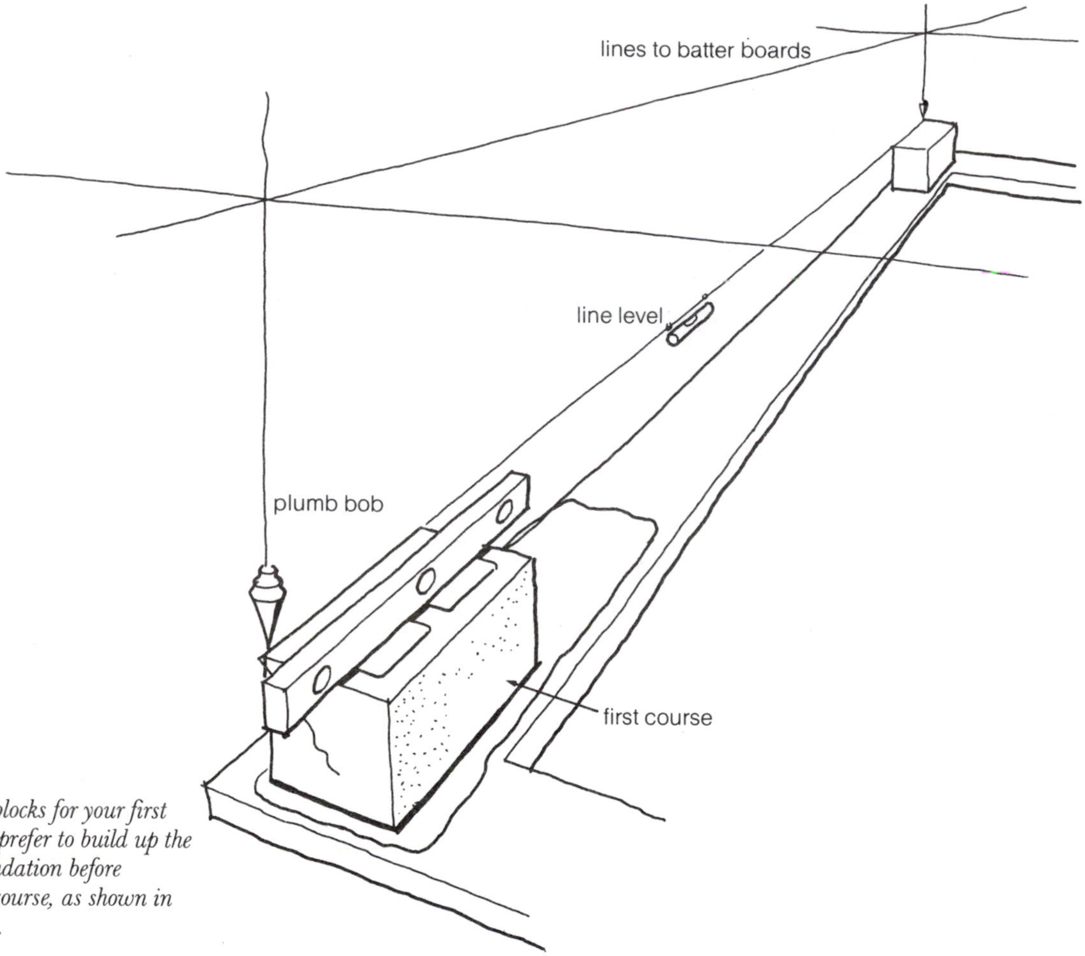

Here's how to set up blocks for your first course. Some masons prefer to build up the corners of a block foundation before completing an entire course, as shown in the photos on page 53.

Mix only enough mortar to use before it hardens (an hour or so), and begin at two of the marked corners. Trowel about ½ inch of mortar under the two corner blocks and, using the string and a line level, adjust them until they are straight and level with each other. After leveling the blocks, you should have a ⅜-inch mortar joint beneath each block.

Smear mortar on the end of a new block and position it against the corner block, again making sure it is level with the first block and straight along the string. Continue laying the first course in this manner. The last block may have to be cut if your joints were too thick or the wall dimensions uneven with multiples of 8 inches. If so, use a mason's hammer or brick chisel to score the block on both sides until it cracks along the desired line.

After finishing the first course, position your strings for the second course and proceed as with the first. Blocks at the corners of the walls should overlap, so that the vertical block joints do not line up. Use a carpenter's level as you build up the corners to insure they are plumb.

Stretch a mason's line from corner to corner, and, to complete the first course, lay the top outside edge of each block to this line.

When installing the last block, in each course, butter all edges of the opening block with mortar and all four vertical edges of the closure block, then carefully lower closure block in place.

Handling the blocks correctly is important. By tipping the block slightly toward himself, the mason is able to see the upper edge of the course below, so he can place the lower edge of the block directly over the course below.

If work is progressing rapidly, mortar scraped from joints may be applied to face shells of block just laid. If there are delays, mortar should be reworked.

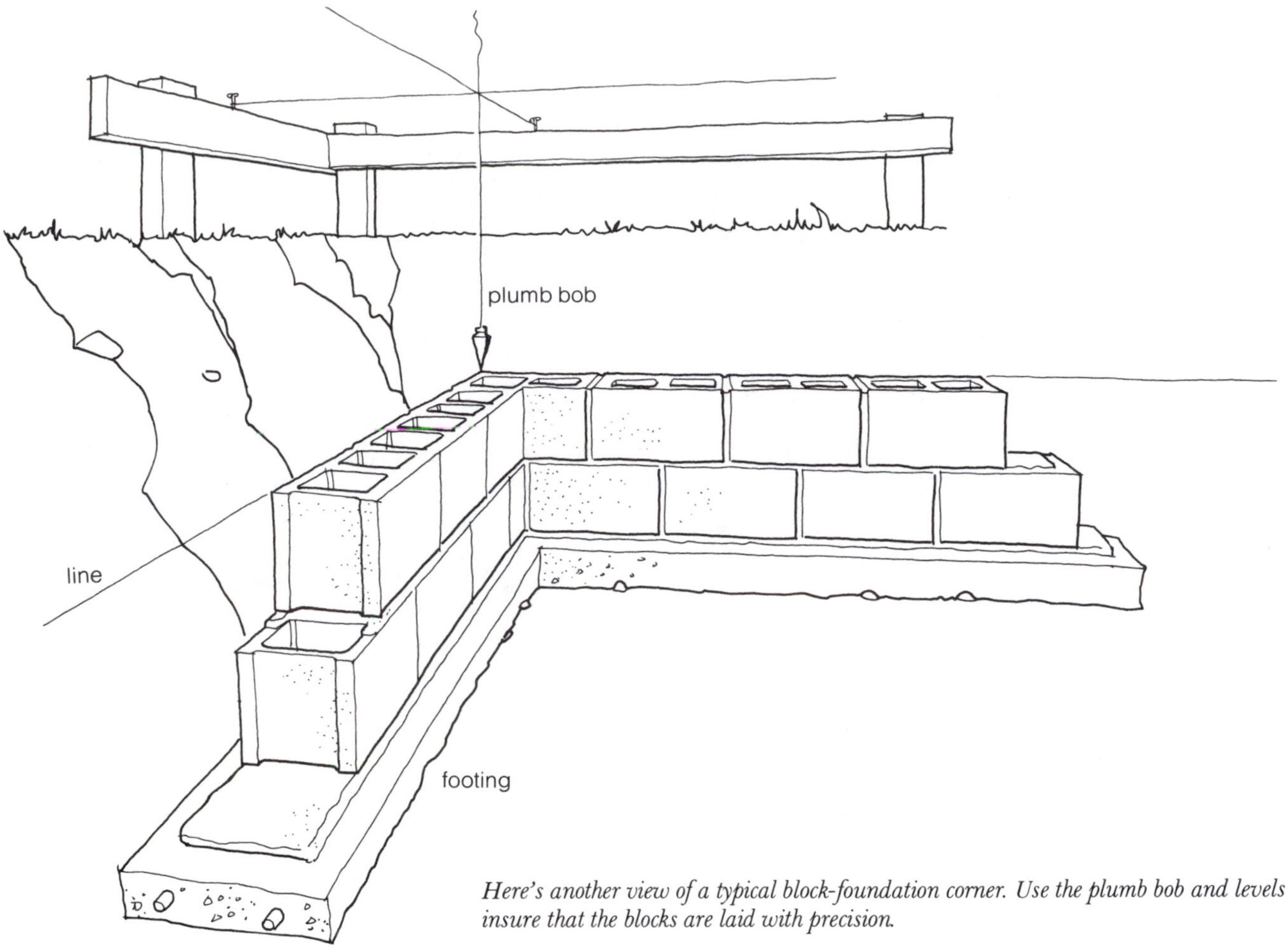

Here's another view of a typical block-foundation corner. Use the plumb bob and levels to insure that the blocks are laid with precision.

Before the mortar dries, use the trowel to strike off excess mortar from the joints. Then smooth the mortar with a jointing tool and create concave recesses along the joint lines. This action packs the mortar joint and helps create a strong, secure wall.

After laying a block foundation, waterproof it with tar just as poured concrete walls are. In northern climates, the walls should also be insulated, and this can be done in two ways. Either apply rigid foam to the outside of the wall or pour loose-fill vermiculite insulation down the block cavities.

Grading

The final step in constructing a foundation is to backfill the foundation trenches or holes and grade the building site. These are important steps for a long-lasting foundation. Proper backfill material and grading will keep water away from the foundation and minimize the possibility of frost cracks and heaving.

Backfilling is simply a matter of placing excavated dirt back against the foundation wall. But, do it carefully to avoid damaging the foundation or insulation. Large rocks should be placed, not thrown, against the wall. This is especially important if the concrete is still green. Use only clean fill. Wood and other organic matter will decompose and leave pockets. If drainage is a problem or you have heavy clay soils, backfill with sand or gravel to protect the wall from freeze expansion.

After the walls have been backfilled and tamped down, do the final site grading. To enhance drainage, make the ground slope away from the foundation. If this is impossible because of an uphill slope on one side, dig a drainage, trench two feet out from the foundation wall to catch surface water and carry it away. During the final grading, use the topsoil that was set aside during excavation. Spread it evenly, seed it to grass and cover it with a thick layer of hay mulch to prevent soil erosion. The grass will grow through the mulch.

CHAPTER 4

Wood Framing

Grant Heilman Photography

TODAY, WOODEN BARNS are built by three principle methods, *platform, post and beam* and *pole* framing. Platform or stud framing is a relatively recent technique dating back to the turn of the century. For the most part, it has replaced other types of framing because of its economy of labor and material. Platform framing uses dimensional lumber, ranging in size from 2 x 4s to 2 x 12s, regularly spaced, to frame walls, floors and roofs. Post and beam framing is an older, traditional method of barn construction, dating back to medieval times. With this method, large beams and posts, such as 6 x 6s and 8 x 8s, form the building's frame. Pole framing is the oldest method of all, dating back to the Stone Age and widely used by Indian cultures in North and South America. Today, pole building is facilitated by machine-cut and pressure-treated poles or posts. Also, it is usually combined with platform framing of floors and roofs.

Framing Methods

Platform Framing

Platform framing is the most widely-used method of wood construction for small barns, sheds and shelters for several reasons. It requires less labor and materials than post and beam framing, and construction details are simpler. Because the individual framing members are relatively light and small, one person can erect an entire building alone. Kiln-dried lumber for platform framing is fairly expensive, but this high cost can be avoided by using native, rough-sawn lumber from local sawmills.

The accompanying figures show details and terminology for conventional platform framing suitable for barns and outbuildings with and without ground-level floors. In either case, the framing begins with *sill plates* laid on the foundation.

If a floor is going in, *joists* are laid on top of the sill plates and then boxed in with *joist headers.* For large buildings, a *girder* is used to support the flooring joists at their midpoint. Girders are usually 6 x 10s or larger beams. The beams and joists form a deck on which subflooring (usually plywood) is laid. *Sole plates* are nailed to the subflooring, and *studs* are nailed on top of the plates to form the vertical support elements of the walls. *Double top plates* are nailed on top of the studs and then the whole process starts again for the second floor. *Rafters* set directly on top

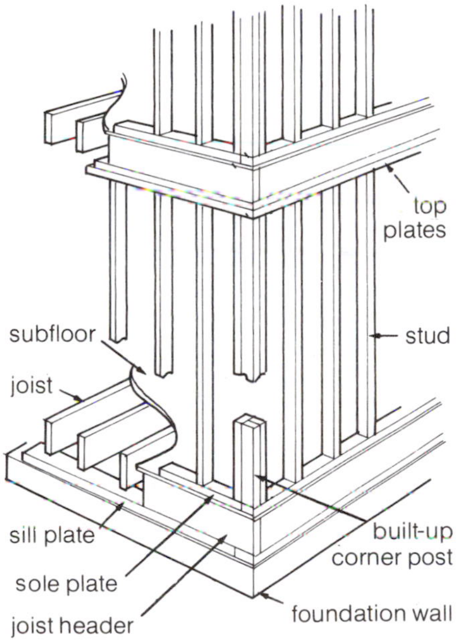

Platform framing with a first floor.

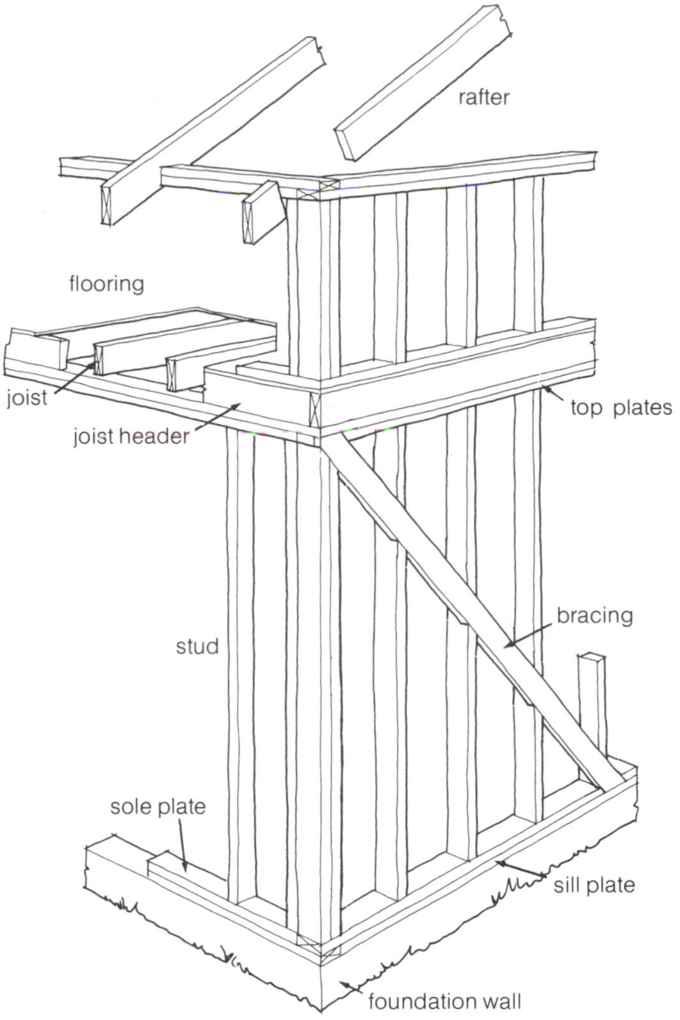

Typical platform framing for a small barn or outbuilding without a first floor.

of the topmost wall plates extend up to a ridge plate to form the top of the roof triangle.

Post and Beam

While post and beam framing requires more materials and skill, it has its advantages. Post and beam structures are extremely strong and durable. If properly joined, framing members can support tremendous loads. Also, they are extremely fire resistant since it takes so long to burn through a solid beam. Another advantage to post and beam framing is its flexibility. Large windows and doors can be framed easily into walls without worrying about headers and other supports because the entire load is carried by the individual posts. Finally, post and beam construction lends itself to a traditional barn raising. After the timbers have been cut to length and notched to fit together, the entire structure can be assembled in one day by a gathering of friends and neighbors.

The *sill beams,* usually 8 x 8s, are the first members to be put in place in post and beam framing. *Girts* are notched into the sills and serve the same purpose that girders do in platform

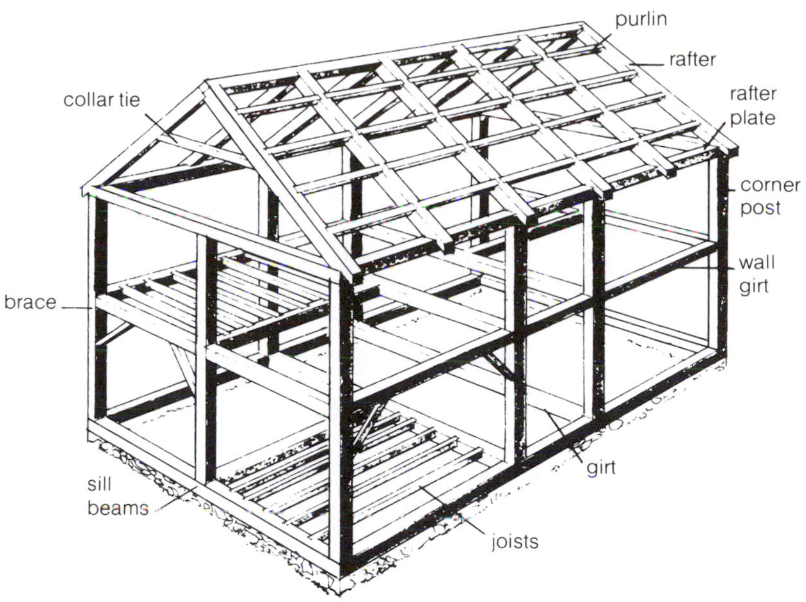

Post and beam structures are distinguished by heavy beams, typically 6 x 6s or 8 x 8s, that provide strength and durability.

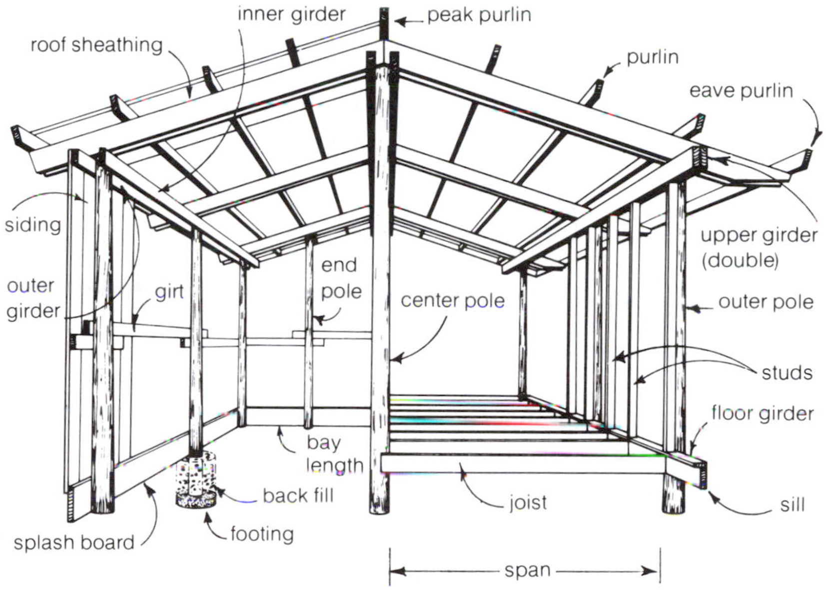

Pole buildings may be framed two ways. The left side shows a widely-used method for barns and outbuildings. The right side is for more finished buildings with suspended floors.

framing. *Joists* between the girts support the flooring. *Wall girts* joining the *posts* at the second floor level are braced diagonally to make the structure rigid. The *rafter plate* supports 4 x 6 or larger *rafters,* and 4 x 4 *purlins* set between the rafters support the roof decking. As in platform framing, *collar ties* keep the rafters from sagging in the middle or spreading out at the bottom.

Pole Framing

Pole framing is much like post and beam construction, using poles or posts set on concrete piers or footings. The beams, however, are replaced by nailed-on or bolted *girts* and girders of dimensional lumber. Floors and roofs are usually framed with dimensional lumber just as in platform framing.

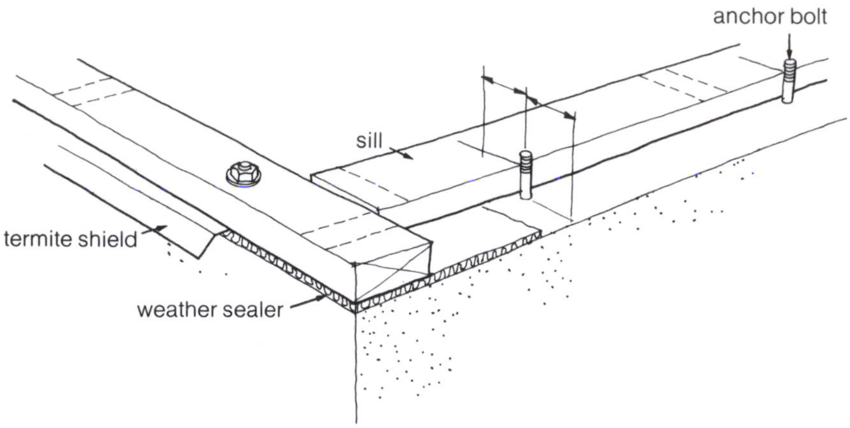

Use a framing square to accurately place anchor bolt holes in sill plate. Dotted lines indicate stud positions.

Proper nail angles for toenailing; use four 8 d nails.

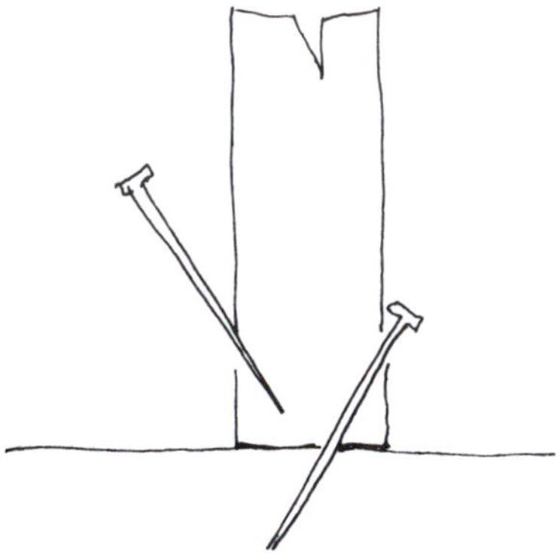

TOENAILING

Toenailing is one of the hardest things for beginning carpenters to learn. This is because toenailing involves holding a piece of wood in place while starting a nail and driving it at an angle. To toenail a stud, first start an 8d nail about 1½ inches from the bottom at approximately a 60-degree angle. Position the stud on the plate and gently but firmly drive the nail into the stud until the nail contacts the plate. If the stud moves, reposition it and then sink the nail fully into the stud and plate. Once the first toenail is in place, the others are much easier to nail since the stud is secured. Never use nails larger than 8d for toenailing studs; anything larger will almost certainly split the wood.

Platform Framing

Because platform framing is a particularly suitable building method for the owner-builder, I discuss it here in detail. Many of the construction details of platform framing may be used in conjunction with post and beam and pole framing, which are covered in general terms at the end of the chapter.

Sill Plates and Beams

After the foundation is in and the concrete has cured for several days, the building frame is started by setting the sill plate or beams. The sills are the building's connections to the concrete foundation.

Use 2 x 8 stock for the sill plate on top of full wall foundations. The 2 x 8 should be flush with the outside of the concrete wall and secured every 4 feet with anchor bolts. To set the sill plates accurately, first cut enough 2 x 8s to cover the top of the foundation and then hold them in place alongside the anchor bolts. The placement of the holes can then be marked accurately using a framing square to find the distance of the bolts from the outside edge of the wall. The holes should be the same distance from the outside edge of the plate. Drill the holes the same size or slightly larger than the anchor bolts and secure the plates with washers and nuts.

For a barn built on a pier foundation, either a built-up 8 x 8 beam or a solid 8 x 8 is used for the sill to carry the building load between piers. A built-up beam is easier to work with, and if properly constructed, is almost as strong as a solid beam. To assemble a built-up 8 x 8, first attach a 2 x 8 plate to the top of the piers and secure it with the anchor bolts. Nail together five layers of 2 x 8s, the length of the building. Make sure, in succeeding layers, to overlap the joints or places where 2 x 8s butt together end to end. This way, the beam will be stronger be-

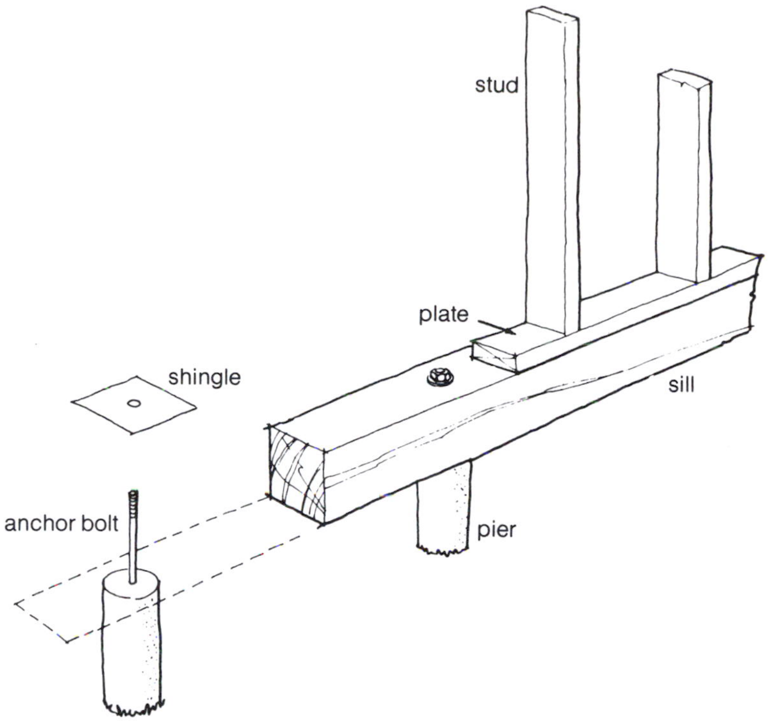

Asphalt shingles help prevent moisture from a concrete pier from seeping into a heavy-beam sill.

Select these nails for framing at points shown.

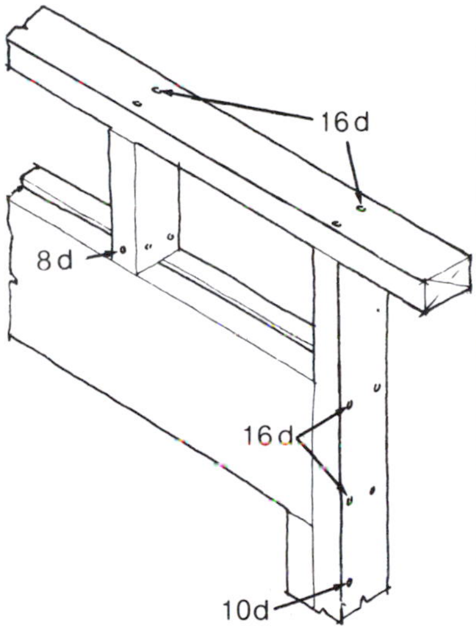

NAILS, NAILS, NAILS

Choosing the proper-sized nail in wood framing is very important. If the nail is too small, it will not hold properly; if it is too large, it may split the wood.

For exterior nailing or working with wet, green lumber, always use galvanized nails coated with zinc to prevent rusting. Galvanized nails also have superior holding power so use them wherever there is a chance regular nails might pull out. For pole building, there are special grooved nails that have even better holding power.

cause one layer's joints will not fall directly above another's. Nail each layer on with three 10d nails every 12 inches.

The built-up 8 x 8 can then be tipped on edge and toenailed into the bottom plate with 8d nails to hold it securely on top of the piers. Where the anchor bolts stick up, the built-up beam will have to be notched to fit over them. Simply drill holes with an electric drill to form pockets for the bolts.

Always separate wooden sills from the concrete foundation walls or piers by a moisture-proof material. Concrete absorbs water and will pick up moisture which will eventually rot the sills. Special foam insulating material is made for residential construction. It separates the plate from the concrete and seals any air gaps as well. For outbuilding construction, I use old asphalt shingles; these provide an excellent and rugged moisture barrier when placed under sills.

Flooring

Flooring joists. If a barn is to have a wooden floor, the next step after setting the sill plates is to lay out and attach the floor joists. (If you plan to have a dirt, gravel or concrete floor, you may overlook the following sections on flooring and proceed to "Wall Framing.") The size and spacing of the joists depend on the amount of weight the floor will carry and the span of the individual joists. Table 4-1 lists the recommended joist sizes and

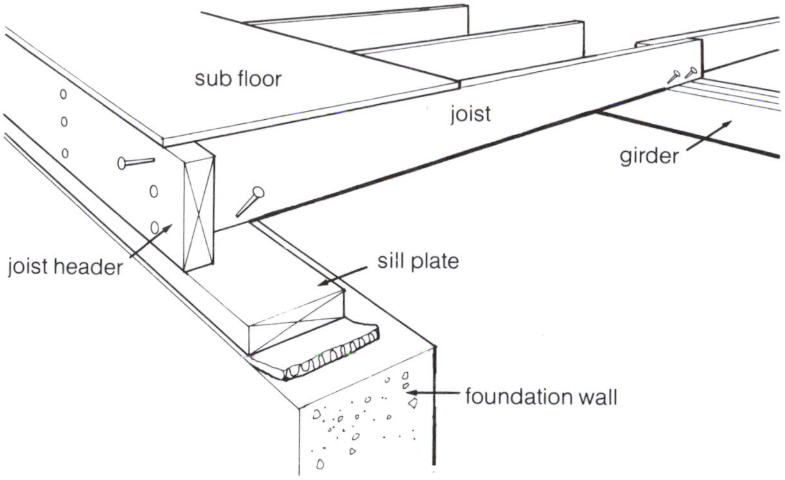

Attach floor joists so their tops are the same height as the joist header. Use a girder to support wide floor spans.

spacing for various spans and load. Normally, floors are built to carry a "live load" of 40 pounds per square foot. For heavy equipment, this should be increased. For more information and assistance, consult your nearest Extension Service office.

Table 4-1 Span of Joists

Span calculations provide for carrying the live loads shown and the additional weight of the joists and double flooring

Size	Spacing	20 lbs. Live Load	30 lbs. Live Load	40 lbs. Live Load	50 lbs. Live Load	60 lbs. Live Load
2 x 4	12″	8′- 8″				
	16″	7′- 11″				
	24″	6′- 11″				
2 x 6	12″	13′- 3″	14′- 10″	13′- 2″	12′- 0″	11′- 1″
	16″	12′- 1″	12′- 11″	11′- 6″	10′- 5″	9′- 8″
	24″	10′- 8″	10′- 8″	9′- 6″	8′- 7″	7′- 10″
2 x 8	12″	17′- 6″	19′- 7″	17′- 5″	15′- 10″	14′- 8″
	16″	16′- 0″	17′- 1″	15′- 3″	13′- 10″	12′- 9″
	24″	14′- 2″	14′- 2″	12′- 6″	11′- 4″	10′- 6″
2 x 10	12″	21′- 11″	24′- 6″	21′- 10″	19′- 11″	18′- 5″
	16″	20′- 2″	21′- 6″	19′- 2″	17′- 5″	16′- 1″
	24″	17′- 10″	17′- 10″	15′- 10″	14′- 4″	13′- 3″
2 x 12	12″	26′- 3″	29′- 4″	26′- 3″	24′- 0″	22′- 2″
	16″	24′- 3″	25′- 10″	23′- 0″	21′- 0″	19′- 5″
	24″	21′- 6″	21′- 5″	19′- 1″	17′- 4″	16′- 0″

This shows maximum safe spans for high-quality wood joists. Values vary for different species of lumber; always check local building codes for exact data.

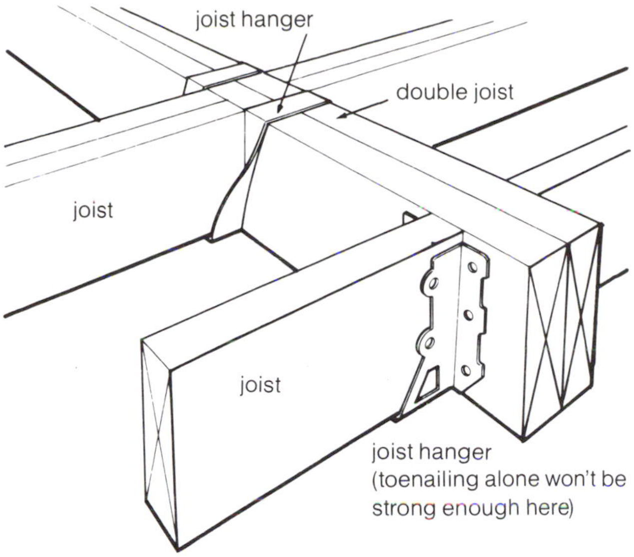

In framing situations like this, or when attaching joists to a beam, use joist hangers.

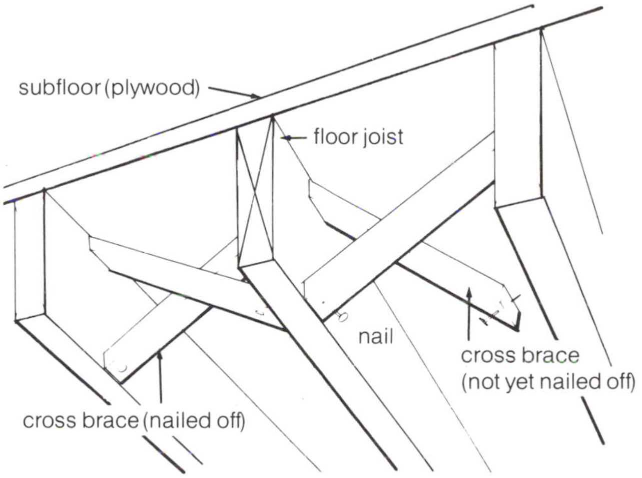

Cross bracing or blocking is recommended for long joist spans. Braces may be cut from 1 x 3 strapping. Blocks may be cut from scrap lumber.

Using a tape and a combination square, mark the position of the joists on the sill plate. The joists will be either 16 or 24 inches on center (o.c.) from the *end* of the wall. This is a modular spacing that allows you to use 4 x 8 sheets of plywood for the subflooring and have a nailer every 4 feet for the edge of the sheets. If you are careful in your layout, the joints of the plywood will land exactly half way on a joist, giving you a ¾-inch nailing surface. If you are sloppy, the plywood will run off the joists and you will have to cut each piece individually, thereby wasting a lot of plywood and time.

After marking the sill plates, cut the joist headers to length and toenail them in place. Then cut the joists to fit between the headers. Nail the joists with three 16d nails driven through the header and several 8d nails toenailed into the sill. Always make sure the *crown* or bow in a joist faces up so the weight of the floor straightens it rather than causes it to sag. Also, joists should be perfectly plumb before they are attached to the header. Check this with a combination square set on top of the joist header.

If the building is too wide for the joists to span the entire width, then put in a girder, a solid or a built-up beam that supports the joists at their midpoint. Girders can be installed on top of the foundation wall. In this case, the joists must be the same width as the girder and fastened to it with metal joist hangers. The girder should be supported by concrete footings or piers placed inside the building.

If you are using a pier foundation with large sill beams, attach the joists with metal joist hangers just as they would be to a girder. Simply lay out the joist marks on the beam, set the hangers, and slip the joists into them. When you are nailing the hangers in place, make sure they will hold the joists flush with the top of the beam.

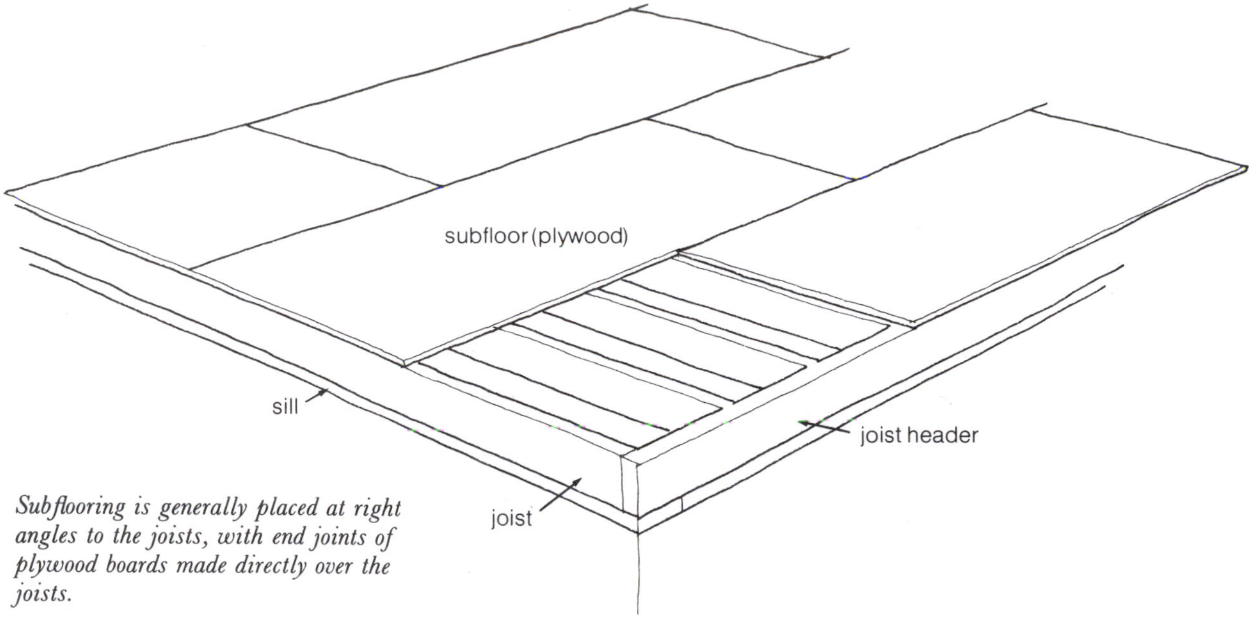

Subflooring is generally placed at right angles to the joists, with end joints of plywood boards made directly over the joists.

For joist spans of 12 feet or more, it is often recommended to brace the joists with blocking or wooden cross braces which keep the joists from twisting. Metal cross bracing is also available that goes on very quickly. Braces or blocks should be set in the middle of the joist span, taking care to preserve the 16- or 24-inch spacing of the joists. Also, if you anticipate having a partition that will run parallel to, but not directly over a joist, now's the time to add blocking that later will be nailed into when the partition is secured to the floor.

Subflooring. After nailing the joists in place, lay down the subfloor. Subflooring can be either plywood, 1-inch boards or 2-inch planks. Plywood is most often used because it is stronger, lighter and goes on faster. Use ⅝-inch plywood for floors that will be covered with another layer of wood for finished flooring. Use ¾- to 1-inch plywood if it is to be the finished flooring.

You can also use rough-sawn lumber for subflooring, a much less expensive choice than plywood. Rough-sawn, 1-inch boards laid across the joists for subflooring and then 1-inch boards laid perpendicular to these for the finished flooring make a durable and economical barn floor. For floors that must support heavy loads, you can use 2 x 6 planks for the subflooring.

Wall Framing

With the floor deck completed, the walls can be framed and raised in place. If the outbuilding has only a dirt or gravel floor, or a poured concrete slab, the walls are built directly on top of the sill plate.

Walls consist of a *bottom sole plates, studs,* and *double top plates*. Rough openings for windows and doors are framed with *headers,* and *jack studs* or *trimmer studs.* Walls can be built with 2 x 4 studs spaced 16 inches on center or 2 x 6 studs spaced 24

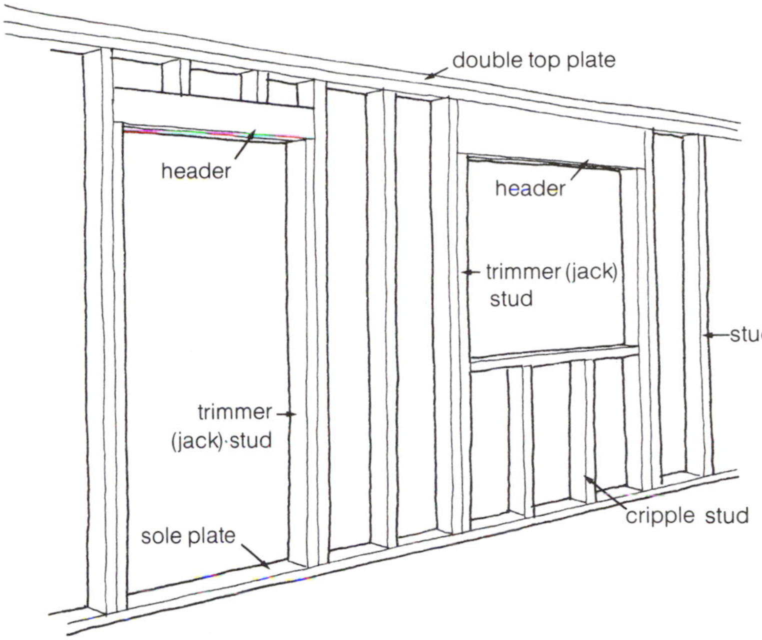

Wall framing includes special studs and headers for windows and doors. For more on laying out walls, see next page.

USING NATIVE LUMBER

If you decide to use low-cost, native, rough-sawn lumber, the first problem you'll encounter is uneven dimensions. One 2 x 6 may be 5¾ inches wide and the next 6¼ inches. If you use this lumber without taking these differences into account, you are likely to end up with uneven walls and floors.

For wall framing, the solution is to line up all the studs flush with the *outside* of the wall plates. This will give you an even exterior wall surface for attaching siding. The inside line of the wall may be a bit wavy, but for barns this is usually not a problem since the studs will be left uncovered. If the interior wall surface is to be finished, it can be evened by attaching horizontal strapping which is shimmed out with shingles.

Laying floor joists is a little more complicated. Always select pieces of oversized stock for the joist headers that are as wide or wider than the rest of the joists. By doing this, the joists will always be the same size or narrower than the headers and won't stick up above them and interfere with the flooring. Simply nail all the joists in place flush with the top of the header. If they are too narrow to sit on the sill plate, slip a shim shingle underneath them until they are supported firmly. This will give you a smooth and level floor deck.

inches o.c. I prefer 2 x 6 studs for barns because there is less labor in cutting and nailing, the walls can be better insulated and the extra cost of the lumber is negligible.

The first step is to find two very straight pieces of lumber for the top and bottom wall plates. If the plates are crooked, your wall will be too. Cut the plates to the length of the wall, or, if the wall is over 16 feet long, cut two or more sets of plates and build the wall in sections. A 16-foot wall, 8 feet high, is about the biggest section two men can lift in place.

Put the plates side by side, and, starting from one end, mark the plates for studs either 16 or 24 inches on center depending on your stud size. One of these plates will be the bottom and the other the top of the wall. Because they are symmetrical, they can be laid out together side by side. With a combination square, go back and make lines across both plates ¾ of an inch on either side of the center marks. These mark the outside edges of the studs.

Rough openings for windows and doors must also be laid out on the plates. These are called rough openings because they are wider than the actual window or door. This extra space leaves room for *jambs,* the pieces of wood that encase a window or door. Usually, rough openings are 1¼ inches larger than the window or door on all four sides to allow for a ¾-inch wide jamb and ½ inch for shims. The shims make the jambs perfectly level and plumb. For preassembled windows and doors that come with jambs attached, leave only a ½ inch on all sides for shimming. Because of the need for accurate rough openings, always select windows and doors and calculate their rough openings before wall framing begins.

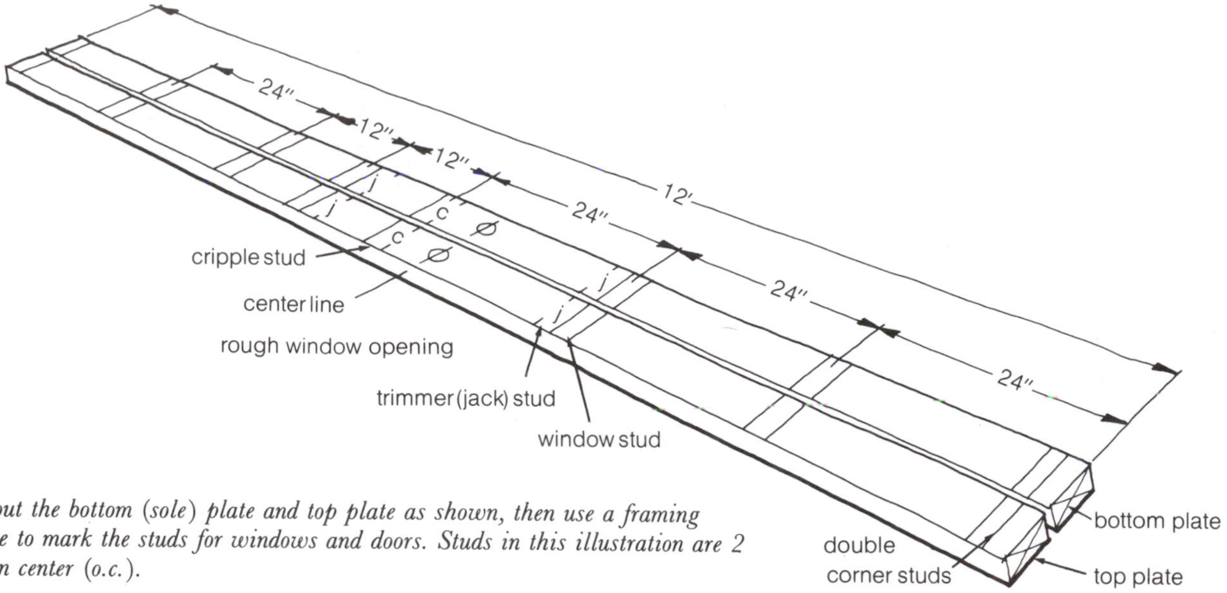

Lay out the bottom (sole) plate and top plate as shown, then use a framing square to mark the studs for windows and doors. Studs in this illustration are 2 feet on center (o.c.).

LAYING OUT WINDOWS AND DOORS

The position of windows and doors should be marked clearly on a building's blueprints or working drawings. Normally, the centerline of the unit and its rough opening are indicated on the drawings. If a rough opening is not indicated, you must get this information from the manufacturer or building supplier. If you are custom building the windows and doors yourself, prepare a detailed drawing showing the construction of the unit, its overall dimensions and how it is to be mounted so you can determine its rough opening.

Rough openings for site-built doors and windows are normally 2½ to 3 inches wider and higher than the unit itself. This leaves room for installing and leveling ¾-inch jambs which form the box the window or door sits in. For large doors that require greater support, 1¼-inch jambs are often used, and this must be taken into account when framing the rough opening. The thickness of thresholds (if used) must also be considered when figuring the rough-opening height for doors. In general, leave ½ inch more on all sides than the width and height of the unit with jambs attached. This will give you room to adjust the unit with shim shingles until it is level.

To lay out rough openings, locate the center of the window or door on the wall plates that are being marked off. Take half the width of the rough opening and measure back on each side of the centerline and make a mark. This marks the inside of your jack studs. Mark another line 1½ inches outside of this and another 1½ inches from that. These lines mark the placement of your jack stud and the window or door stud. The lines for the jack stud should be marked with a large J to remind you that this is not a full stud. Any of the regularly-spaced stud marks (either 16 or 24 inches o.c.) that fall within the rough opening of windows should be marked with a C to remind you that these are cripple studs that only extend up to the window sill. It is important that these cripple studs be placed at the regular stud interval to serve as nailers for the wall sheathing.

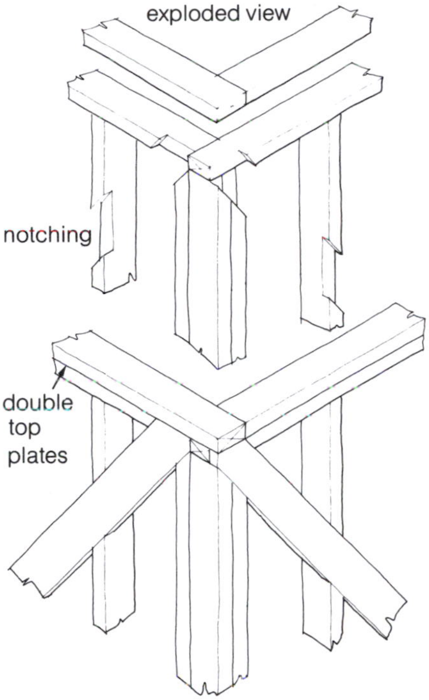

Diagonal bracing may be notched into studs to add strength. This is recommended for walls without plywood siding.

Stud assembly. After laying out the plates, build the walls on the platform or ground and then lift them into position. Cut the studs to the correct length and then secure them with two 16d nails through the top and bottom plates. The rough openings should be framed with: *headers,* which carry the load over windows and doors; *jack studs,* which support the header; and *cripple studs,* which support the rough sill. Headers can be made from two pieces of 2 x 6 or 2 x 8 lumber, depending on the load to be carried. One piece is nailed flush with the inside, and, the other, with the outside of the wall. Nail the headers well into the top plate and the two side studs with 16d nails. Then support the headers with jack studs.

When the framing is complete, brace the wall before lifting it into place. All stud walls must be braced to keep the studs at right angles to the plate and to make the wall a rigid structure. If the walls are going to have plywood sheathing or siding, this will act as a brace. If they are not covered with plywood, diagonal 1 x 4 bracing must be nailed across the studs.

First, square up the wall section using a tape to measure the diagonals. Use a sledge hammer to tap the corners of the wall until the diagonals are equal. If you use plywood sheathing, attach this to the outside of the wall to brace it. Remember to leave enough overhang on the bottom of the wall so the plywood will hang down and cover the top of the foundation by at least an inch. If you are using diagonal 1 x 4s, nail these directly to the *inside* of the wall so they won't interfere with the exterior siding. Or, if both sides of the wall are to be sheathed, cut notches in the studs for the 1 x 4s so they are recessed.

Erecting walls. Once braced, the wall is raised into position. Usually this is a two- or three-person job, since the wall must be held level while temporary braces are set to hold it upright. Line up the bottom wall plate with the outside edge of the building and secure it with 16d nails. Drive these through the plate and into the subflooring and joists or sills below. While one person holds the wall plumb using a carpenter's level, another can attach 2 x 6 braces to the edge of the wall. Arrange the braces so they run out at a diagonal to stakes in the ground. This temporary bracing holds each wall section plumb until all the sections are built and tied together with the top double plate.

Build, raise and brace the other wall sections in the same manner. When laying out the side walls, be sure to take into account the width of the front and back walls which the side walls butt against. Add the width of the front wall when marking off

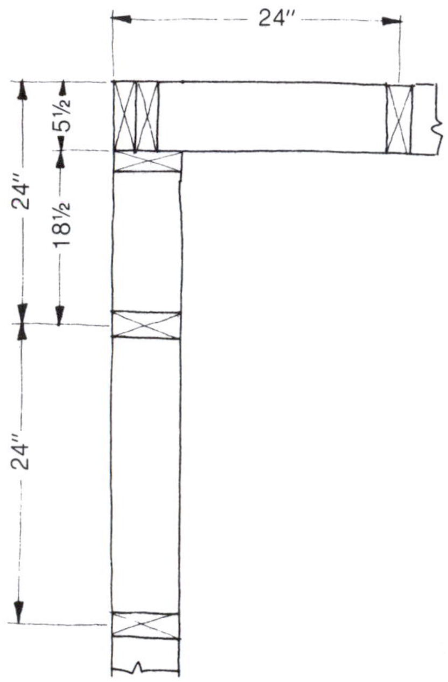

Sidewall layout for 2 x 6 studs (1½ x 5½ inches actual dimensions), spaced 24 inches o.c.

the side wall plates so that the second stud from the end is 16 or 24 inches o.c. from the outside of the building, not the end of the plate. This will keep the sidewall studs on center for plywood sheathing. In conventional residential framing, two studs and blocking are often used for corners. One advantage of this design is to provide needed interior nailing surfaces for sheet rock. However, this is usually unnecessary for small barns and outbuildings.

When all the wall sections have been built and braced, attach the top double wall plate. This top plate should overlap any joints in the lower plate and overlap the corner joints. This will tie the wall sections together securely and give a rigid box structure that can carry the second floor or roof rafters.

Partitions. Normally, interior wall partitions are made of 2 x 4 stock and built in the same manner as outside walls. The plates are laid out, the studs nailed in place on the ground, and then the wall raised into position. If the wall runs at right angles to the floor joists, simply nail the plate into these through the subflooring. If the wall runs parallel to the joists, but not directly on top of one, it is necessary to block between the joists under the wall to support the plate. This, of course, must be done before the subflooring is put down.

Partitions for pole barns or barns built without a floor can be erected in at least two ways. Either poles can be driven into the ground to support a fence-like wall or concrete piers can be poured inside the barn to support load-bearing partitions.

With the first-story walls up, the second-floor joists and subflooring can be built if there is to be a hayloft or storage area above the first floor. If the barn has only one story, then the roof rafters are framed on top of the double plate.

The Right Roof

The choice of a roof style depends on the nearby architecture, the climate and the building's function. The most common type is the *gable* roof which is widely used because of its simplicity and adaptability to any climate. A gable roof has two equal pitches that meet at a center ridge. The *shed* roof has a simple continuous slope from the front of a building to the back. It is most often used for animal stalls, woodsheds or other narrow buildings that require high front and low back walls. *Salt box* and *gambrel* roofs are two other designs common to barns and

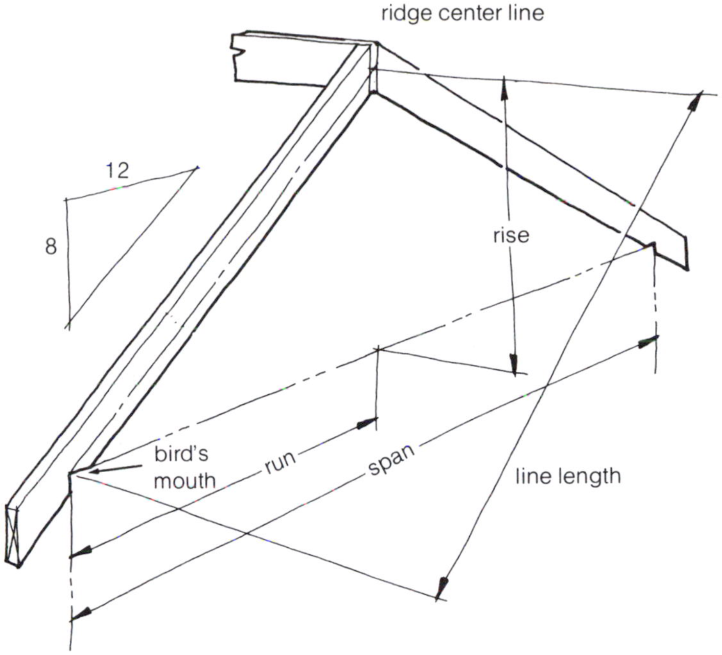

The roof pitch is expressed as a ratio of the vertical rise to the horizontal run; in this case, the ratio is 8:12. The line length is the distance from the center of the ridge to the outside edge of the bird's mouth cut.

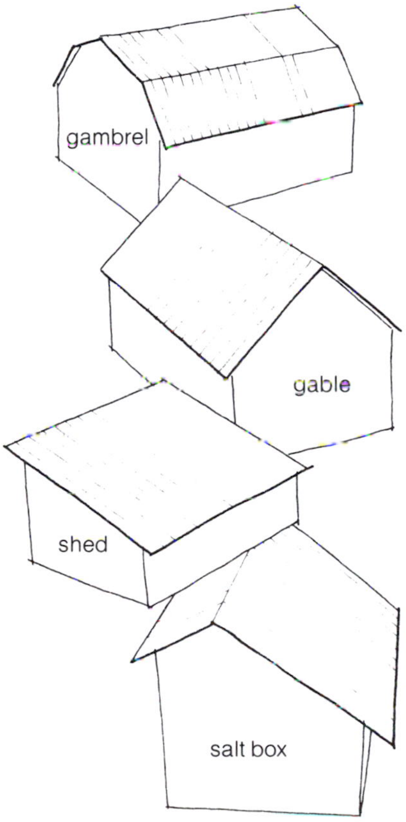

Four common roof styles are the gambrel, gable, shed and saltbox.

outbuildings. The gambrel provides more useable space than a shed or gable roof. The salt box design is popular in New England because the long, sloping roof provides a lot of protection from winter storms.

The most important factor in roof design is its pitch. The pitch determines how much room is available for storage and how well the roof sheds rain and snow. The pitch is the ratio of vertical rise to horizontal width. Thus, if a gable roof rises 8 feet vertically from the top of the wall plate to its peak and is 12 feet wide from the outside wall to its centerline, it is called an 8 in 12 roof.

For northern climates, an 8 in 12 pitch is a good angle to keep excessive snow loads off the roof. For southern climates, lower pitches are acceptable. The minimum pitch that can be covered with asphalt shingles or corrugated metal roofing is 3 in 12. At lower pitches than this, continuous roll roofing or built-up tar roofing must be used to keep it waterproof.

Gable Roof Framing

To frame a gable roof, first determine the length of the rafters. This is often done by a method called "stepping off" or using a rafter square to mark off the units of rise and run on a rafter. With this method, you can quickly and accurately determine the length of the rafter and the position of the rafter cuts. There are three special cuts: the *plumb cut* at the top of the rafter which rests against the ridge plate; the *bird's mouth* which allows the rafter to sit on the top of the double top plate; and the *tail*

STEPPING OFF RAFTERS

The most accurate way to mark rafters for cutting is to use a rafter square and "step off" the rafter. This means using the square to represent the rise and run of the roof pitch. Since the square forms a right triangle, like the imaginary rise and run of the roof, you can use the scale on the small blade, called the *tongue,* to represent the roof's rise, and the scale on the large blade, called the *body,* to represent the run.

For this example, we'll mark off a rafter with a total run of 12 feet 3 inches, a rise of 6 feet 1½ inches, and an overhang of 1 foot 8 inches. Dividing the rise by the run, we see that this is a perfect 6 in 12 roof.

First, select the straightest piece of rafter stock you have to use for the pattern. With the rafter up on sawhorses, locate the 12-inch mark on the body of the square and position it against the edge of the rafter at one end. Locate the number for the unit of rise, in this case 6, on the tongue of the square and position it against the edge. Draw a line along the back of the tongue. This marks the top plumb cut at the centerline of the ridge. When this is cut you will have to subtract half the thickness of the ridge plate from this line to compensate for the thickness of the plate.

To begin stepping off the rafter, first measure off the odd unit of run, in this case 3 inches. With the square in the original position, measure off 3 inches on the body and make a mark on the rafter. Slide the square down the rafter, holding the 6 in 12 position, until the back of the tongue is on the 3-inch mark you made. Now mark a line along the back of the tongue and body.

Once you have marked off the odd unit of run, you are ready to move the square down the rafter in full 12-inch increments to mark off the remaining 12 feet of run. Simply move the square down the rafter in the 6 in 12 position until the 6-inch mark on the tongue lines up with the previous 12-inch mark on the body. Step the rafter off 12 times in this manner to measure the 12 feet of run. Your last mark will indicate the entire length of the rafter from the centerline of the ridge to the outside of the wall plate.

To mark the bird's-mouth cut that will sit on the wall plate, turn the square upside down and position the 12-inch mark on the body at the top of the last plumb line and the 6-inch mark on the tongue on the rafter's upper edge. The horizontal line along the bottom of the body marks the bird's-mouth cut.

Finally, two more steps can be made with the square in its upside down position, one full 12-inch step and another 8-inch step. This will give you the 1-foot, 8-inch overhang for the eaves. Make the tail cut, then return to the ridge and make the top plumb cut. Remember to shorten the rafter, one half the thickness of the ridge plate.

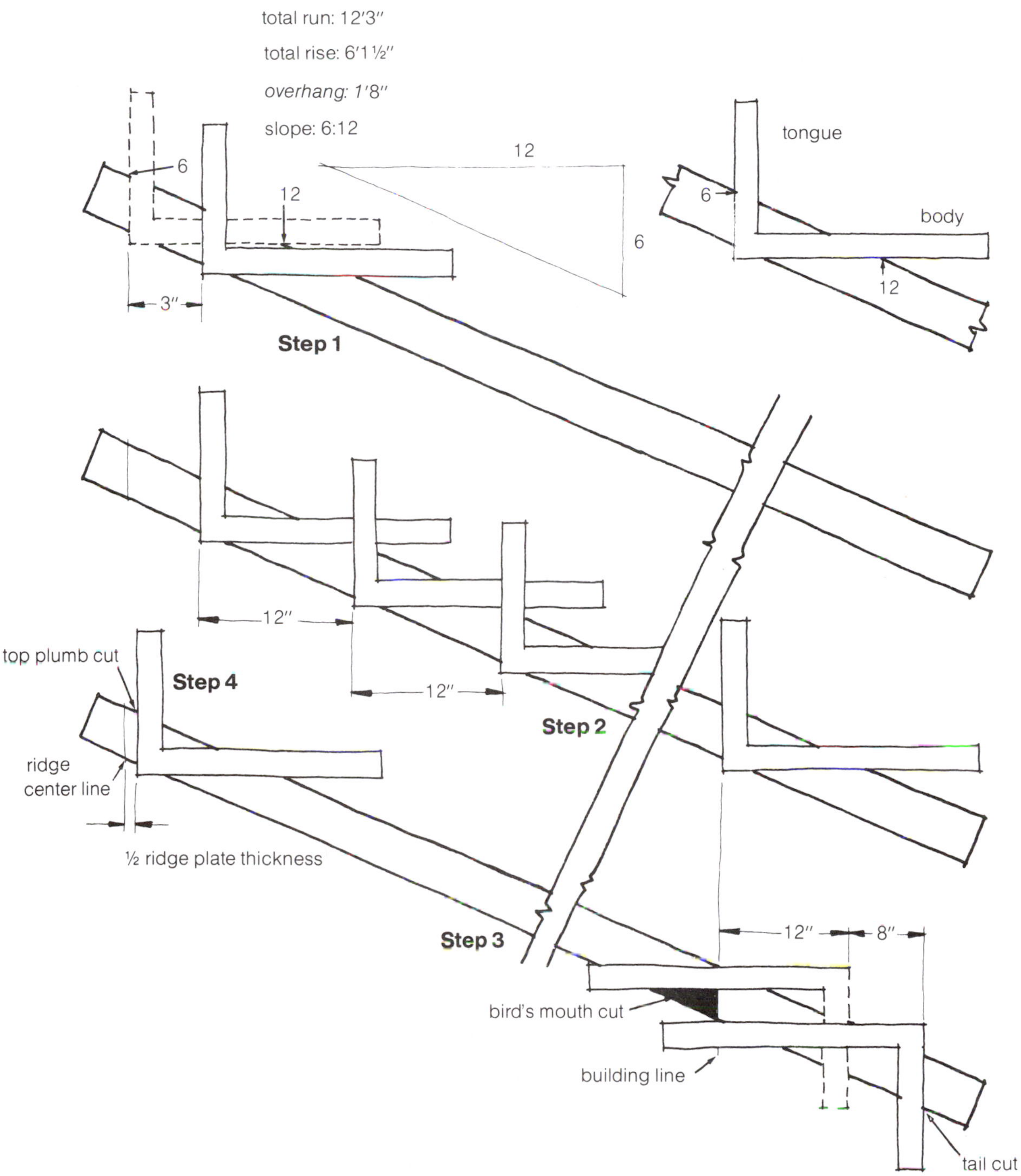

Stepping off rafters; 1) mark off odd unit, in this case, 3 inches from ridge center line; 2) step off 12-inch increments until building line is reached; 3) reverse square to mark bird's-mouth cut, overhang and tail cut; and 4) return to top of rafter to make top plumb cut.

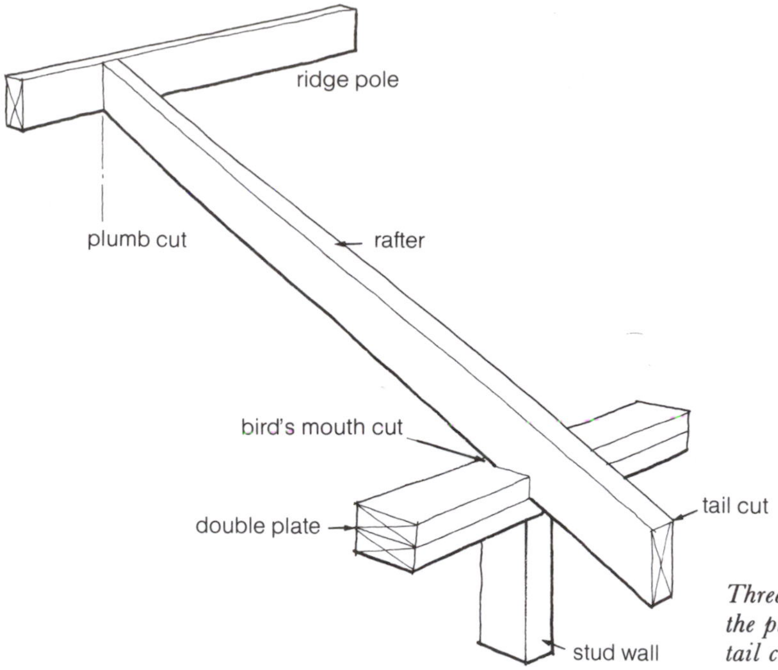

Three special cuts are made in rafters: the plumb cut; the bird's mouth; and the tail cut.

cut which forms the outside edge of the building eaves. The tail cut can either be a plumb cut or a square cut depending on how you want the eaves to look.

Before cutting the rafters, decide how large the building eaves or roof overhangs are going to be. The primary purpose of eaves is to protect the siding of the building and the foundation from rain coming off the roof. For this purpose, they should extend from the building at least 1 foot. Eaves can also screen out the summer sun, but let the winter sun at its lower angle shine through windows. To achieve this, build the eaves larger than normal; extend them 2 or 3 feet out from the building. The optimum distance depends on the local sun angles.

After you determine the roof pitch and overhang, cut a *pattern rafter* to use as a template for cutting the others. Test this pattern rafter by holding it against a temporary ridge plate held at the proper height, to see if the length is right and that all the cuts fit. (The temporary ridge plate need only be a short piece of 2 x 8 material held by hand at the center of the wall.) If the rafter does not quite fit squarely against the ridge plate or the wall plate, adjust the cuts until it does by a process of trial and error. When you are certain the pattern is correct, cut four rafters from it.

Also cut a section of ridge plate from the same width stock as the rafters. The ridge plate must extend the full length of the roof line, but it is easiest to put it up in 10- or 12-foot sections.

Next, mark positions for the rafters, every 16 or 24 inches o.c., on the ridge plate and on the top of the wall plates starting from the outside edge of the building. Mark these with double lines, 1½ inches apart, just as you would wall studs.

Assembling the roof. When both plates have been marked off, attach the two end rafters to the ridge plate on the *ground.* Use

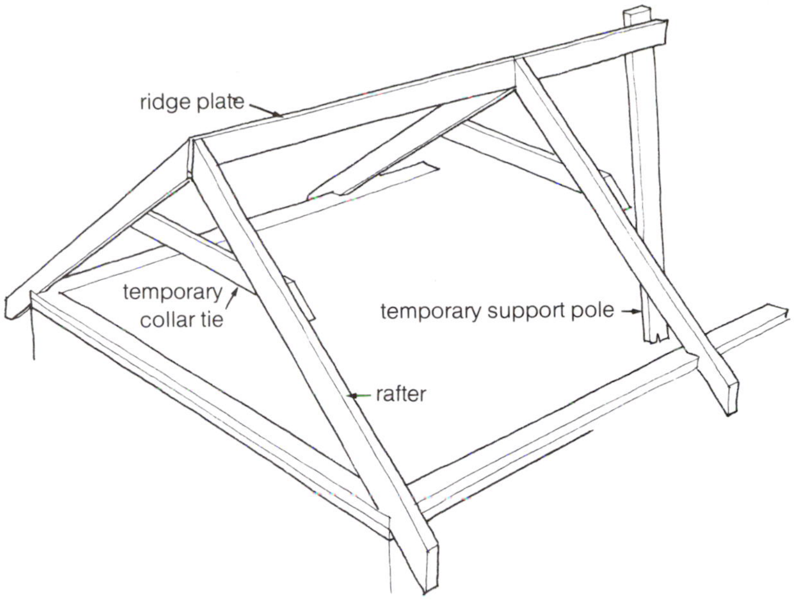

Use a temporary support pole to hold up a ridge plate section and rafters. When all rafters are in place, remove the temporary collar tie at the gable end and replace it with gable-end framing.

16d nails driven through the plate to attach the first rafter and then toenail the second onto the plate with 10d nails. Adjust the spread of the rafters until they span the width of the building exactly and lock them in this position using a collar tie nailed across them temporarily. Then with the help of two other people, hoist the rafters on top of the wall plate and fasten the ridge plate level with a support pole.

Toenail the end rafters to the end of the wall plate using 10d nails. With the ridge plate level, attach a set of rafters to its other end. The ridge plate is then self-supporting and the rest of the rafters can be cut and nailed in place. Follow the same procedure for the next section of ridge plate until the entire roof is framed.

If you don't have time to put the roof deck on immediately, brace the rafters with 1 x 4s. Simply tack long strips of 1 x 4 or other scrap lumber diagonally across the roof rafters on both sides. If metal roofing is planned, you can install 2 x 4 nailers instead of temporary bracing. Install the nailers horizontally across the rafters, spaced 24 inches o.c.

As soon as the rafters are set and before any weight is put on them, attach collar ties to keep the weight of the roof from bowing the rafters in or pushing the building walls out. Usually, collar ties are cut from 2 x 6 or 1 x 8 stock and attached to at least every third pair of rafters. Don't use permanent collar ties on the gable ends.

To finish off the gable-end framing of a small barn, shed or shelter, remove temporary collar ties and install vertical studs. The accompanying illustration shows two ways to fit the studs to the gable end: notching; and cutting the stud at an angle to fit beneath the rafters.

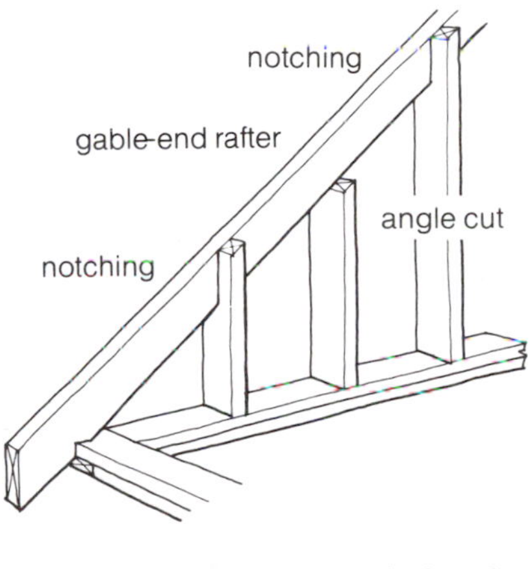

Gable-end framing, with vertical studs cut or notched to fit beneath rafters.

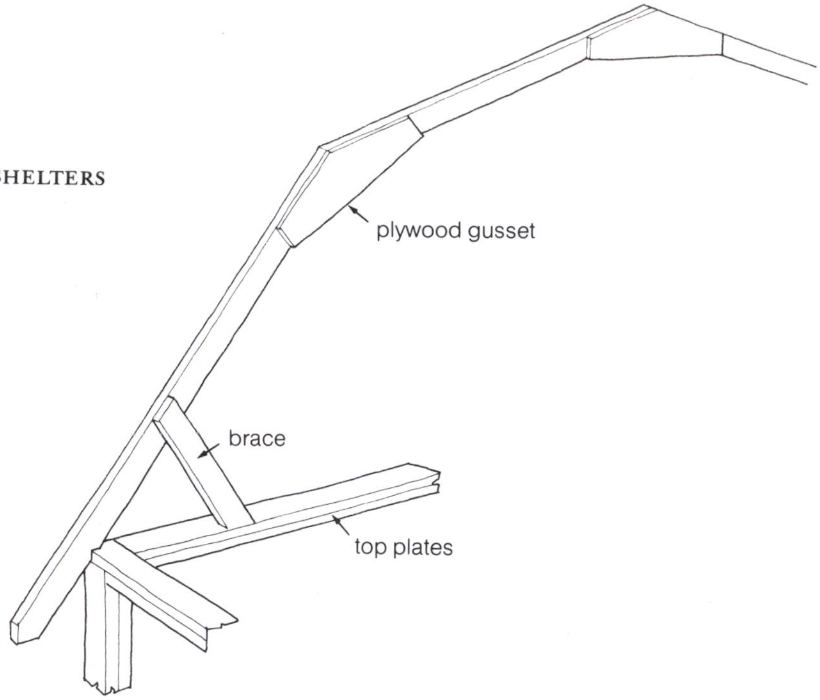

Gambrel roofs have two sloped pitches and framing members held together with braces or plywood gussets.

Shed, Gambrel, Salt Box Roofs

Shed roofs are built in much the same way as gable roofs except they only have a single pitch. Make two bird's-mouth cuts in each rafter so it fits over the front and back wall plates. Eave headers are sometimes attached to both ends to lock the rafters together and to finish off the overhangs. This type of roof is much simpler to frame and install than a double-pitched roof, but because of its low pitch and design limitations, it is normally suitable only for small sheds and outbuildings.

Gambrel roofs are often selected for barns because they provide more head and storage room. They have two sloped pitches; usually, the top is about 30 degrees and the bottom 60 degrees to the horizontal. Each is framed with two members held together with braces or plywood gussets.

Salt box roofs are similar to gable roofs in a structural sense. However, there are differences: the salt box has short rafters for one roof pitch and long rafters for the other; usually, the long rafters are formed by overlapping two long boards; and the long rafters are often supported at their midpoints by a girder or interior wall. (For an example of a salt box, see p. 146.)

Trusses

When optimum utilization of the space immediately beneath the roof is not essential, you can use roof *trusses.* Trusses are quite strong and, in many cases, very economical because they can be built of lumber that's smaller than normal rafters. You

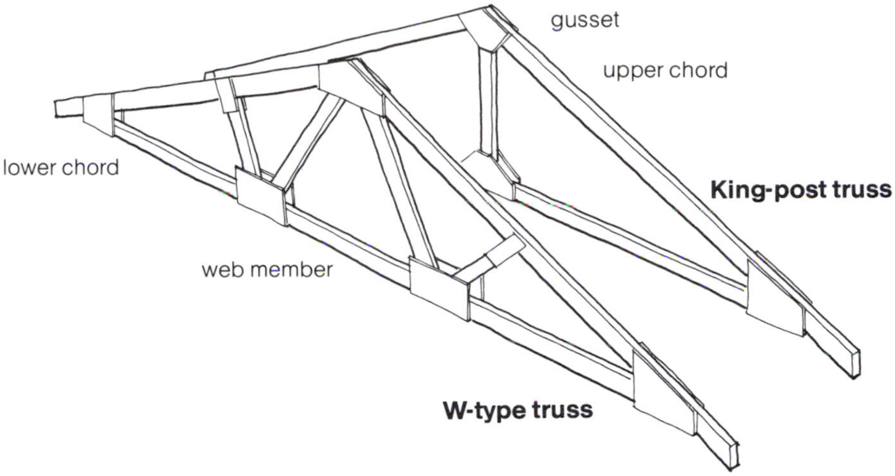

King-post and W-type are two standard trusses; if you assemble these yourself, space nails in the gussets about 3 inches o.c.

can order factory-built roof trusses to fit almost any roof pitch and design. And these preassembled units can cut roof framing time down to several hours. If you decide to build your own trusses, follow an engineered design that will support anticipated roof loads. Be sure to make accurate cuts and to fasten all joints with metal web plates or plywood gussets and the correct number of nails.

When trusses are installed, they must be tied together temporarily with 1 x 4 strapping to keep them from blowing over before the roof deck is put on.

Post and Beam

Post and beam, or timber framing as it is commonly called, has been the traditional method for building barns for centuries. Only in the past 100 years has it been replaced by platform framing because of rising material and labor costs. In many parts of the country, however, timber framing is making a comeback due to the availability of low-cost native timbers and a revived interest in craftsmanship. A brief introduction to timber framing follows.*

The type of wood selected for timber framing varies from one part of the country to the other. Oak is the popular hardwood, valued for its strength and durability. It also has excellent working qualities for joinery. Spruce, hemlock, and pine are the two most widely-used softwoods. They are easier to cut and chisel than oak, but not as strong. Often other types of wood are chosen simply because they are locally available at low cost.

The principal challenge of timber framing is the joinery of posts and beams. Only standard wood-working tools such as saws, chisels and drills are necessary to make the joints, but to do the job properly, requires patience and attention to detail.

* For further reading on post and beam framing, see Benson, Ted with Gruber, James, *Building The Timber Frame House,* Charles Scribner's Sons, New York, 1980.

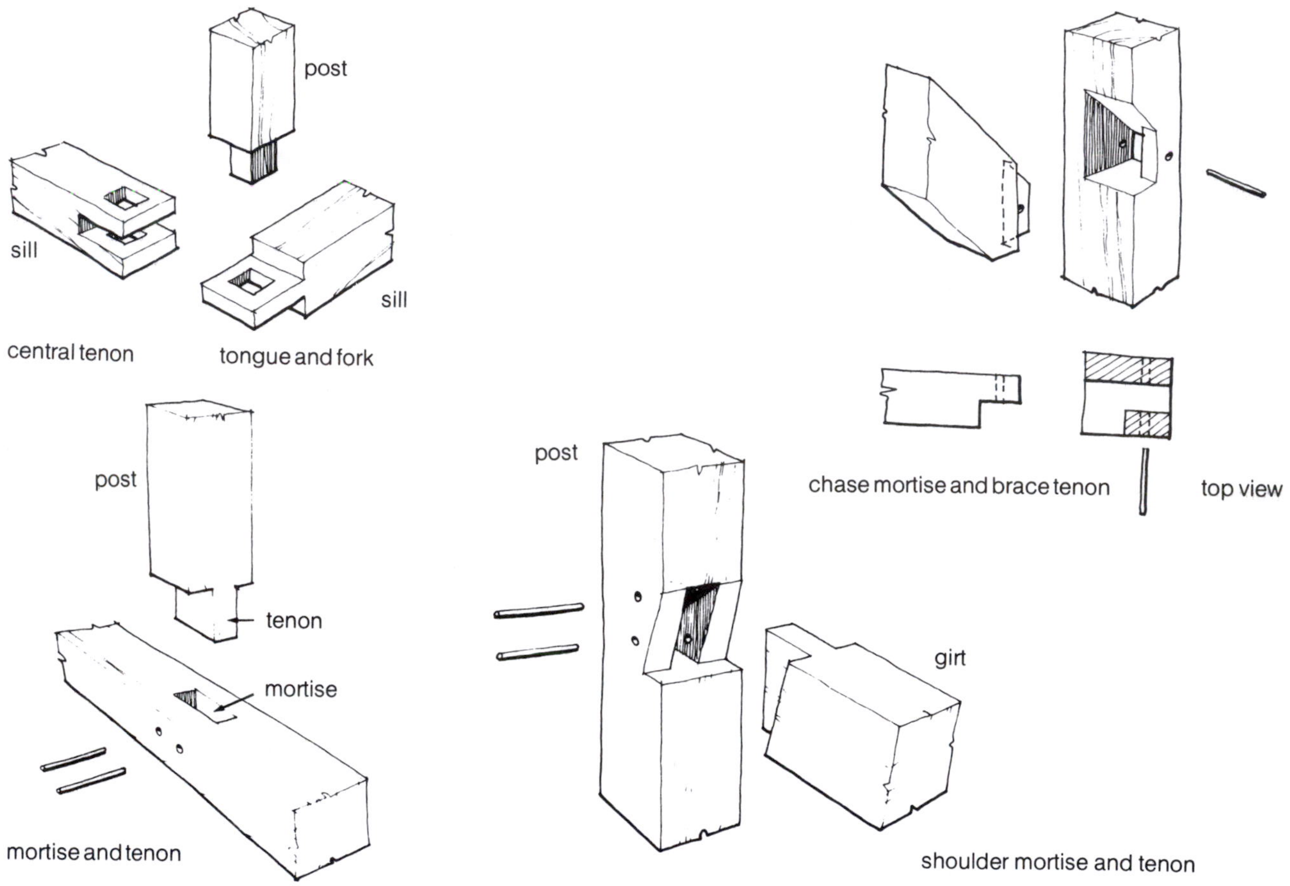

Four principal joints used in post and beam construction.

Four principal joints are used in timber framing though there are many variations of these for special situations. A *tongue and fork* with a *central tenon* is the simplest way to lock adjoining sill plates and the corner post together. A *mortise and tenon* is the standard joint for setting posts into the sill. A *shoulder mortise and tenon* is used to fasten the girts. It is important to note that the shoulder carries the weight of the girt with the full thickness of the beam while the tenon anchors it without undue stress. A *chase mortise and brace tenon* is used for corner braces that keep the post and beam structure rigid.

The process of timber framing is much like platform framing. Once the sills are in place on top of the foundation, the walls (or *bents,* as they are traditionally called) are lifted into place as a unit. Girts are set across the first-and second-story floors, and then rafter beams are set as pairs. The two principal differences from platform framing are that the beams and their joints are

MORTISE AND TENON JOINERY

The mortise and tenon joint is a standard timber framing joint that can be made with four basic woodworking tools: a hand saw, chisel, drill and square.

To cut the mortise for a post and sill beam joint, first use a combination square to lay out where the post and its tenon will sit on the beam. Usually, the tenon runs the width of the post and is 1½ inches or more in thickness. The tenon always runs parallel with the beam so pegs can be driven through both to lock them together.

Drill out the mortise using an electric drill or hand auger with a bit that is slightly smaller than the width of the tenon. Be careful to drill squarely into the beam so that the walls of the mortise are plumb. Use a stop guide on the drill bit to accurately measure the depth of the mortise. It can be a little deeper than the tenon, but it can't be shallower.

After removing most of the waste from the mortise with a drill, use a sharp chisel to even out the sides of the housing and to scrape the bottom clean. Don't overdo it. The tenon must fit snugly into the mortise and a few too many passes with the chisel can make for a loose joint. Once the mortise is finished, drill two holes in the side of the beam through the mortise for the wooden pegs. These holes should be at least 1½ inches up from the bottom of the mortise.

Now cut the tenon. Square off the bottom of the post to the depth of the mortise. Then

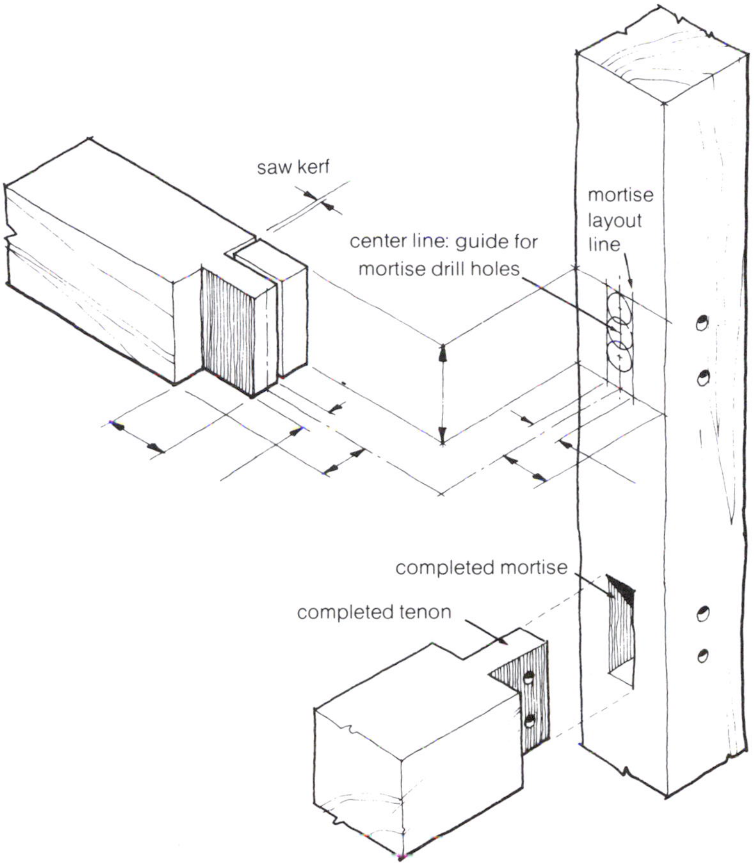

Use a square to mark beams for tenon and mortise; note corresponding dimensions shown here.

mark the outline of the tenon on the end of the post and running back up the sides to meet the square lines. Use a circular saw or hand saw to cut along the square lines that run with the tenon. Cut only to the depth where the tenon begins. Make cuts 1 inch apart and to the proper depth across the two faces of the beam that will be removed to expose the tenon. The waste wood can now be removed with a wide chisel using the depth of the saw cuts as a guide.

After chiseling out the tenon, test fit it in the mortise and make small adjustments as necessary to get a good fit. With the tenon in place, mark the peg holes on the tenon with a pencil stuck through the mortise holes. For barn construction, the normal peg size is ¾ to 1 inch in diameter. For a tight fit, the holes in the tenon and mortise are slightly offset so that the peg draws the tenon into the joint. This is referred to as "drawboring." The tenon should be drilled ⅛ of an inch off center toward the top of the post so that the pegs will be forced in a shallow S curve that helps pull the tenon into the mortise.

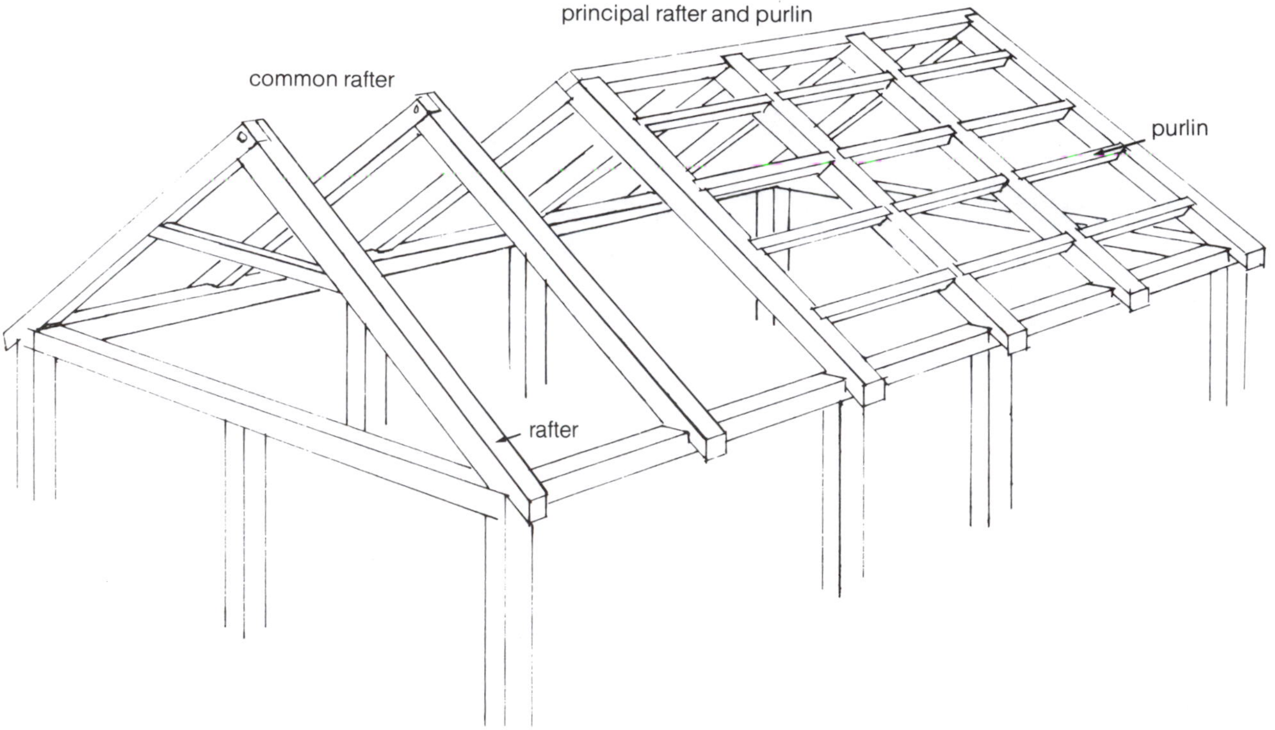

In timber framing, two basic roof systems are used: the common rafter *and the* principal rafter *and* purlin.

precut and test fitted on the ground and then the entire frame is assembled before any subflooring or platforms are put down.

Two types of roof systems are used in timber framing, the *common rafter* and the *principal rafter-and-purlin.* The common rafter system uses 4 x 6 or larger rafters set 3 to 6 feet apart with the roof sheathing laid perpendicular over the rafters. The principal rafter-and-purlin system uses larger rafters, 6 x 6 or bigger, spaced 6 to 9 feet apart with 4 x 4 purlins running between them. The roof sheathing is then laid parallel to the rafters over the purlins.

After the timber frame has been assembled, usually with a small army of people, the walls are studded in and floor joists set using regular dimensional lumber. This is necessary to hold the siding and the floor boards and results in the additional materials required for timber framing. Windows and doors are framed in the same way as platform framing except that load-carrying headers are not required since the building load is completely carried by the wall girts. All that is necessary is a single plate above and below to frame in the rough opening.

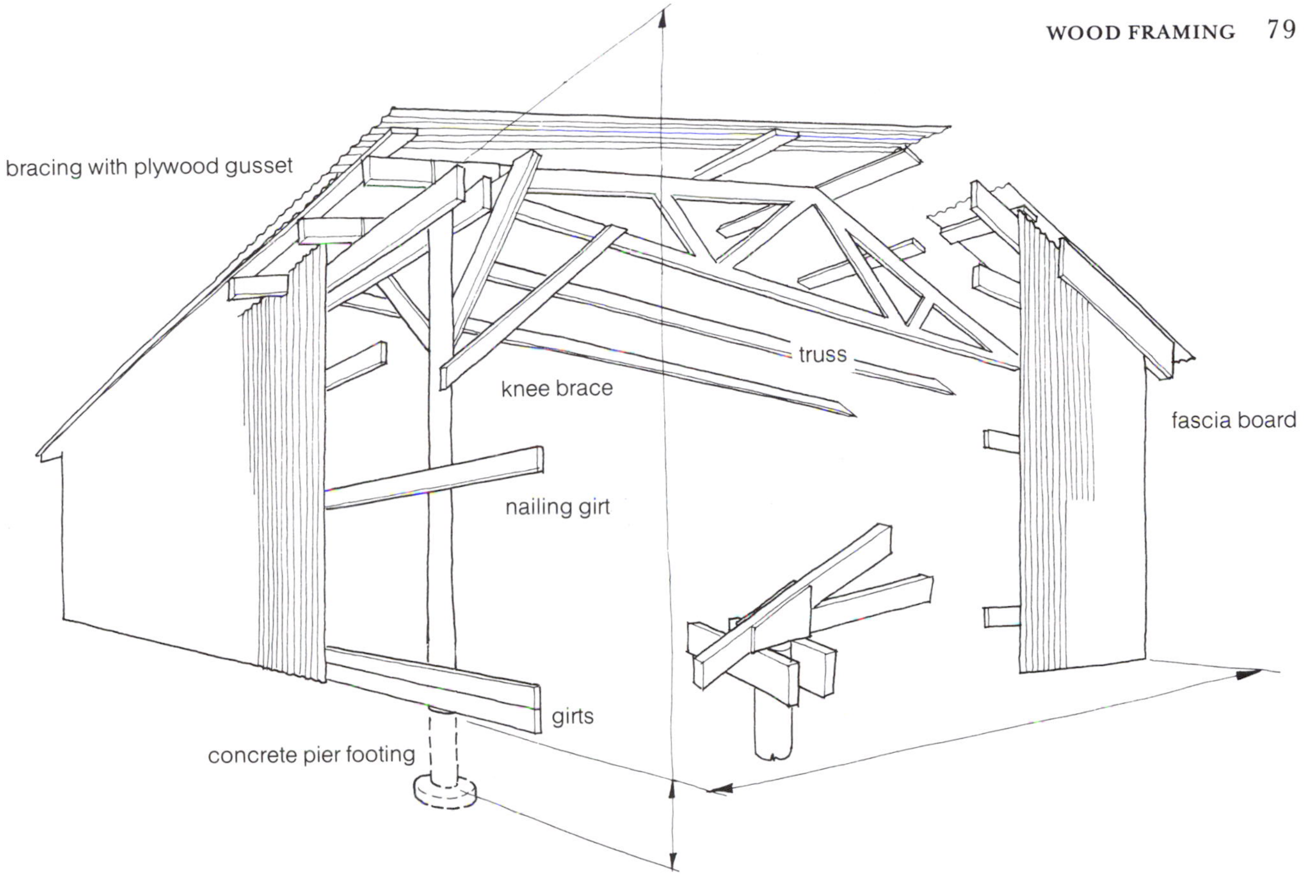

Basic framing pieces of a pole building are shown in this cut-away view. Although this framing method is known as pole *building, posts are often substituted for poles.*

Pole Framing

Simplicity and economy are the main advantages of pole building. This type of framing is ideally suited for small barns and outbuildings; it enables quick, low-cost construction and requires fewer materials than any other framing method.

Poles are available in many sizes and lengths ranging up to 60 feet or more. Common building poles are under 30 feet in length and range from 4 to 8 inches in diameter. If they are to be set in the ground, they must be pressure treated to last more than a few years. Pressure-treated, southern yellow-pine poles are readily available in all parts of the country. Second-hand poles can also be bought at half price from utility companies and salvage operations. Posts in sizes such as 4 x 6 or 6 x 6, are also suitable.

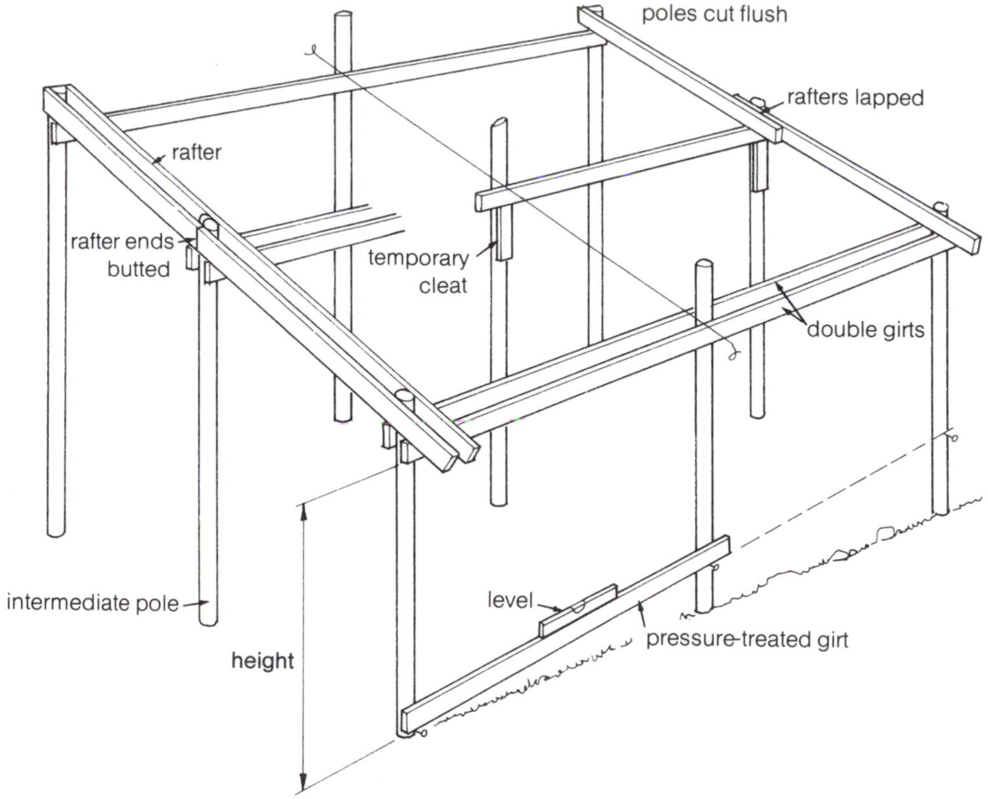

Make sure girts are level. Use the first girt as a guide for measuring the height of the building. Line up girts on intermediate poles with a string. It may prove helpful to rest girts on temporary cleats before securing with nails. Compound rafters may be butted or lapped as shown.

The first step in pole building is to set the poles in the ground on concrete footings or on concrete piers with anchor brackets (see p. 47). Poles can be spaced from 4 to 16 feet apart depending on their size and the load they must carry. Normally, the poles will extend from the foundation right up to the roof line. Large poles can either be raised by hand with several people or raised with a tractor and rope. Once they are up, they must be held plumb while braces are attached to keep them in position.

When all the poles have been raised, girts are nailed onto them horizontally every 2 feet to carry the siding. These are normally 2 x 4s or 2 x 6s depending on the span between poles and the weight of the siding. Bottom girts in contact with the soil should be absolutely level and pressure treated or coated with a preservative.

The top members that carry the rafters are usually double girts made from 2 x 8s or 2 x 10s placed on the inside and outside of the poles. These are bolted together through the poles or nailed in place with 20d spikes. The height of the top members is measured from either the top or bottom of the bottom girt.

Wall bracing is absolutely necessary in pole buildings. Braces are run from the upper girt to the poles and are often reinforced with plywood gussets. These braces are the only things that give the walls any lateral strength if metal or board siding is used. Plywood siding helps brace the building, but is not a substitute for adequate diagonal bracing.

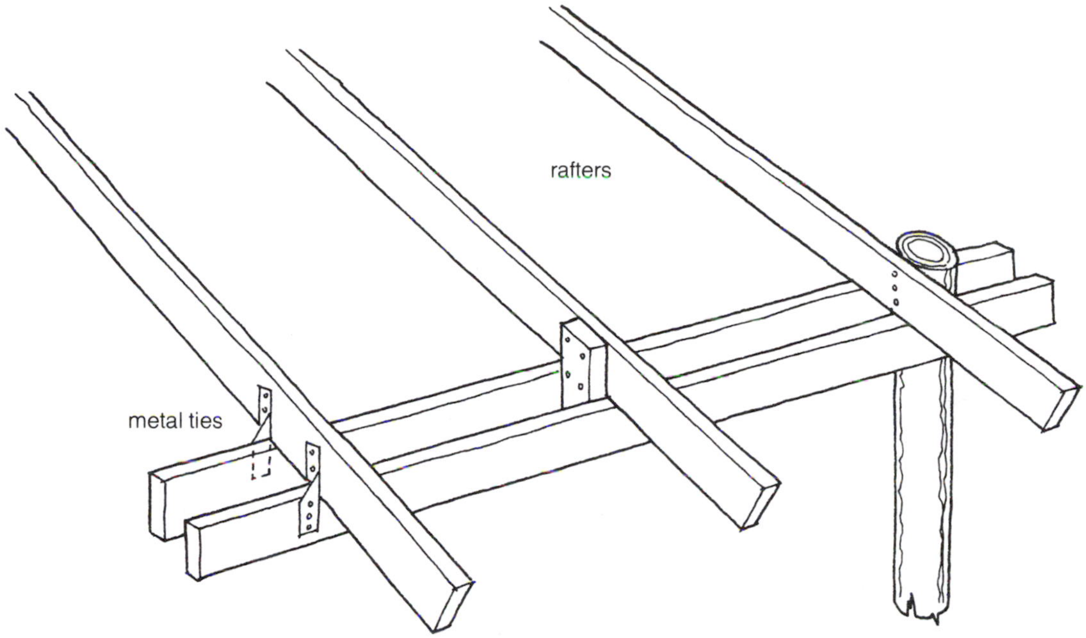

You can attach rafters to girts with metal ties or wooden blocks. Sometimes, it's convenient to simply attach the rafter to a pole.

With the walls framed and braced, you can assemble roof rafters or trusses on top of the double girts. You can cut and assemble rafters in the same manner as described in the section on platform framing. Or, to simplify things, omit the bird's-mouth cuts, and, after placing the rafters on top of the double girts, secure them with wooden or metal ties. And whenever possible, secure them to adjacent posts. If you use rafters rather than trusses, be sure to attach collar ties so the outside walls will not be pushed out.

In a pole barn, you can build floors at any level by using wall girts to support the joists. Attach joist headers to both the ends of the joists and the poles.

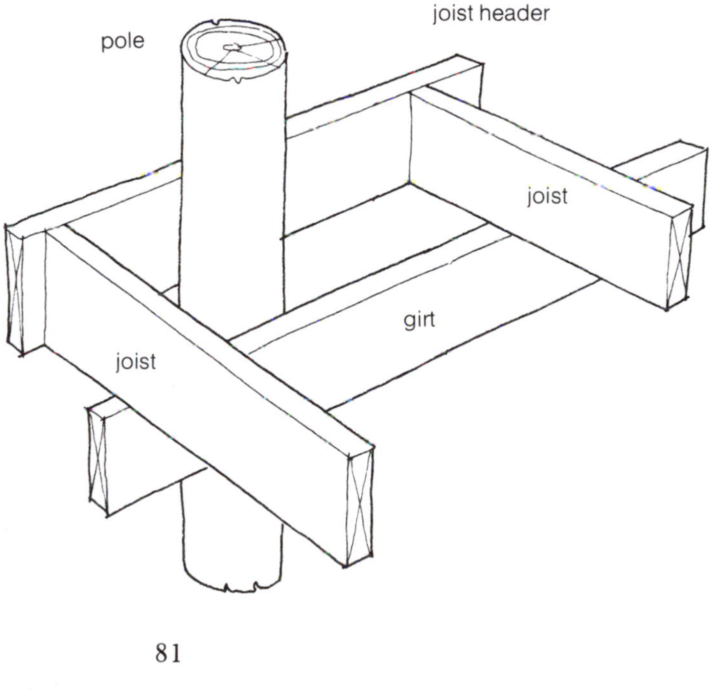

To install a floor at any level of a pole barn, simply support the floor joists on the wall girts.

CHAPTER 5

Siding and Roofing

Weyerhauser

MANY SIDING AND roofing materials are available for barn and outbuilding construction. When you make your choice, consider not only the cost, durability and appearance, but also the all-around suitability of a particular material. For instance, galvanized-steel siding might be the cheapest and the most weather-resistant siding you can find. For animal housing, however, it might be a poor choice because it is dented easily.

In another situation, you might decide that native board and batten siding is the ideal, low-cost solution for your garage. But, your house and all the others in the neighborhood are covered with painted clapboards. The juxtaposition of the two different sidings would be unsightly and might detract from the value of your home. Despite the added cost, it would probably be better to use clapboards on the garage to match the style of the house and neighborhood.

Two other important items to consider before installing siding, especially if you are planning a livestock barn or outbuilding, are ventilation and insulation. Discussion of these topics follows in Chapter 6, pp. 110–113.

Siding

Native Lumber

Rough-sawn, native lumber is an attractive, low-cost siding material; it may be installed vertically to achieve one of four principal styles (see p. 84). *Board and batten,* the traditional barn siding, consists of 6-inch or wider boards butted edge to edge and 1- to 2-inch battens which cover the gaps between the boards. In a variation of this style, *batten and board,* the battens are placed behind the boards, not over them. Another variation, *board on board,* dispenses with the battens altogether; the boards are simply overlapped. Finally, there is *shiplap siding.* In this case, the siding boards have ¾-inch half-lap joints. When overlapped, these joints create a weathertight seal.

Pine, spruce and hemlock are three softwoods commonly used for board siding. Pine and spruce are preferable because of their better weathering characteristics. Oak, a hardwood, is occasionally used for siding, especially in the south where it is plentiful. If oak is used, it must be installed while green. Once it has dried, it is nearly impossible to drive a nail through it. Cedar is

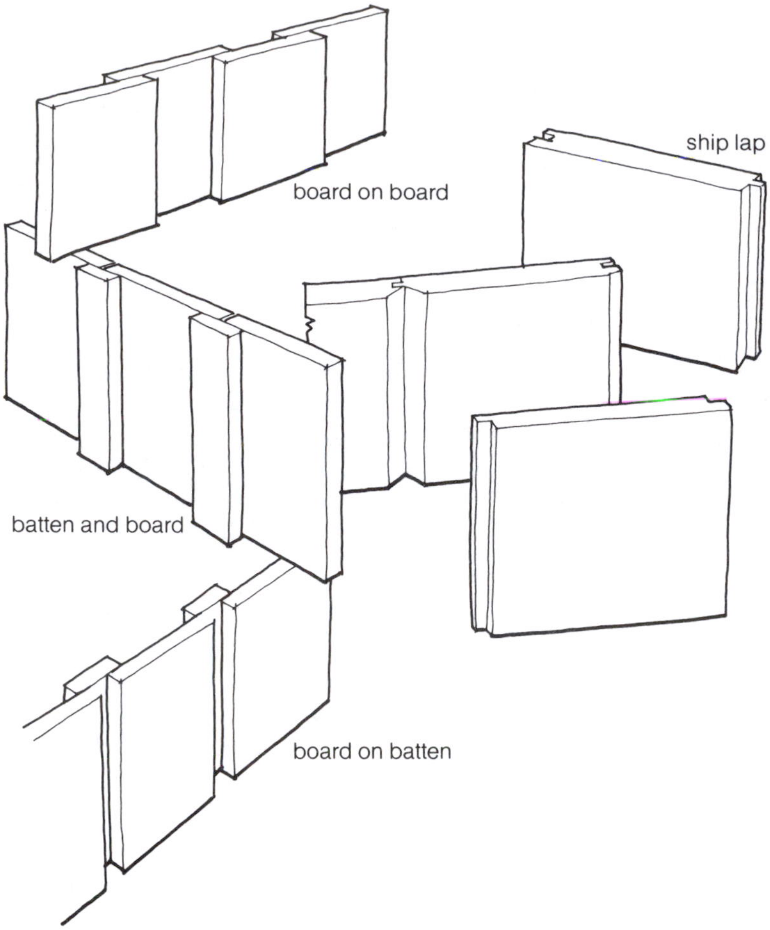

Native-lumber siding may be purchased in several styles, including batten and board, board and batten, board on board and shiplap.

INSTALLING BOARD AND BATTEN SIDING

Board and batten is an attractive and economical choice for barn siding. Select high-quality boards because the siding determines the tightness and appearance of your barn. Reject boards with loose knots and visible cracks.

To install the siding boards, first set a level string line across the side of the building about 1 inch below the bottom of the sill plate or beam. Draw it taut by attaching it to two stakes driven in the ground at either end of the building. This string marks the bottom of the siding and keeps it straight and level.

Start in one corner and attach the first board using 8d galvanized nails driven into the horizontal strapping or blocking. Make sure the board is plumb and that the bottom is just above the string, almost touching but not quite. The next board should be butted against the first, the bottom lined up and then nailed in place.

Because the widths of rough-sawn boards sometimes vary from end to end, make sure each board is fairly plumb and not simply butted against the edge of the adjacent board. Don't worry about small cracks between the boards; these will be covered by the battens.

When all the boards are installed, cut battens to the desired width on a table saw and nail them over the joints between the boards. Also, nail battens over the corners of the building to seal the joints there. (Use 8d or 10d nails.)

another excellent siding material that has the best weathering characteristics of any commonly-available wood. Unfortunately, it is also very expensive in most parts of the country. See Table 5-1 for additional wood weathering characteristics.

Rough-sawn boards can be milled to any width and length. Normally, mills turn out 6-, 8-, and 10-inch wide boards; lengths are in increments of 2 feet, starting at 8 feet. You can also buy boards in random widths, sometimes at a savings over dimensional lumber. I prefer 8-inch boards since they are wide enough to cover a wall quickly, but not so wide as to warp or crack.

Installing board siding. To install vertical board siding, there must be horizontal wall framing members to provide a nail base. For pole construction, use girts to carry the siding. For stud construction, nail 2 x 4 blocks between the wall studs or 1 x 3 strapping horizontally across the studs. I prefer to nail strapping across the studs every 2 feet o.c. because there is less cutting, and strapping goes on faster than individual blocks.

Where a single board will not reach the top of the wall, such as on the gable end of a barn, another board must be installed. There are three ways of joining the boards together. The sim-

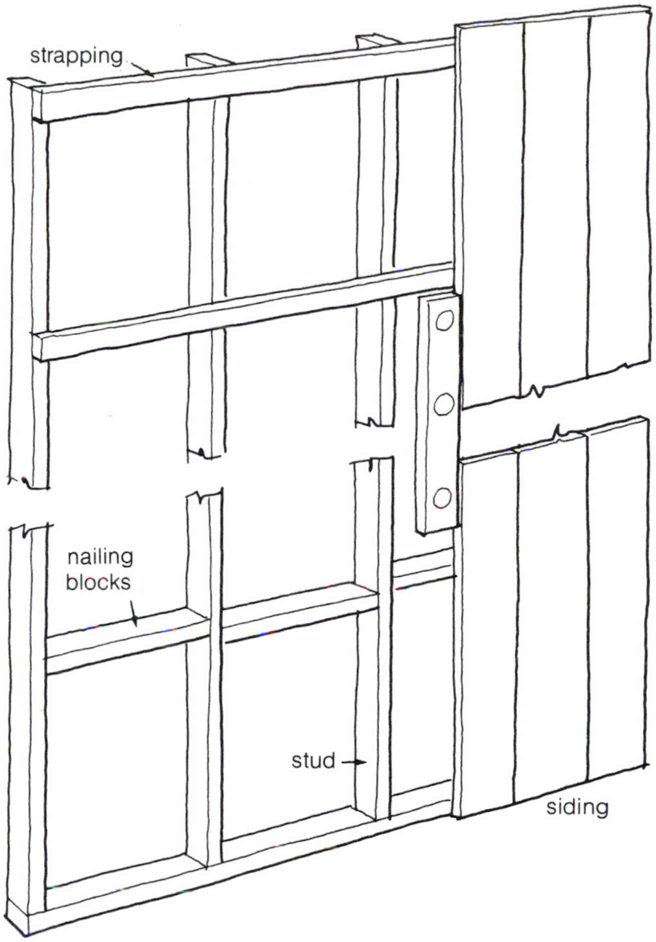

When installing vertical board siding, make sure each piece is plumb. With standard platform framing, you can nail siding to blocks or strapping. With pole barns, nail the siding to girts.

Table 5-1 Weathering Characteristics of Wood*

Softwoods	Resistance to Cupping 1 = Best 4 = Worst	Conspicuousness of Checking 1 = Least 2 = Most	Ease of Keeping Well Painted 1 = Easiest 4 = Hardest
Cedar, Alaska	1	1	1
Cedar, white	1	—	1
Redwood	1	1	1
Pine, eastern white	2	2	2
Pine, sugar	2	2	2
Hemlock	2	2	3
Spruce	2	2	3
Douglas fir	2	2	4
Hardwoods			
Beech	4	2	4
Birch	4	2	4
Maple	4	2	4
Ash	4	2	3
Chestnut	3	2	3
Walnut	3	2	3
Elm	4	2	4
Oak, white	4	2	4

* From *Wood Handbook,* Agriculture Handbook No. 72, United States Department of Agriculture.

ESTIMATING SIDING MATERIAL

Estimating the amount of siding material you need is not always easy. You must calculate the square footage of wall area to be covered, adjust for differences in nominal and actual dimensions of the siding boards, account for window and door openings and adjust for waste and errors.

For plywood siding, the task is fairly easy. Simply measure the square footage of your walls by multiplying their length times height, then add up all the surfaces. Divide this figure by 32, the number of square feet in a 4 x 8 sheet of plywood, and you'll have the number of sheets that you need. To calculate the area of the triangular gable ends, multiply the width of the building by the height of the triangle from the top wall plate to the peak and divide by 2. Don't subtract any square footage for windows or doors, as these will usually be covered with the plywood and cut out later. Large garage doors or other openings should be subtracted, of course. When you arrive at a final figure for the number of sheets you need, add 10 percent to cover for waste in cutting and errors.

To calculate the number of boards you will need for vertical board siding, use the same method. First, find the wall area and divide this by the number of square feet an individual board will cover. For example, if you are using 1 x 8 rough-sawn siding, 8 feet long, each board contains 5.3 square feet. To cover a 30 x 8 wall, you would need about 45 boards (240 ÷ 5.3). Make sure you use the actual width of the board and not its nominal size. A rough-sawn 1 x 8 is actually 8 inches wide, but a planed, kiln-dried 1 x 8 is only 7¼ inches wide. When you arrive at a final figure for the number of boards you need, always add 10 to 15 percent for waste and bad wood.

Estimating bevel siding is perhaps the most difficult task. You buy bevel siding by the nominal board foot, so you must convert this figure into square foot coverage. First, determine the overlap on the siding. If the siding is to have 3 inches exposed to weather, then each row will only cover a 3-inch wide section of wall. The minimum-width siding you could use for this would be 6-inch nominal-width siding which would actually measure 5½ inches wide. Because each row will only cover half the width of the siding, 2 board feet of siding will be needed to cover one square foot of wall area. A formula for figuring this would be:

$$\frac{\text{nominal width of siding}}{\text{inches to weather}} \times \text{sq. ft. of wall} = \text{bd. ft. of siding}$$

Add ten percent to this figure for waste.

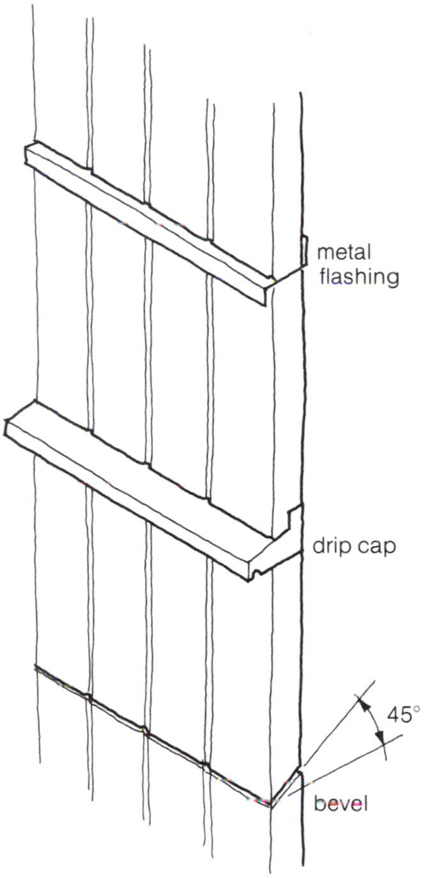

Three ways to join siding boards are with metal flashing, with a drip cap or with a 45-degree bevel joint.

plest is to use metal flashing. Or you can cut a 45-degree bevel joint so the top board overlaps the bottom one. This ensures that gaps for wind and water won't develop between the ends of the two boards. Always make this joint on top of a horizontal nailer so the ends of both siding boards can be fastened securely. Another way to join boards is to use a drip cap which overhangs the bottom boards and sheds water from the joint.

Metal Siding

Corrugated sheets of aluminum or galvanized steel are popular barn and shed sidings. Both metals are long lasting and require almost no maintenance. They are available in a full range of colors and with different protective coatings.

I prefer galvanized steel to aluminum because it is generally sturdier and less expensive. Metal siding's biggest disadvantage is that it can be dented easily. If used for large-animal housing, the insides of stalls and walls must be protected with wooden rails to keep the animals from kicking or rubbing against the siding.

Two common widths for galvanized siding are 32 and 38 inches. When the pieces are overlapped, these two widths cover 30 and 36 inches, respectively. A standard length is 8 feet, though the material can be special ordered to any length in 1-inch increments.

Cutting corrugated metal. Despite the fact that it is metal, aluminum and galvanized steel siding and roofing can be cut easily on the job. All that you need is a circular saw with a metal-cutting blade that can be bought at any hardware or building-supply store.

When cutting metal, always wear safety goggles to protect your eyes from metal bits thrown out of the saw. Also, leather gloves are a good precaution since freshly-cut edges can be ragged and sharp.

To mark for a cut along a piece of metal siding, use a chalk line. With it, you can snap a colored line that is easily visible. Pencil marks are nearly invisible on the gray surface of sheet metal.

When you are cutting metal, hold back the blade guard on the circular saw with your left thumb so that it doesn't catch on the corrugations. Slowly, but steadily, move the saw, taking care not to force it as you cut. The blade will send sparks flying, so don't cut around flammable materials or gases.

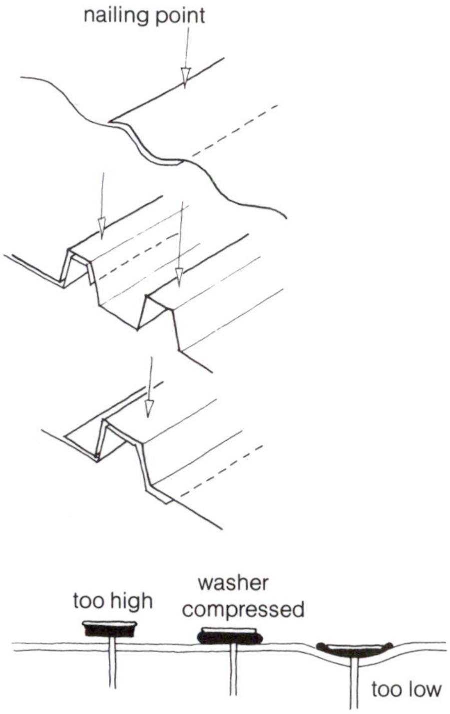

Arrows indicate nailing points for various metal sidings. When installing this material, follow manufacturer's instructions. Drive nails until washers are slightly compressed.

After cutting for a couple of hours you might notice that the blade seems smaller than when you started. If you have an abrasive blade, you're right, it is smaller. The blade cuts by abrasive action and grinds itself away with every revolution. I prefer a more durable, non-ferrous metal cutting blade.

Installing metal siding. Like vertical board siding, metal siding requires horizontal wall strapping or nailers for support. In most cases, the sheets are installed vertically and overlapped according to the manufacturer's instructions. When installing the first sheet on the corner of the building, *make sure it is absolutely plumb.* Otherwise, all the following sheets will run off.

Galvanized siding is fastened with hot-dipped, galvanized, ring-shank nails with lead or neoprene washers. Galvanized hex head screws are also used. These can be driven with an electric drill with a hex head bit. Aluminum siding must be fastened with aluminum ring-shank nails not galvanized nails. Dissimilar metals will react with each other causing corrosion and oxidation.

Nail the siding at least every 8 inches into the girts or strapping set 2 feet o.c. Drive the nails through the top of the corrugations and into the wood until they just compress the washer. If they are not driven far enough, the siding will be loose and rattle in the wind; and water will get in under the washers. If they are driven too far, the siding will collapse leaving an unsightly dent. Always tack each sheet in position with just a couple of nails and check it for squareness before driving all the nails. Once driven home, the nails are very hard to remove without damaging the siding.

Trim pieces are available for finishing the corners of the building and around the edges of windows and doors. Install these according to each manufacturer's instructions.

Plywood

Plywood sheets are available in a variety of textures and patterns for siding. T 1-11, a commonly-used type, has a rough-sawn surface and grooves every 4 inches to imitate rough-sawn boards. Another choice, regular A/C exterior glue plywood, has a smooth surface for painting and is less expensive. (For more on plywood characteristics, see Table 2-1, p. 15.)

Plywood siding goes on quickly and is very strong and light; you can apply it directly to the studs without any other sheathing, strapping or bracing. Cost is its main disadvantage. Also,

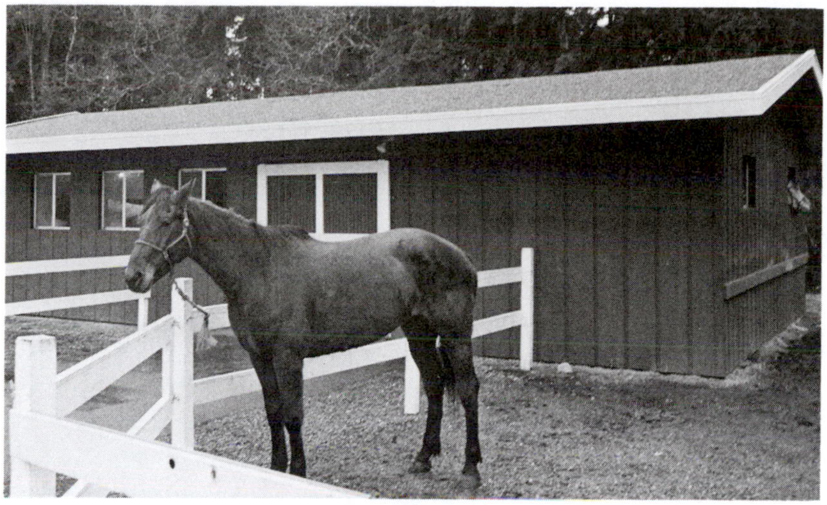

T 1-11 plywood siding has a rough-sawn surface and grooves, a good texture for the exterior of small barns and outbuildings.

applying the large sheets is a two-person job, whereas only one person can apply board siding.

The normal size of a plywood panel is 4 x 8, though it can be special ordered in longer lengths. For most outbuilding construction, the plywood should be ½ or ⅝ of an inch thick. Normally, it is installed vertically to eliminate horizontal seams. Secure plywood with 8d galvanized nails spaced 12 inches apart along the studs and 6 inches apart along the 4-foot edges of the sheets.

If you wish to decorate the plain plywood siding or cover the vertical seams, you can nail 1½-inch battens over the plywood either every 8 inches or every 4 feet at the seams. Horizontal seams, such as on the gable end of a building where a second row of plywood must be installed to reach the peak, should be flashed with an aluminum drip cap. The aluminum flashing keeps water from entering the joint as it drips down the siding.

When installing plywood sheets around windows and doors, don't bother to precut the sheets for the rough openings. Simply nail the siding right over the opening, then cut out the openings with a saw. A small electric chain saw is ideal for cutting out the opening from the inside of the building. Use the rough opening frame as a guide. A circular saw will also work, but first mark the area to be cut out on the outside of the plywood since the saw must be used from the outside. To do the marking, drive nails through at the corners from the inside, then join the nails with a chalk line.

Bevel Siding and Shingles

Two horizontally-installed wood sidings are bevel siding and shingles. Bevel siding boards, usually cut from spruce or cedar, come in widths 5 inches and larger. When overlapped in horizontal rows, they form a weathertight seal. Shingles, also made from spruce or cedar, are overlapped in the same manner, but in wider rows.

Both bevel siding and shingles require a nailing base of either

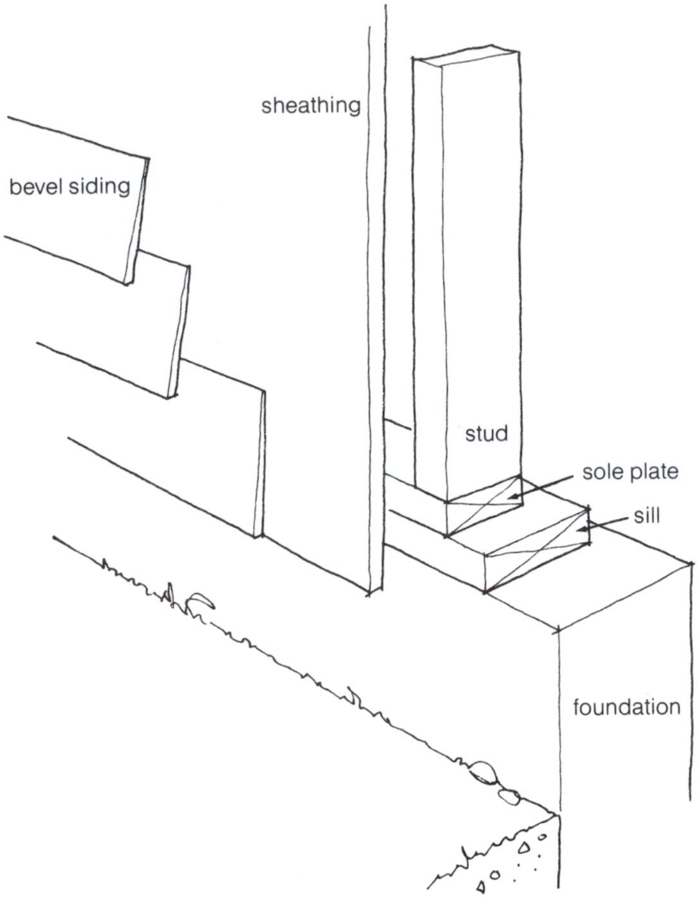

Bevel siding and shingles are installed similarly. Note that the plywood sheathing overhangs the foundation slightly to create a tight seal.

boards or plywood sheathing. Combined with the high cost of the siding, this makes for an expensive and labor-intensive protective covering. Bevel siding and shingles are extremely durable, however, and are still widely used in residential construction. These sidings are also often selected for outbuildings where architectural standards or the severity of the climate dictates their use.

Install both sidings from the bottom up, overlapping each course by about half the width of pieces. For example, install bevel siding with 3 inches exposed "to weather." Use 4d galvanized nails to fasten the siding to the sheathing, nailing through the siding's thick bottom edge.

In addition to wooden bevel siding, manufactured hardboard and plywood siding are available in strips 12 to 16 inches wide. These go on like bevel siding, but are much less expensive and require less labor.

Oil Stains and Paints

All wood siding should be protected from the weather. Paint, a time-tested preservative, requires periodic and expensive maintenance. Every few years or so, it must be scraped, washed and replaced with a new coat. Clear linseed oil and oil-based stains are beginning to serve as replacements for paint because they require less labor to apply and maintain. Stains protect wood just as well as paint and they can be reapplied without scraping and preparation.

Roofing

Like siding, roofing materials come in many shapes and sizes. Galvanized-metal roofing is perhaps the most common choice because it is durable, inexpensive and easy to apply. Asphalt shingles are also used where it is important for appearance to match existing shingled roofs. Shingles, however, require more labor and must go on over a roof deck of plywood which adds to their expense. On roofs with extremely low pitch, asphalt roll roofing or built-up tar roofs are used to form a watertight seal.

ESTIMATING ROOFING MATERIALS

Roofing materials, except metal sheets, are measured by the square. A square is a 10- x 10-foot area or 100 square feet.

Asphalt shingles are sold in bundles, each containing enough shingles to cover ⅓ of a square. Thus nine bundles would cover 3 squares or 300 square feet of roof area. Shingles are also measured by their weight, the heavier ones being more durable; 235-pound shingles are common for most roofing applications, meaning a square of these shingles weighs 235 pounds. When ordering shingles, simply measure the square footage of both sides of the roof and divide by 100 to determine the number of squares you need. Add about 5 to 10 percent to this figure for starter courses, ridge caps and waste.

Double coverage roll roofing comes in 3-foot wide rolls that are 36 feet long. When installed with a half lap, each roll covers a net area of 50 square feet or one-half a square.

Some materials you will need for shingling are tar paper and roofing nails; 15-pound tar paper comes in 3-foot wide rolls that are 144 feet long. One roll will cover 4 squares. Galvanized roofing nails, 1¼ inch, are used for new roofs, and you'll need about 2½ pounds per square of roof area.

Galvanized metal and aluminum roofing come in different widths from different manufacturers, but two common widths are 32 inches and 38 inches. When properly overlapped, these widths will cover a net width of 30 and 36 inches, respectively. To figure the square foot coverage of metal roofing, simply multiply the net width by the length of the sheets.

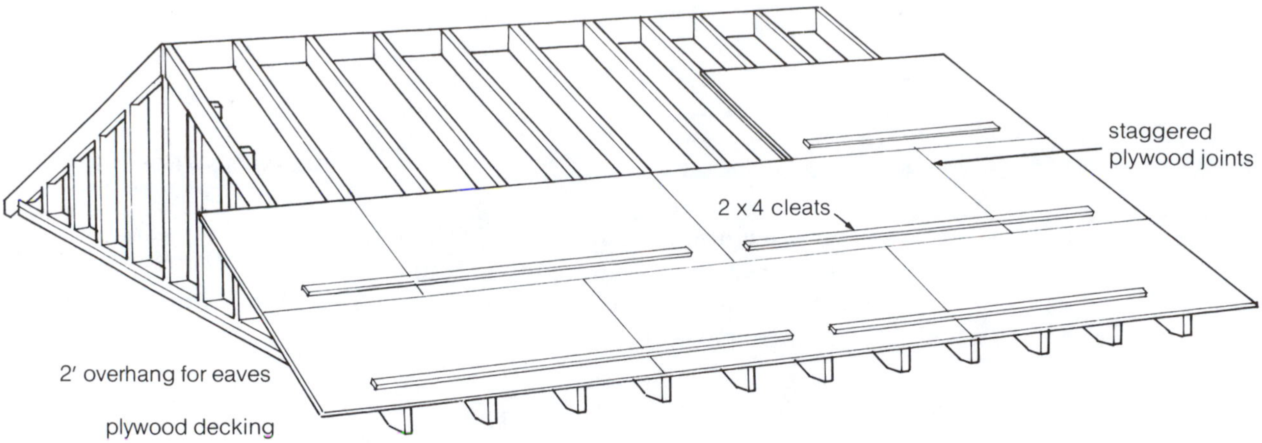

How to install plywood decking with overlapped joints and overhangs for eaves.

Plywood decking and eaves. Before the roofing goes on, the roof deck must be put down and the eaves framed in. (Eaves and overhangs are desirable to help protect barns, sheds, shelters and outbuildings from the weather. But, fully-enclosed eaves are not always necessary; fascia boards, soffits and other finishing details may be omitted when air tightness is not required.) For shingles and roll roofing, a full plywood or flakeboard deck is needed. Extend the roof deck beyond the two end rafters, the width of the eaves.

Here's how to put down a plywood deck. Starting at the bottom, run 4 x 8 plywood sheets horizontally across the rafters for greatest strength. Make sure the plywood is flush with the bottom of the rafters and overhangs the end walls of the building. Stagger the joints between pieces of plywood so they don't all line up on the same rafter. The first row of plywood is set using ladders and standing on the top of the wall plate. Next, nail 2 x 4 cleats across the plywood and into the rafters to give footholds as you work up the roof.

After the plywood has been nailed in place, the gable-end eaves must be framed to support the overhanging deck.

First, cut two sets of *fly rafters* which form the outside edges of the two gable-end eaves. Fly rafters are exactly like the normal rafters except they don't have bird's-mouth cuts and they are ¾ of an inch longer because there is no ridge plate to butt against. Normally, fly rafters are cut from 2 x 4 or 2 x 6 stock depending on the width of the fascia board which will cover them.

With the fly rafters cut, haul them up on the roof. With one person on the roof and another on a ladder, hold a fly rafter in position under the edge of the decking. Drive 16d nails through the decking to secure the rafter in place. The fly rafters should butt against one another at the roof peak and be even with the tail cut of all the other rafters.

Once all four fly rafters are nailed to the decking and suspended temporarily, they should be blocked against the building every 2 feet o.c. These blocks, called *lookouts,* should be toenailed into the end rafters and nailed to the fly rafters with 10d

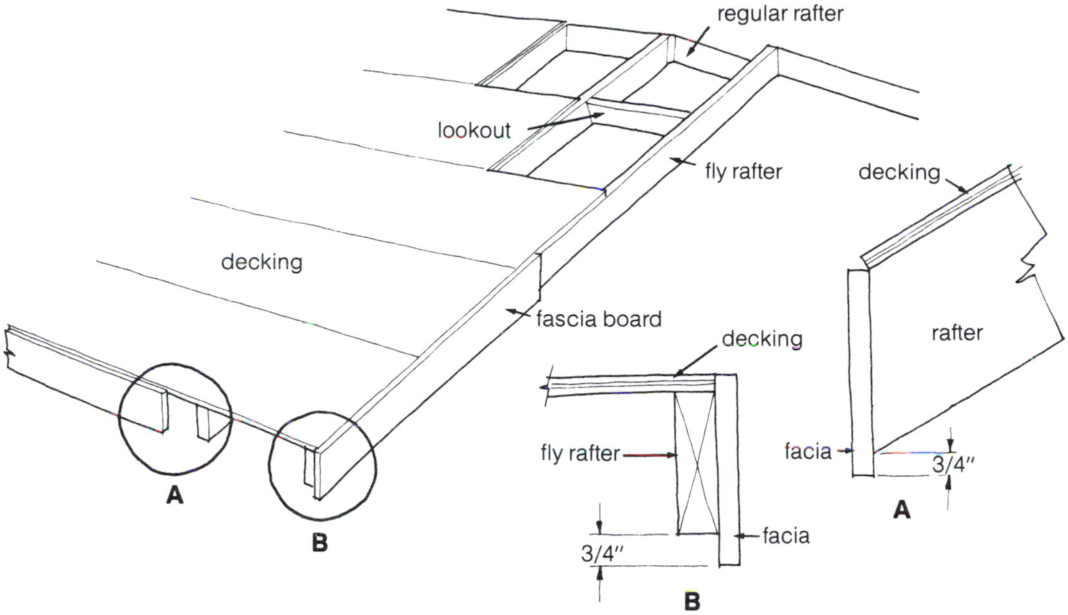

After decking has been applied, fly rafters, lookouts and then fascia boards may be added.

nails. When the blocking is in place, the eave overhangs will be strong enough to support the roofing and to work on. The fascia boards can now be applied to the fly rafters and over the ends of the main rafters. The top of the fascia boards should be flush with the top of the roof decking. Also, they should be wide enough to hang down ¾ of an inch below the soffit boards which will be nailed up under the lookouts. Once the fascia boards are on, you are ready to start roofing.

Metal Roofing

Like metal siding, metal roofing consists of long rectangular panels installed vertically. All that is needed to support the metal is 2 x 4 strapping running across the rafters every 2 feet o.c. The first row of strapping should be even with the bottom edge of the rafters and overhang the ends of the building the width of the eaves.

After the eaves are framed in and the fascia boards attached as previously described, nail the metal sheets in place. Starting at the bottom corner of the roof, place the first sheet square with the rafters and overhanging the fascia boards by at least 1½ inches on the sides and bottom. This overhang will keep water from dripping onto the fascia boards. Then complete the bottom row of roofing, overlapping the sheets according to the manufacturer's directions and keeping a constant 1½-inch overhang on the bottom. If more than one row of sheets is required to reach the peak of the roof, simply overlap the rows like shingles. For steep-pitched roofs (6 in 12 or more), a 4-inch overlap

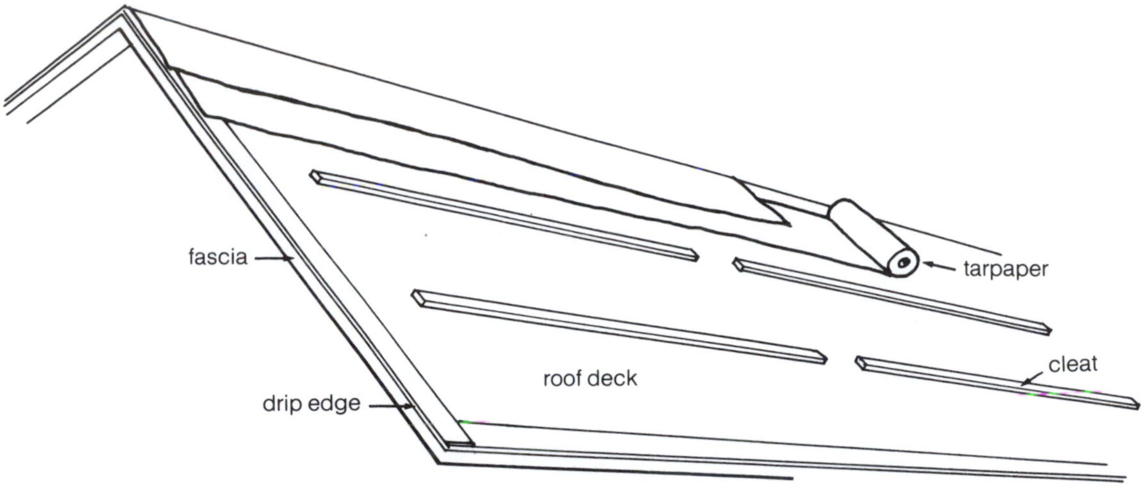

Before installing asphalt shingles, first attach a drip edge, then lay down tar paper. Underlay the paper, from the top down, to avoid having to repeatedly attach and remove cleats.

will seal out water; for lower pitched roofs, use an 8-inch overlap.

When both sides of the roof have been covered, nail a ridge cap over the peak to seal the joint there. Metal ridge caps are adjustable to conform to any roof pitch.

Corrugated fiberglass panels are also available that you can use in conjunction with metal roofing to let light into a building. Install these just as the metal sheets with an overlap on each edge. Though these panels are fairly expensive, they allow natural daylight into the barn, thereby cutting your electrical bill.

Asphalt Shingles

Asphalt shingles come in several different styles and a wide variety of colors. They must be installed over a ½-inch plywood or flakeboard roof deck covered with 15-pound tar paper and edged with galvanized metal drip edge.

When the plywood deck is in place, frame the gable eaves and attach the fascia boards as previously described. After the fascia boards are on, nail a galvanized drip edge to the edges of the roof deck. The drip edge, an important feature, keeps water off the fascia and gives a straight line for the roof edge.

Next, staple 15-pound roofing felt (tar paper) to the plywood. Start at the top of the roof, not at the bottom as is commonly recommended, so you can use the 2 x 4 cleats to walk on. Roll out the first course at the peak and staple it down except along the bottom. Move down 3 feet and roll out the next row, slipping it up under the top row to give a waterproof overlap of about 4 inches. It requires a little more patience to "under lap" the tar paper this way, but it saves the time and work of having to first remove all the cleats, reattach them as you install the tar paper, and then remove them once again. If you install the tar paper from top down, you only need to attach the cleats once and remove them once as you work down the roof.

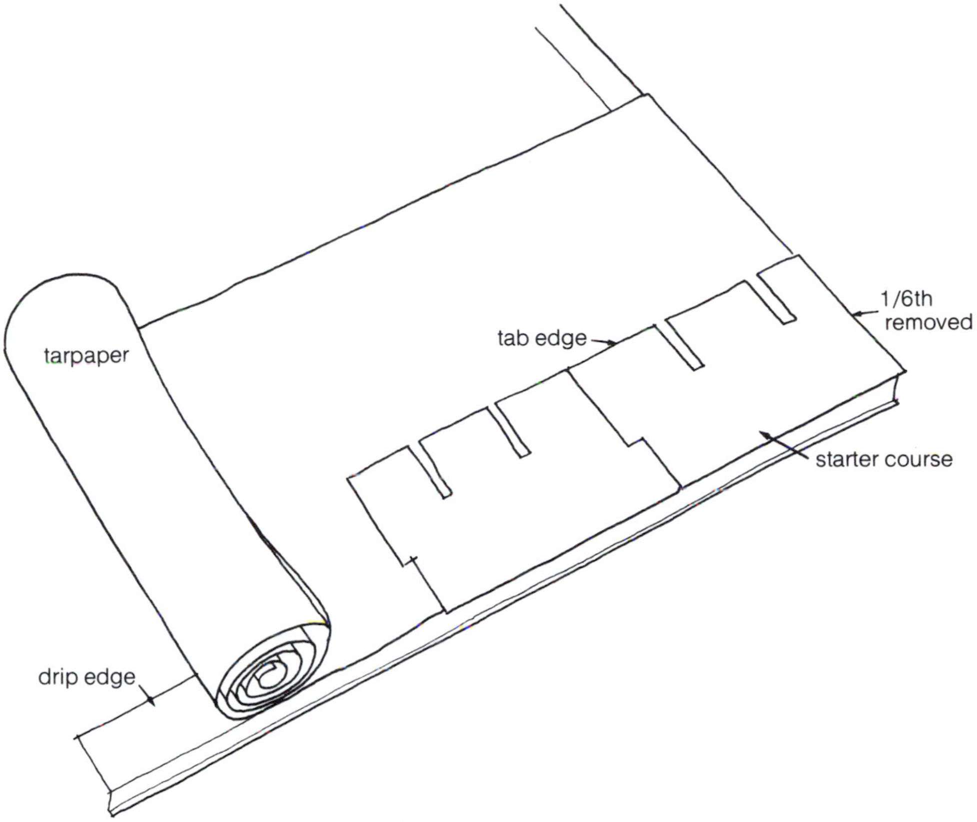

After laying down the tarpaper, begin a starter course for three-tab shingles with the tabs pointing up and 1/6th of the first shingle removed.

With the tar paper installed, you are ready to shingle. Some tools you will need are a straight edge (a rafter square is good for this); a utility knife for cutting shingles; and a chalk line for marking shingle courses. You will also need roof jacks which are wooden brackets for holding the staging if the roof has more than a 4 in 12 pitch. You can attach these brackets to the roof without damaging the shingles to create a secure work platform. Usually, you can rent them from tool rental shops or building suppliers.

To install asphalt shingles, follow manufacturer's instructions. The first course of shingles is called the *starter course.* Install this course of shingles with the tab edges pointing up toward the peak of the roof. One-sixth of the first shingle must be cut off so there won't be an overlap of the tab slits of the next course. When cutting shingles, always cut from the back so the knife blade won't be dulled by the pebbled front surface. Nail the shingles with four 1¼-inch galvanized roofing nails.

Nail the first "exposed" course directly over the upside down starter course, using full shingles. To start the next course, use a shingle with ⅙th of one end removed. The bottom of the shingle should fit up against the tab slits on the underneath course. For the next course, cut off ⅓ of the first shingle, then cut off ½ of the shingle that starts the next course, and so on. The result is that several courses have been started. Once these are com-

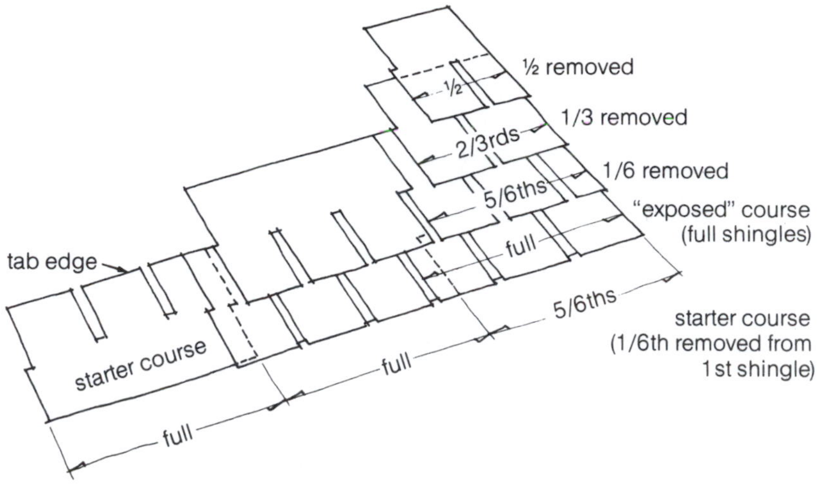

When installing asphalt shingles, follow manufacturer's instructions, or trim shingles in a sequence like this to avoid overlapping slits.

pleted across the roof, start the process over again. As you move up the roof, use a chalk line and tape measure to keep the rows straight and horizontal.

When you get to the ridge, don't cut the last course off even with the peak. Bend it over to the other side of the roof and nail it down. When both sides have been shingled, a ridge "cap row" is nailed in place. This consists of shingles cut vertically into three 12-inch pieces. These are folded over the ridge sideways, leaving the normal amount "to weather." Work along the ridge *toward* the direction of the prevailing winds, so the overlaps face away from the wind and driving rain.

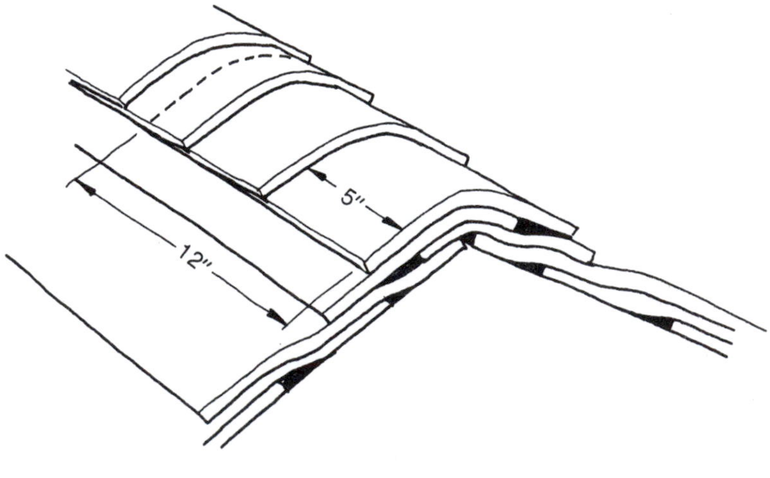

At the ridge, the last courses of shingles are bent over so they overlap, then a "cap row" of 12-inch shingle pieces is placed as shown.

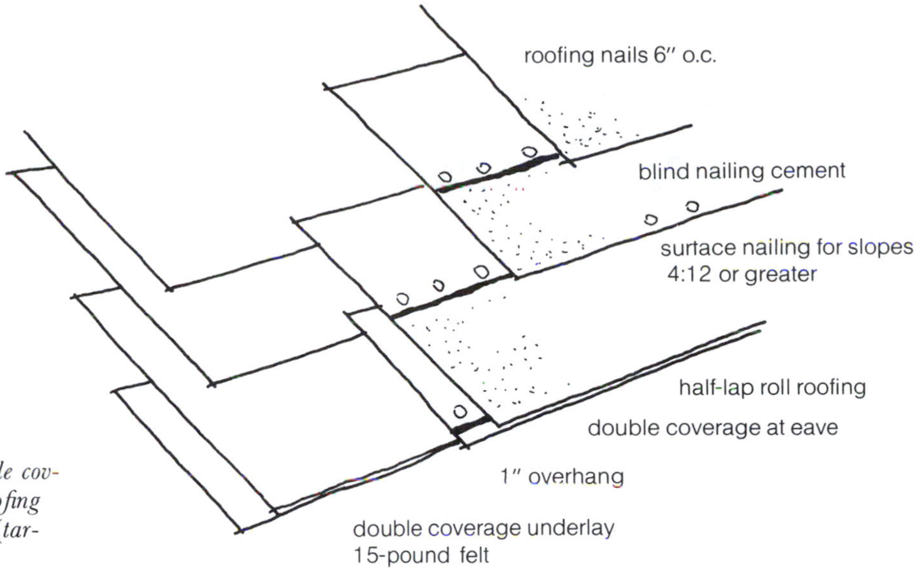

Roll roofing is installed with double coverage at eaves. And beneath the roofing is a double layer of 15-pound felt (tarpaper).

Roll Roofing

For shallow-pitched roofs with less than a 4 in 12 slope, use asphalt roll roofing applied on top of 15-pound felt (tar paper). Staple the tar paper to the roof deck.

Roll roofing comes in 3-foot wide rolls that cover 50 square feet each, when applied as double coverage (with a half lap on each row). Unroll horizontally across the roof and apply strips from the bottom up. Overlap each strip half its width by the row above it and fasten with roofing nails along its top and bottom edges. Spread a sealing compound called *blind nailing cement* under the bottom edge of each row to seal the nail holes where they penetrate the roofing.

For extra protection on shallow-pitched roofs, tar the roll roofing with asphalt roof coating. Apply the coating with a tar brush. The coating hardens into an impervious surface within a few days.

CHAPTER 6

Windows, Doors and Finishing Details

Borg

WITH A ROOF over a building, the rush of construction is over. The building is protected from the rain so you can slow down the pace of your work and concentrate on the windows, doors and finishing details. The type of windows you select for your outbuilding will depend on your budget and carpentry skills. Factory-built, pre-hung windows are available that can be simply nailed in place. These are easy to install and weathertight, but expensive. If you have a limited budget, you may want to make your own windows out of fixed panes of glass or recycled window sashes. Homemade or recycled windows require extra carpentry work and may not be as air tight as factory units, but you can easily save $50 or more per window.

Windows

Four principal types of windows are used for small-barn construction. The most common is a *double-hung window* with two sashes that slide up and down in tracks. On old windows, the sashes are divided into "lights" (small panes) by wooden muntins. On modern double-hung windows, the muntins have been replaced by wooden or plastic grills mounted over a large single pane of glass to give the impression of small panes. This allows for easier cleaning of the glass and the use of double or triple glazing for greater energy efficiency.

Awning windows have only a single sash that opens out at the bottom. These are usually less expensive than double-hung windows and more energy efficient. They come with either a roller opening mechanism or sliding-friction hinges to hold them open at any desired angle.

Sliding windows are available that have either two moveable sashes or one fixed sash and one moveable one that slides to the side over the fixed one. These are available with both wood and aluminum frames. These units are ideal where wall height is a limiting factor or where windows must be placed high on a wall to keep them away from animals.

Barn sash is another type of window widely used for outbuilding construction. It is simply a fixed pane of glass in a frame mounted in a fixed position.

Windows should be concentrated on the south side of a building to capture solar heat and to avoid winter winds. While the primary function of windows is to let in light, they serve other purposes as well. At strategic points, windows should be operable to help with summer cooling and cross ventilation. Not all

sliding

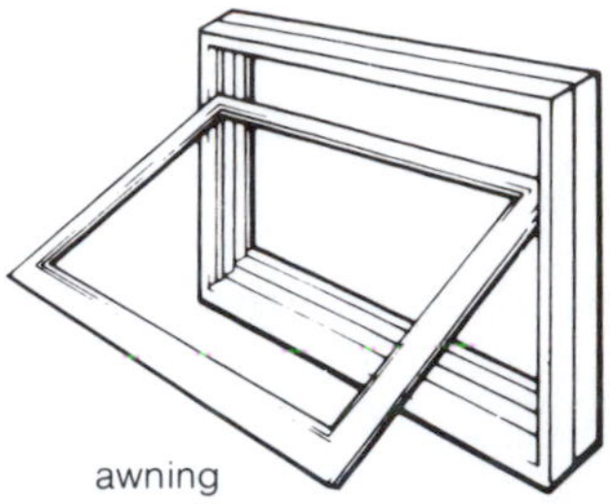
awning

barn sash

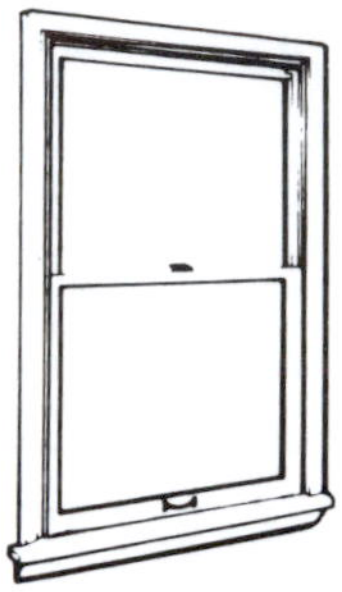
double hung

Windows typically used for small barns and outbuildings include the sliding and awning type, the double hung window and ordinary barn sash.

windows need to be operable, however. Fixed panes of glass and barn sash are less expensive, easier to install and more energy efficient than operable windows.

Installing Windows

Pre-hung units. Pre-hung windows are installed after the sheathing is on, but before the finished siding is in place. If there is no sheathing (if, for example, the plywood beneath board siding has been omitted), install the windows over the siding.

An important measurement of windows is the width of the jambs. The jambs must be wide enough to extend from the outside of the sheathing (or siding if there is no sheathing) to the inside of the finished wall. This will allow window casings to be nailed onto the edge of the jambs to finish the window trim. You can order windows with different-width jambs to fit different walls. Or, you can order or make jamb extensions to extend the inside of the jamb. If the inside of the building will not be finished with wall paneling and trim, then the width of the jamb is not as critical.

To install pre-hung windows, first staple 15-pound felt tar paper in a strip about 12 inches wide around the window rough opening and over the sheathing. (Omit this step if the window

Wherever possible, use tarpaper to reduce air infiltration. (Illustration shows tarpaper partially installed.) Shims can hold the window so it is perfectly level and square.

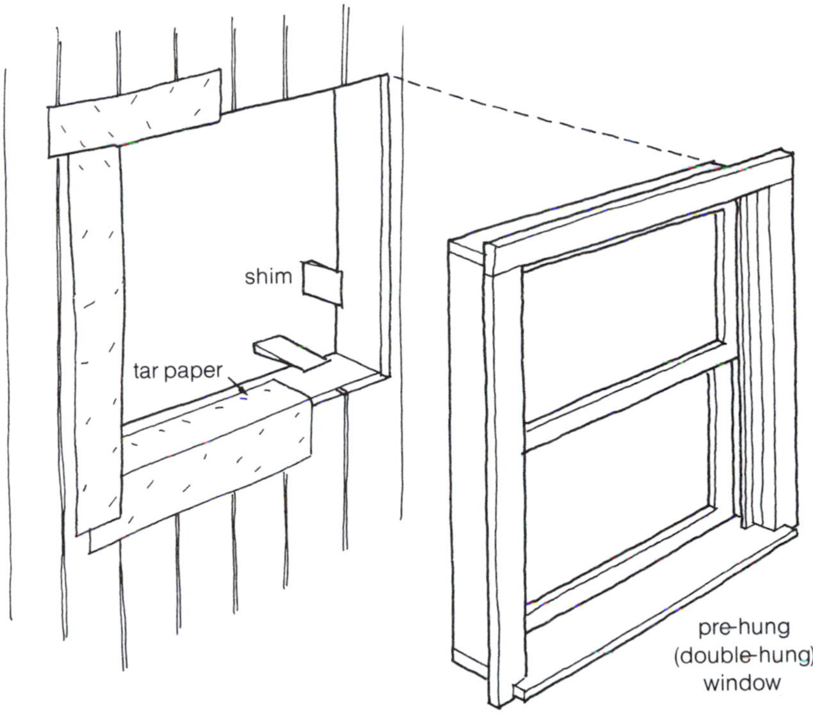

is being installed on siding and there is no sheathing.) Then run a bead of caulking on the tar paper or siding, under the places where window casings will be nailed to the *sheathing*. This will help minimize air infiltration. Hold the unit tightly against the building and have a helper wedge shim shingles under the window from the inside of the building until the window is level and at the proper height. When the unit is level, adjust the side jambs with shims until the window frame is perfectly square. You can check this by either using a rafter square on the inside of the jambs or checking the diagonals of the window opening to make sure they are equal.

When the unit is level and square, drive a nail part way in each corner to hold it in place. Check again for squareness, and if it hasn't moved, nail the window permanently through the top and side casings. Use 12d galvanized casing or finish nails driven securely into the framing of the rough opening.

Recycled windows. Recycled window sashes are ideal for barns and outbuildings. You can rebuild sashes into double-hung, awning or fixed windows at a fraction of the cost of factory-made units. It is best to try to get a set of similar windows from a single building so that they will all match. If the units are in good shape, you can remove entire frames in one piece from the old building and simply reinstall them in your rough openings. If the frame is rotted or cracked, you can still salvage the sashes and either rebuild the frame and tracks or install the sashes as fixed units.

A horizontal, sliding window is the easiest operable type to build from old sashes. The first step is to build and install a window frame that will fit in your rough opening. The frame

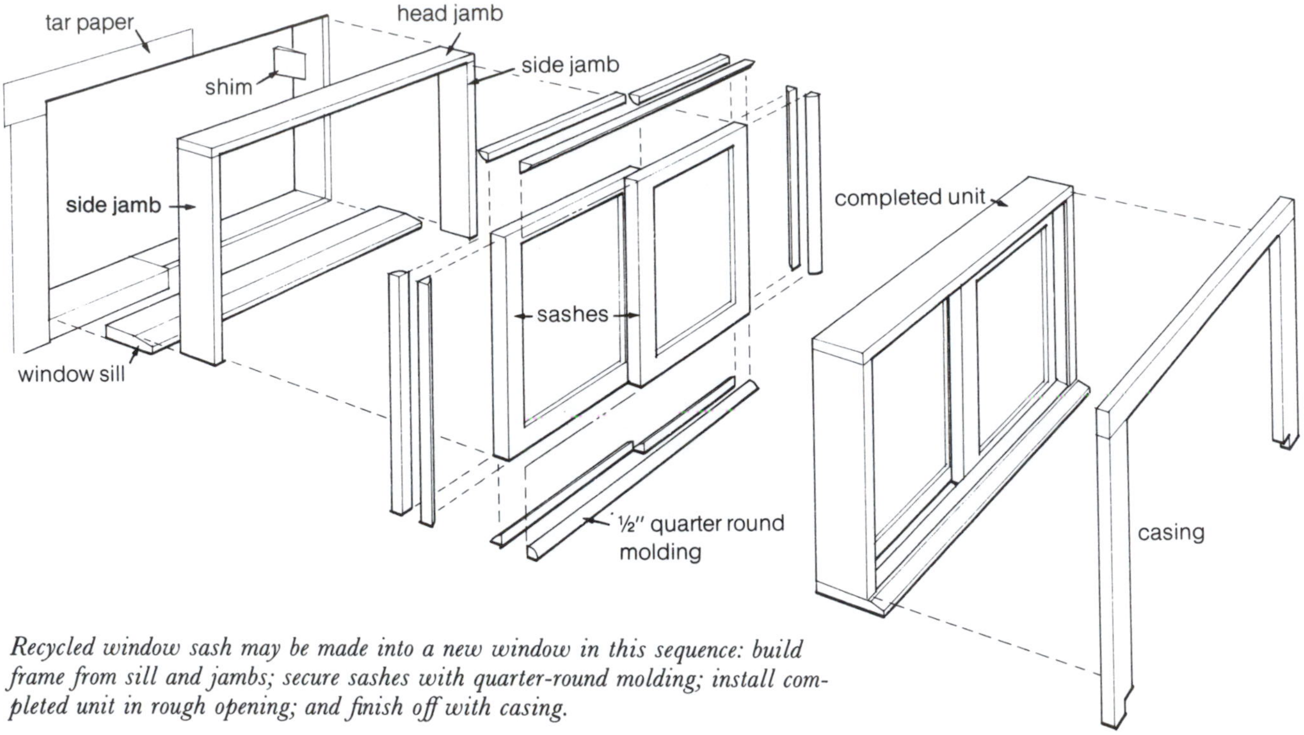

Recycled window sash may be made into a new window in this sequence: build frame from sill and jambs; secure sashes with quarter-round molding; install completed unit in rough opening; and finish off with casing.

also must have the proper dimensions to hold the sashes. To make the frame, use a clear piece of 2 x 8 or 2 x 10 for the window sill and 1-inch stock cut to the proper width for jambs.

First, cut and install the sill. It should overhang the siding by 1½ inches so that it projects out beyond the window casing and allows water to drip free of the siding. Bevel the part that extends out beyond the siding at a slight angle so that it sheds water. Level the sill and nail it in the rough opening with 12d galvanized casing nails (about 8 inches o.c.).

Cut and nail the side jambs and the head jamb to complete the frame. The finished height of the window frame should be ⅛ of an inch taller than the height of the sashes so they can slide easily. And the width should be 1 inch less than the combined widths of the sashes so that they overlap. Use shims behind the jambs to get the proper dimensions and to make sure the frame is square.

When the frame is nailed securely, mount the two window sashes. To do this use ½-inch quarter-round molding to form tracks for the sashes to slide in. Lay the tracks so the inner window sash is fixed and the outer one slides by it.

Awning windows are also simple to make from old sash. First, a sill and jambs are installed as outlined before to make a window frame. Make the frame exactly to the size of the window sash. The sash can then be hinged either to the inside or the outside edge of the top (or bottom) jamb so it swings either in or out. Use pieces of ½-inch quarter round for stops to keep the

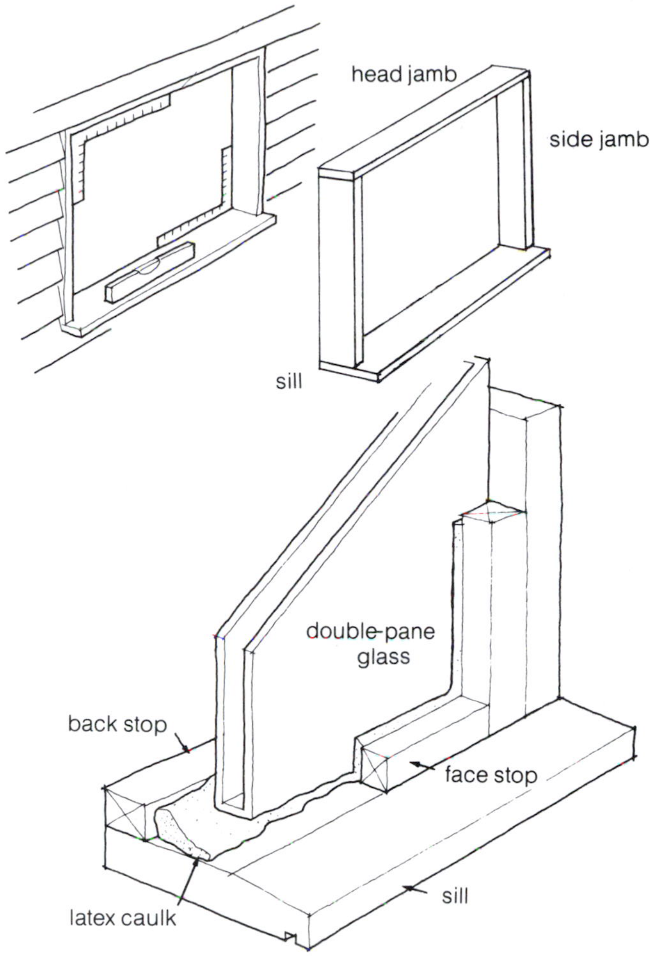

Fixed glass is easy to install in a rough opening; simply follow the sequence of steps outlined at left.

INSTALLING FIXED GLASS

Fixed glass is double-pane or insulated glass, set in a casing to make a window. "Fixed" means the window is not operable. Here's how to install fixed glass:

1. Build a frame for the glass that is ⅛-inch wider and higher on its inside dimension than the piece of glass. Make the sill from 2 x 8 stock with a beveled outer edge to shed water. Make the side and head jambs from 1 x 6 pine.
2. Install the frame in the rough opening, insuring that the sill is level and that the frame is perfectly square. Use shim shingles driven between the frame and rough opening to adjust its squareness. Secure the frame with 12d galvanized casing nails driven through the jambs.
3. Attach a back stop to the rear of the sill and jambs. The stop can be made from a small strip of wood or quarter-round molding and nailed with 4d finish nails.
4. Apply a good bead of latex caulk or a neoprene gasket strip around the window frame against the back stop. This should form a weathertight seal and expansion joint for the glass.
5. Place the glass pane or insulated glass unit against the back stop. Press it inward until it makes a tight fit along the back stop.
6. Nail the face stop in position to secure the glass. Be careful not to angle the nails into the glass.

window in position when it is closed. To hold the window open, you can use a variety of arrangements. A short stick, a hook and eye screw attached above the window or friction side hinges are all possibilities. All you need to keep the window closed are a hook and eye screw.

There is no reason that every window in a barn should be operable. After you have provided for adequate ventilation with moveable windows, use fixed panes or barn sashes that are cheaper and easier to install. Barn sash can simply be mounted in a window frame with quarter-round molding and caulking to make it air tight.

If you want large areas of glass for either solar heating or more light, consider using custom-made pieces of Thermopane glass. Thermopane is two layers of plate glass fused to a thin metal frame. It is also called insulated glass and is readily available from glass suppliers, made to your specifications. Thermopane units can be mounted directly in framed window openings just as barn sash is.

If you are considering passive solar heat for your livestock barn, an inexpensive option is using replacement glass panels

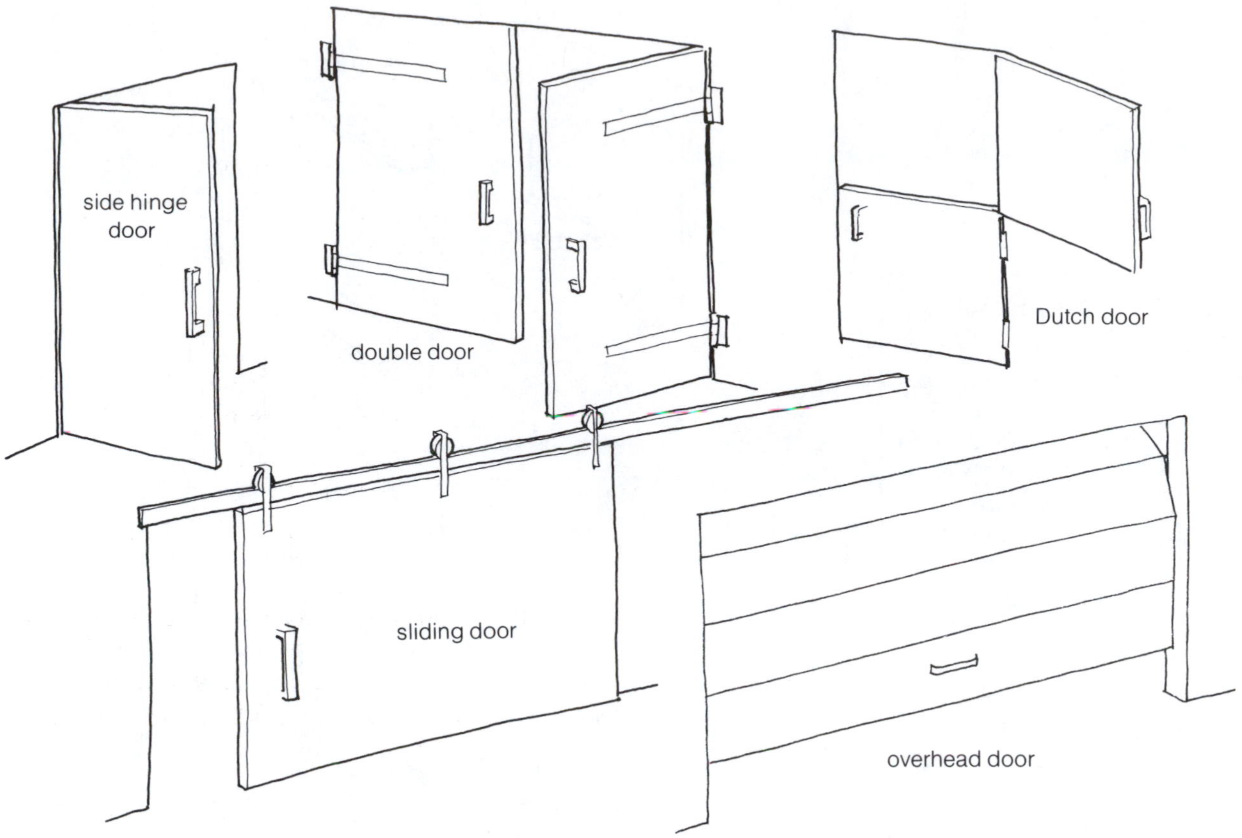

Doors suitable for small barns and outbuildings include the side hinge door, double door, dutch door, sliding door and overhead door.

for sliding doors. This is sometimes called patio glass. This glass comes in ⅝-inch Thermopane, and in several standard sizes including 34 by 76 inches. An extra advantage of these units is that the glass is usually tempered to resist breakage. On a square-foot basis, they are inexpensive, energy efficient and easy to install.

Other options for solar heating are fiberglass and plastic glazing panels. Most of these are not transparent so you can not see through them, but they let in almost as much sunlight as clear glass and can be bought at a fraction of the cost. They are also stronger than glass. Solar hardware suppliers have complete listings of these glazings along with installation information.

Doors

Several types of doors are used for outbuilding construction. The *side-hinge door* is used for both interior and exterior doors that need only be large enough to allow a single person through. These are available in solid wood and insulated steel (exterior doors) and in lightweight hollow core units (interior doors). The normal dimensions for exterior hinged doors are 36 inches wide by 80 inches high. Interior doors are often narrower.

Sliding doors are often used for large barn openings to accommodate tractors and oversized equipment. Usually, these doors are made on the job site out of rough boards, metal or plywood. They are hung on an overhead track with rollers so they can slide to one side of the door opening.

Overhead doors or garage doors are sometimes preferred to sliding doors because they take up less room and are out of the way when open. They are made of hinged panels that are spring mounted in a curved steel track.

The *Dutch door* and the *double door* are two variations of the side-hinge door. The Dutch door has a separate top and bottom panel so that either can be opened while the other is closed. These are used with horse stalls so the horse can have light and air through the opened upper part while remaining confined by the lower panel. The double door is simply two hinged doors mounted side by side to form a larger opening. Usually, these are made on the site out of boards or plywood and mounted with large T-hinges.

Installing Doors

Hinged doors can be bought pre-hung, complete with jambs, casings and threshold. Install pre-hung doors in the same manner as window units. Prepare the rough opening with 12-inch strips of tar paper and a bead of caulk around the outside. (Again, omit the tar paper if the door is to be installed on siding and there is no sheathing.) Tilt the door unit into place; plumb and square it, using shim shingles; and drive 12d casing nails through the side jambs to secure the door.

If you are using old recycled doors or new ones that are not pre-assembled, you will have to frame in the rough opening and hang the door yourself. The door frame is built just like a window frame except a door threshold, if used, replaces the window sill. Normally, thresholds are made from fir or oak to withstand the constant wear of boots and shoes. You can buy them pre-cut from building suppliers or make them from 1½-inch stock. Install the threshold first and then the side and head jambs.

When the door frame has been built and installed, mount the door hinges about 6 inches from the top and bottom of the door, and for extra heavy doors, add a third hinge in the middle. Notch the leaves of the hinges into the side jamb and into the door so the door fits tightly against the jamb when it is closed. Use a sharp chisel for this job. The surfaces of the hinge leaves should be flush with the surfaces of the jambs when you

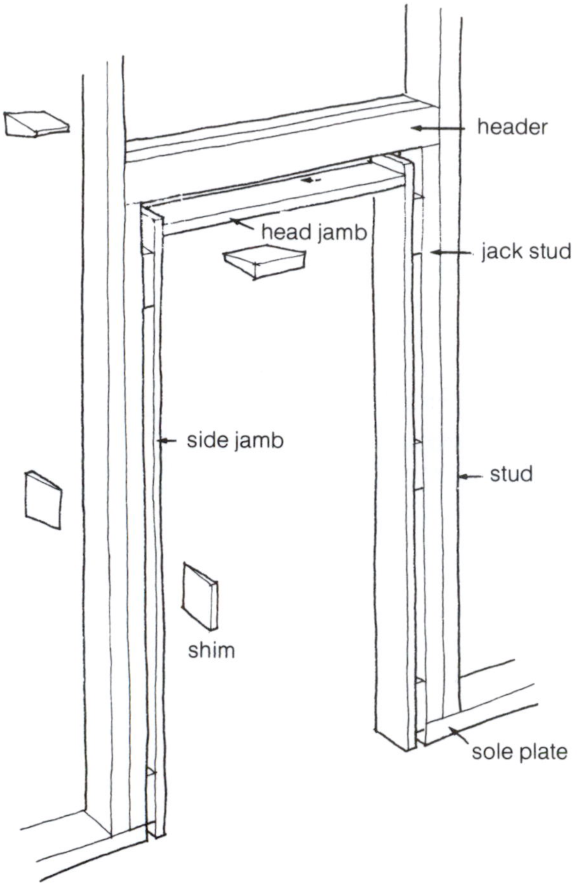

Major parts of the door frame are the side and head jambs. Thresholds are often omitted for small barns and sheds.

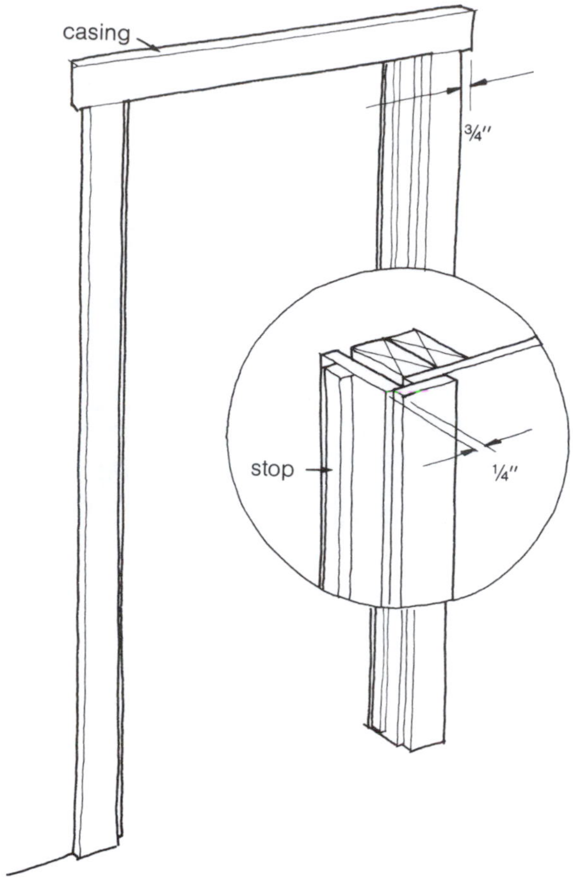

Apply stops to the side and head jambs, and cover the framing with casing.

finish. To make fitting easier, pull the pins out of the hinges so you can work with the individual leaves.

When the outer leaves have been notched and installed on the door jamb, place the door in the frame in its closed position. With a ¼-inch shim under the door for clearance when it swings, mark the position of the hinges on the side of the door. Remove the door from the opening and chisel out mortises for the other leaves and mount them on the door. If you have measured accurately, you can now hold the door in place, slip the hinges together and reinsert the pins.

After the door is hung, apply *stops* to the side and head jambs to keep the door from swinging too far when it is closed. You can buy stops in a variety of styles with different edge moldings. For outbuildings, make them from 1 x 3 spruce or pine stock ripped in half. With the door in the closed position, nail the stops in place so they make a tight fit against the door. On exterior doors, you can add weatherstripping to the stops to give an airtight seal.

To finish off the door, nail casings around the inside and outside of the door to cover the jamb and frame installation. Normally, casings are made of 1 x 4 or wider pine boards nailed over the edge of the jambs and the siding. The casings are set back from the edge of the jambs by ¼ inch on all sides and the top piece of casing overhangs the sides by ¾ inch on each side. You can also set casings flush with the jambs and omit top overhangs. It is a matter of preference.

Fit the tops of doors and windows with drip caps to keep water from getting in behind the casings. Make drip caps from a piece of crimped aluminum or a beveled piece of wood set on top of the window casing. Even a large bead of caulk is often satisfactory if the window or door is partially protected by the eave overhang.

Homemade Doors

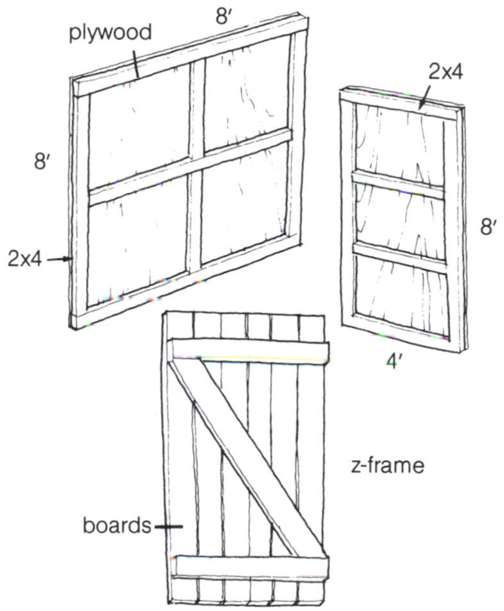

Simple home-made doors can be fabricated from 2 x 4s and plywood or rough-sawn boards backed by a Z-frame.

An inexpensive alternative to buying solid-wood doors is to make your own out of plywood mounted on a 2 x 4 frame. First, make a 2 x 4 frame of kiln-dried lumber to fit the size of your door opening. Nail the 2 x 4s together on edge, so that the thickness of the frame is 1½ inches, not 3½ inches. The frame should have cross members at least every 4 feet to support the plywood skin.

Make sure the frame is square before attaching the plywood to it. Then fasten the plywood to the frame with 6d galvanized nails.

You can make doors for stalls and pens for small livestock from single sheets of plywood mounted with hinges; ⅝- to ¾-inch plywood is the minimum thickness that can be used alone without warping.

Doors must be sturdy enough to hold up to the rough wear and tear an outbuilding receives. A flimsy door or one that is hung on undersized hinges won't last long. If ⅜-inch plywood facing isn't sturdy enough for your purposes, increase the thickness of the plywood faces or use rough-sawn, 1-inch boards.

If you use boards for the door facing, you must brace the frame so it has lateral rigidity. The easiest way to do this is to make a Z-frame.

Sliding and Overhead Doors

Sliding and overhead doors are good equipment entrances to barns. You can build sliding doors using a wooden frame covered with plywood or board skins. The door has rollers bolted to its top that slide on an overhead steel track.

When designing a sliding door, consider its size and weight carefully. Most overhead tracks are designed to carry a maxi-

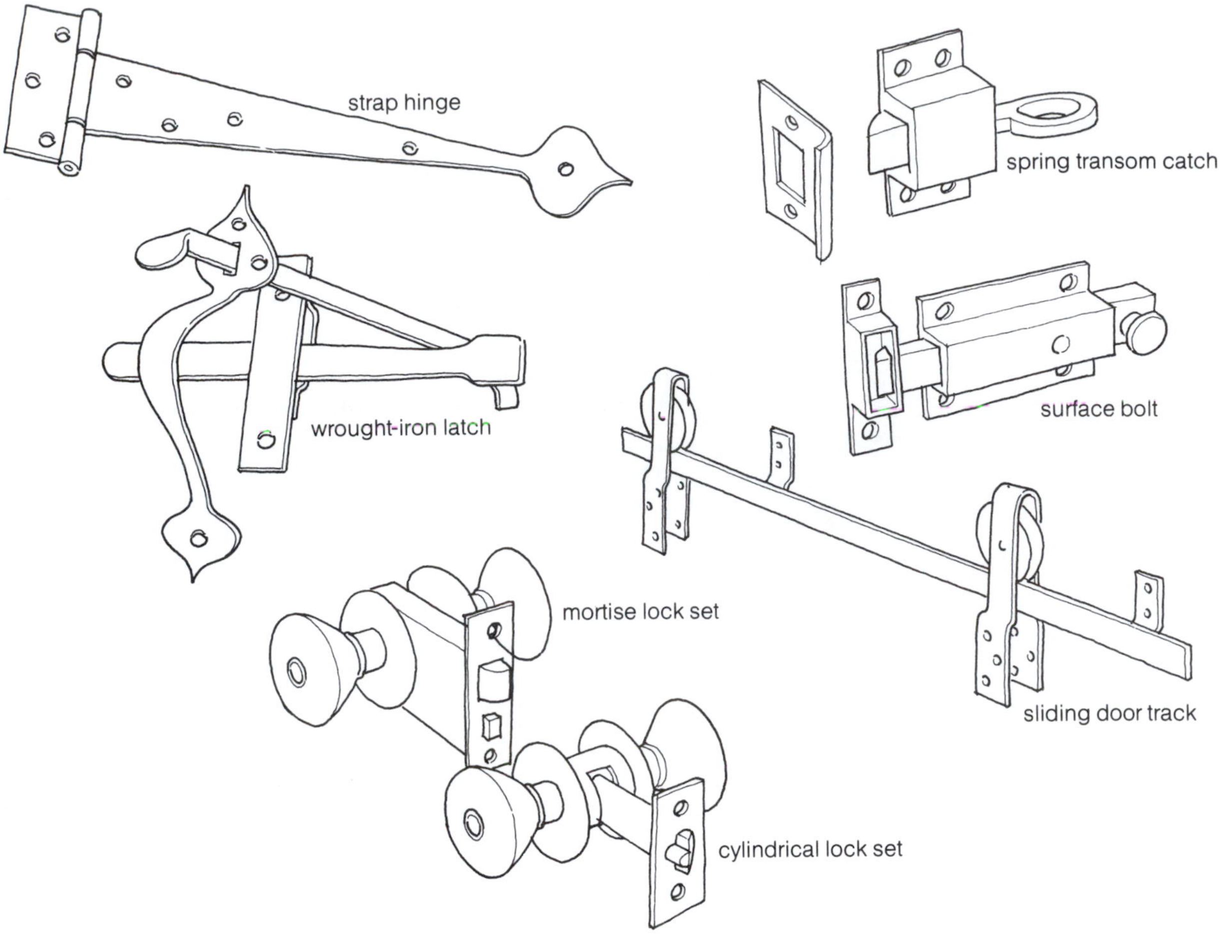

Door hardware, suitable for small barns, sheds and shelters, includes these various pieces. Simple latches or hooks are also often adequate.

mum weight of 300 pounds. A 10- x 8-foot sliding door made with rough-sawn framing and covered with 1-inch boards might put this carrying capacity to test. It would also take three people just to stand it up and put it in place. One way to minimize the door weight is to use kiln-dried framing members and plywood skins.

Another problem with large, horizontal sliding doors is figuring out where they will slide to when they are opened. Keep in mind a 10-foot sliding door mounted inside a 20-foot wide barn won't open all the way. Also, when the door slides along the inside of the wall, this wall surface must be free from obstructions and partitions. One way to minimize the area large sliding doors take up is to break them into two pieces that either slide open in opposite directions or overlap and slide to the same side. When built with two halves, sliding doors are also easier to mount and operate. Obviously, doors mounted on the outside would be encumbered by few obstructions. But, to prevent rusting, their mounting hardware should be protected from the weather.

If the barn layout restricts or interferes with sliding doors, overhead garage doors are an alternative. Because of the com-

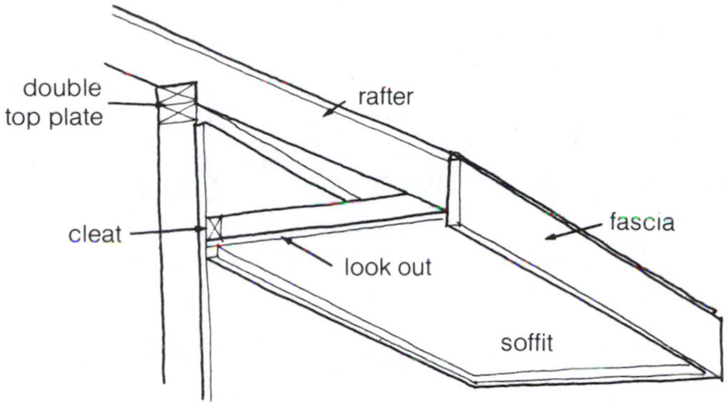

Horizontal Soffit

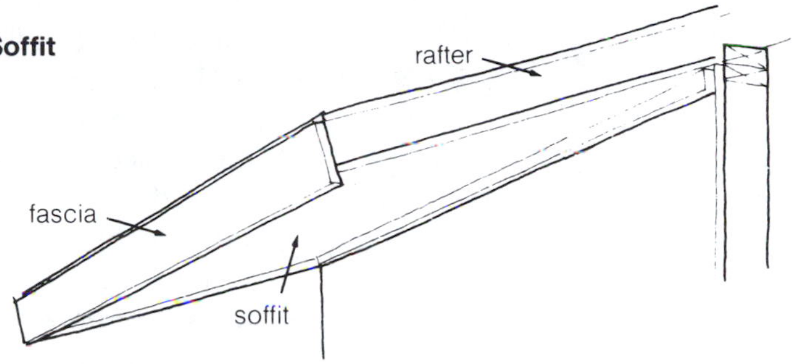

Sloping Soffit

If you plan to finish off the eaves of your outbuilding, you can do so with horizontal or sloping soffits.

plicated construction of the door panels and runners, it is not practical to build these yourself. Contract with a local overhead door company to install a completed unit or buy one with the necessary hardware from a building supplier and install it yourself.

Door Hardware

There are many different types of door latches and locks ranging from burglar-proof lock sets to simple sliding bolts.

Finishing Details

Trim

After the windows and doors have been installed, only the eaves are left to close in. On barns that are used primarily for storage, the eaves are left open to allow ventilating air to pass through. For livestock barns or others that must be kept warm, however, the eaves should be closed in. (In southern climates, this may not be necessary. In any case, be sure to provide adequate ventilation.) Closing the eaves will help keep out insects and pests such as squirrels.

Since the fascia boards have already been attached, all that is left to do is to attach the soffit boards to the underside of the eaves. Soffit boards, either single pieces of ⅜-plywood or 1 x 6 or 1 x 8 boards, can be attached horizontally or on a slope. (If

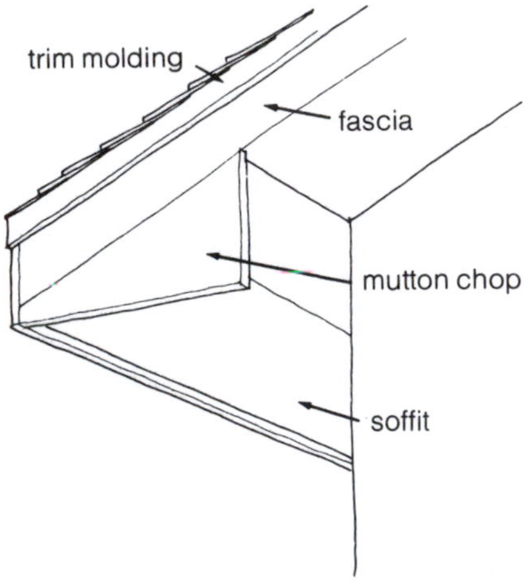

Mutton chop or eave return are names for the end piece that joins the horizontal soffit and vertical soffit board.

added air circulation is desired, you can install soffits with vents.)

If the soffit is to be sloping, simply attach the soffit board(s) to the underside of the rafters, butting the edges tight against the fascia board in front and the barn siding in the back. Secure the soffit with 8d galvanized nails. On the gable ends of the building, nail the soffit up under the lookouts.

If you are using the horizontal soffit detail, you need to attach lookouts to the ends of the rafters as horizontal nailers for the soffit. The lookouts can be 2 x 4s nailed into the side of each rafter, flush with the bottom of the rafter and extending horizontally back to the siding. Nail them into the rafters with 10d nails, and with 8d nails toenail the other end to a cleat mounted on the siding. Then attach the soffit boards underneath the lookouts.

A horizontal soffit requires an eave return to join the two different soffit planes where they come together at the corners of the building. Eave returns can be ornate, complicated structures, especially on older residential houses, but for a barn they can be quite simple. Your basic task is to join together two boards that are meeting at different angles.

Ventilation

A controlled environment is important to the health and productivity of farm animals. Adequate ventilation can help remove summer heat, moisture, dust and toxic gases; when these accumulate, mold, dampness and disease result.

The simplest way to ventilate a barn is to have well placed windows that provide cross ventilation. Consider prevailing winds carefully when laying out the windows and interior spaces of a barn. But, do not locate windows so livestock receive direct drafts. Drafts on confined horses and cows will cause pneumonia.

Rising hot air also induces ventilation. By installing air vents in the roof, such as cupolas or gable-end vents, hot air can be exhausted naturally from the building. As the hot air escapes, it pulls in fresh, cooler air through the ground floor windows. This process, called *thermosyphoning,* occurs naturally whenever the air inside the barn is heated.

Natural ventilation can also be assisted by electric exhaust fans placed in a cupola or wall vent. Fans can assist cooling by pulling cool night air through a building and lowering its temperature. In the morning, shut off the fans and close up the building to keep out the hot daytime air.

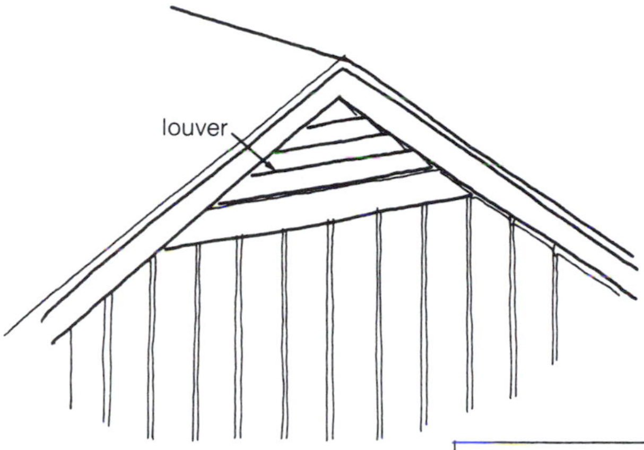

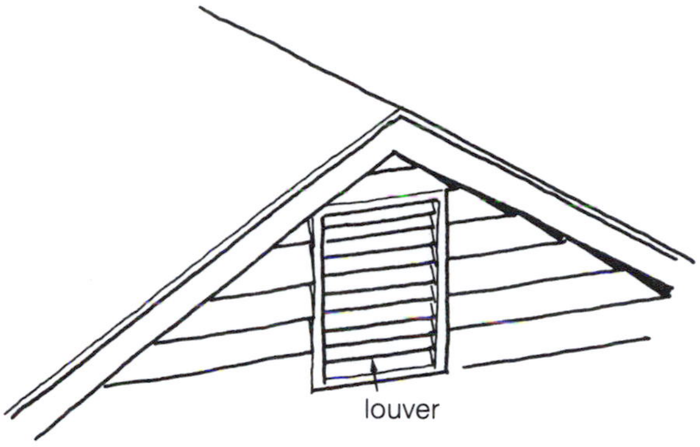

Gable-end vents such as these assist air circulation in small barns and livestock buildings. Vents should be screened inside to keep out insects.

VENTILATION FOR LIVESTOCK

Ventilation in livestock barns is critically important to the health and well being of the animals. Ventilation is necessary for cooling, to remove disease pathogens and to control moisture.

A common rule of thumb for livestock buildings with average animal density is to provide for 4 air changes an hour during the winter and 40 air changes an hour during the summer. Thus for a 30-by 20-foot barn with an 8-foot high ceiling, you would have to change 4,800 cubic feet of air every 15 minutes in the winter. This would require a ventilation system capable of handling 320 cubic feet per minute (cfm) of air. In the summer, you would need ten times this rate or 3,200 cfm.

To meet the dual requirements of summer and winter ventilation, you can either buy a single exhaust fan with variable speeds or buy two fans, a small one for winter and a larger one that can pick up the extra summer load. A fan is rated by the cfm so that you can determine how much air it will move. Never buy a fan solely on the basis of its horsepower or blade diameter; these features may have little relation to its air-moving capacity.

The fan size must be matched to the air inlet's area if the fan is to work at maximum efficiency. A good rule of thumb is 18 square inches of air inlet for every 100 cfm of summer fan capacity. This will give you an air velocity of 800 cubic feet per minute. During the winter, part of the air inlets can be closed off to compensate for the reduced air flow.

Allowing four air changes an hour during the winter can put a real strain on your heating system and pocketbook, especially in northern climates. One possible solution is to install an air-to-heat exchanger. A heat exchanger works by extracting the heat value of the air that is being exhausted from the building and transferring it to cool, fresh air that is entering the building. Heat exchangers are widely used in residential construction today, and for livestock operations in areas with large heating requirements and high fuel costs, this may be an economical solution.

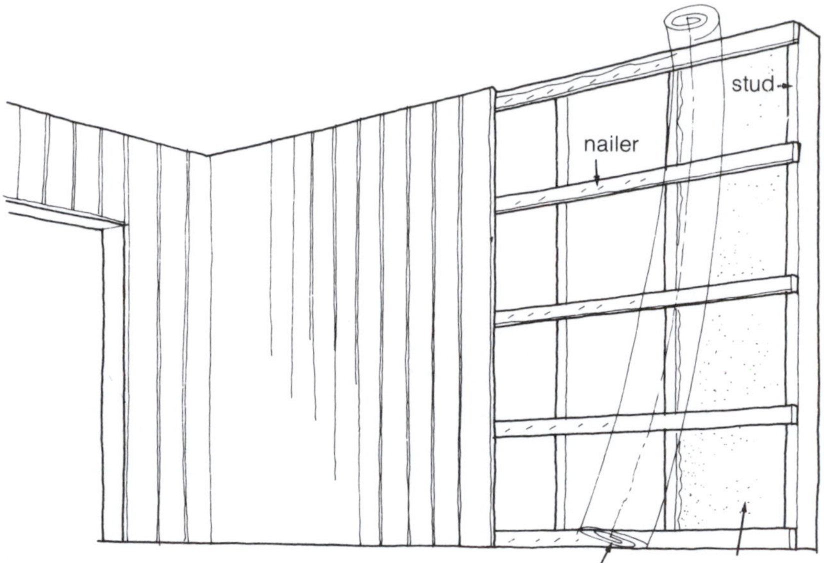

Place the vapor barrier behind the paneling to prevent moisture from entering the walls and damaging the insulation.

Insulation

Heated livestock barns, tool shops or garages should be well insulated to maximize comfort and minimize fuel bills. Insulation is easy to install, and it is one of the best investments you can make. (For more information on insulation, see Appendix C.)

The minimum amount of insulation recommended for heated buildings is 3½ inches of fiberglass in the walls and 6 inches in the ceiling. For northern climates, the minimum standard is 6 inches in the walls and 12 inches in the ceiling. Keep in mind that the entire building doesn't always need to be insulated; only those rooms that are going to be heated or cooled to maintain fairly constant temperatures.

Insulation value is measured by a number called the R factor. The higher the R factor, the better the insulation. Fiberglass, 3½ inches thick, has an R of 11; 12 inches has an R of 38. Fiberglass comes in batts or rolls, 16 or 24 inches wide, to fit between studs and ceiling joists. It also comes with an aluminum-foil backing, a kraft paper backing or unfaced. I would recommend using kraft faced or unfaced fiberglass as it is the most economical per R value. Don't bother with rigid-foam insulation, except for insulating concrete walls and foundations, as it is very expensive.

Always cover insulation with a *vapor barrier* to keep moisture out of it. A good vapor barrier is 4 mil polyethylene plastic that comes in 8-foot and wider rolls. This is stapled over the *inside* of wall studs and ceiling joists to keep moisture which originates inside the building from entering the walls. Never put a vapor barrier on the outside of a building or insulation; this would trap moisture in the walls and rot them. Exterior building surfaces should be able to "breathe." Even a layer of tar paper over the wall sheathing is a bad idea as it will trap moisture in the building.

To protect the insulation and vapor barrier from puncture, some type of interior wall sheathing is necessary. One possibility is building board, an inexpensive wall covering made from compressed cardboard. It is available in several different brands and thicknesses. Plywood is a more rugged alternative, but also more expensive.

If you are building an insulated and heated barn, ventilate the roof to avoid ice build up and moisture problems. If there is an attic space that has insulation in the floor and is unheated, install gable-end vents at both ends of the building. These vents allow air to flow beneath the roof, keeping it cold and venting moisture.

If the rafters are insulated, then install soffit vents and ridge vents; these will allow air to flow up between the rafters and out the peak of the roof. Leave at least 1 inch of air space between the insulation and the roof deck so the air can travel freely. Premade aluminum soffit and ridge vents are available that provide the correct amount of ventilating area.

CHAPTER 7

Wiring

Grant Heilman Photography

As a general practice, electrical wiring, whether for the home or an outbuilding, should be left to the professionals. If you want to do some of the work yourself (and find that it is legal to do so), you will probably need to secure a wiring permit before work starts and have an authorized inspector approve the work when it is completed. Many communities permit you to install new circuits, but require that a licensed electrician make the final connection at the service entrance panel. This is a good way to have your work inspected for safety; at the same time, you will be saving the electrician's fee for mounting the boxes and routing the wires.

Check local codes carefully to insure that you follow recommended wiring standards and requirements.* Don't skimp on materials or time; many barns and outbuildings have been lost because of hasty wiring jobs. And a hayloft aflame with animals below is a nightmare.

Before embarking on a wiring project, it's a good idea to have a basic understanding of electricity, and the terms used to describe it.

The quantity of current flowing through a wire of a specific diameter is expressed as *amperes*. If the wire is too small for the amount of current it must carry, it will overheat and trip the circuit breaker or blow a fuse.

In order for electricity to travel through wiring, it must be under pressure. This electrical potential is measured in *volts*. Most modern homes receive 240 volts of electric power. This allows the installation of 240-volt equipment such as electric water heaters, ranges and dryers, as well as standard 120-volt lights, outlets and small appliances.

Wattage, the amount of power flowing at a given point, is determined by multiplying amps times volts. Every electrical device has its rated power usage printed or stamped on it. Some have the wattage on an attached nameplate or tag. By adding up the wattage used by the appliances on a given circuit, you can determine the electrical load of that circuit.

For example, on a 20-ampere, 120-volt circuit you would have a maximum of 2,400 watts available (20 amps x 120 volts = 2,400 watts). Leaving a safety margin, you would plan to only have 2,000 watts of load on this circuit. If this was to be a circuit for infrared heat bulbs for warming pen areas, you would divide the wattage of each bulb into the designed circuit load to determine how many lamps could be on this circuit. If

---

* Also see: Summers, Wilfred I., technical consultant, and Ross, Joseph A., editor, *National Electrical Code Handbook* (Quincy, Massachusetts: National Fire Protection Association, 1981).

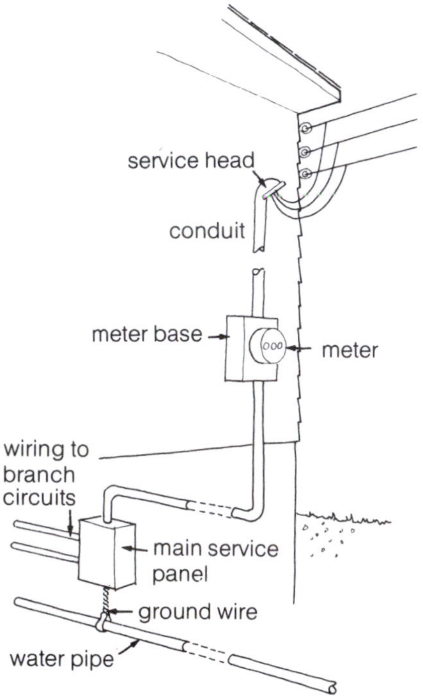

Wiring can be run directly from the basement main service panel to an outbuilding, or from the main service panel to a subpanel at an outbuilding.

the lamps are 300 watts each, you could attach six of them to this circuit bringing the total load to 1,800 watts, well under your design load.

Determine Your Power Needs

Before doing any wiring, examine your existing electrical service and determine whether this power is sufficient for your new electrical requirements. Electrical utility company personnel can help you make this determination. Also, after reading this chapter, prepare a wiring schematic, showing routes, wire sizes and the location of all fixtures.

Typically, a 100- or 150-ampere main service panel is sufficient to service a three-bedroom home, as well as a small outbuilding such as a nearby shed or chicken house. Each outbuilding should be on a separate power take-off from the main service panel, and each should have its own disconnect switch, so power can be turned on or off at the outbuilding.

If your main service panel has adequate capacity, but there is no room for a new circuit from the panel, you must install a sub panel for your outbuilding. If the main service panel has *in*adequate capacity, or if you plan a large outbuilding with a heavy electrical load, have the power brought from a utility pole directly to a new meter and main service panel at the outbuilding. If two or more outbuildings with heavy electrical loads are planned, it might be best to bring the power from the utility pole to a centrally-located yardpole and meter; then feed the power from there to service panels at the individual buildings.

Installation of the yardpole and incoming power lines to the meter should be done by electrical contractors or utility-company personnel. If the wiring is done from a yardpole, locate the meter and pole conveniently and centrally between the home and outbuildings. This routing system is better suited for larger operations with several buildings because it assures that no building is too far from the main service. Wiring from a main to a sub panel is recommended for a home and single outbuilding.

In either case, the feeder lines to the outbuildings can be located above or below ground. Use underground lines whenever possible, and plan the trench carefully so as not to interfere with underground plumbing. Overhead lines can be damaged by water, ice, windstorms or falling trees. They also can be snagged by tall farm equipment. In sum, they are an unsightly nuisance.

SAFETY PRECAUTIONS

•Don't attach wires to service panels before disconnecting power.
•Make sure the circuit you plan to work on is dead. Test it before making any repairs or connections.
•Don't wire buildings while standing on wet or damp surfaces. Even though the circuit you're working on is off, you may inadvertently come in contact with another live circuit or wire.
•Be sure to check with local codes before beginning any wiring projects.
•Be sure all equipment and wiring is properly grounded.
•Use pliers and other electrical tools with insulated handles.
•Don't staple or smash through cables.
•Protect underground cable with conduit wherever damage might occur.

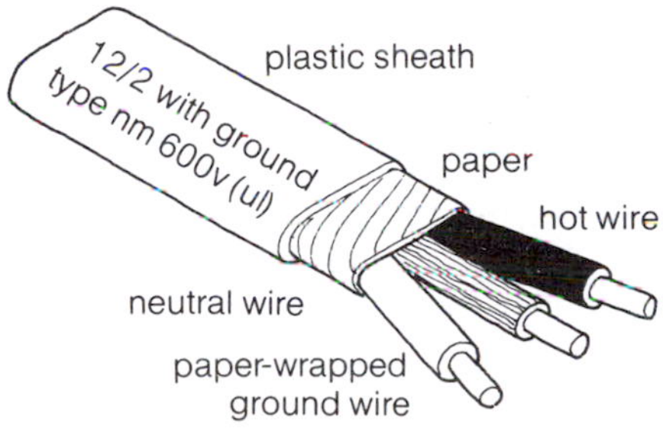

Twelve-gauge type NM cable with two conductors (12/2).

Check local codes for the types of underground wiring approved in your area. Bury the wires 12 to 18 inches below the surface and enclose them in conduit wherever they may be subject to damage.

Service Panel

Electric utility personnel will install the electric lines that run from the utility pole to the outside of the building or yardpole. In most cases, a homeowner or electrician then must continue the wiring to the meter and service panel.

The service panel box distributes power to the various branch circuits. It also contains the main disconnect switch for the entire wiring system as well as circuit breakers for each individual circuit.

Each circuit includes a series of outlets and switches, and each is protected by a circuit breaker at the panel. Whenever a circuit is overloaded, the circuit breaker will trip. When the problem has been remedied (usually by decreasing the number of appliances simultaneously using the circuit), power can be restored by flipping the breaker back on.

A three-wire, 30-ampere service panel (7,200 available watts) is usually sufficient for lighting a small barn. This service provides only limited capacity for lighting and a few smaller appliances. Larger barns and confinement buildings require at least a 100-ampere (24,000 available watts) panel; some may require 150- and 200-ampere service for full use of electricity for major appliances.

Remember that when planning your wiring system, extra capacity is important because the size of the service entrance and connecting wires limit the amount of electricity you can use at one time. With extra capacity, you avoid the fear of overloading the wires and tripping circuit breakers. Ask electric-utility company personnel for advice on the service panel best suited for your buildings.

The Wire

All wiring materials and installations must meet specifications set down in local codes. Most interior lighting and receptacles for barns and outbuildings are installed with plastic-sheathed cable, stapled or strapped every three feet to beams or other frame members.

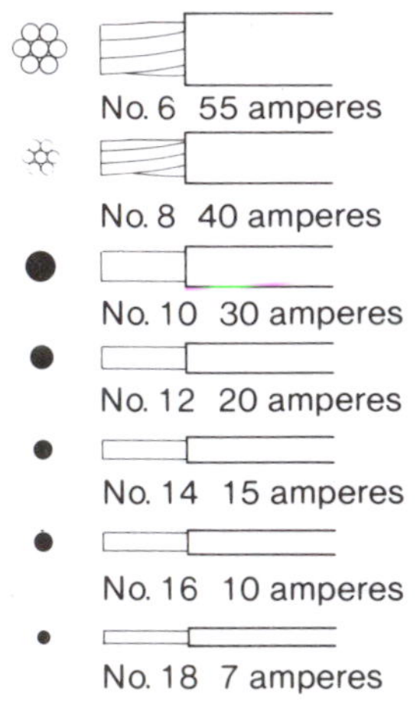

The smaller the wire number, the greater the diameter and ampacity.

Type-NM (non-metallic, sheathed) cable, commonly called *Romex,* is suitable for most outbuildings because its plastic sheathing protects it from moisture. Type NM-cable is made of a tough outer sheath which covers two or more plastic-insulated copper conductors and a bare copper grounding conductor. The ground wire is essential to your safety. It provides a conductor from the electrical system, either directly or through other conductors, to the ground.

Wire is also rated by numbers; the lower the number, the larger the diameter of the wire and the more current it can carry. To determine the right size of wire to use for a circuit, you must know the length of wire in the circuit and the maximum load from appliances and lights. Table 7-1 gives wire sizes for various loads and distances to the load center.

For example, the table shows that No. 12 wire, which has a

Table 7-1 Wire Size Required*

(computed for maximum of 2-volt drop on two-wire 120-volt circuit)

Load Per Circuit	Current 120-volt Circuit	Length of Run (Panel Box to Load Center)—Feet																	
Watts	Amps	30	40	50	60	70	80	90	100	110	120	130	140	150	160	170	180	190	200
500	4.2	14	14	14	14	14	14	12	12	12	12	12	12	10	10	10	10	10	10
600	5.0	14	14	14	14	14	12	12	12	12	10	10	10	10	10	10	10	8	8
700	5.8	14	14	14	14	12	12	12	10	10	10	10	10	10	8	8	8	8	8
800	6.7	14	14	14	12	12	12	10	10	10	10	10	8	8	8	8	8	8	8
900	7.5	14	14	12	12	12	10	10	10	10	8	8	8	8	8	8	8	8	6
1000	8.3	14	14	12	12	10	10	10	10	10	8	8	8	8	8	8	6	6	6
1200	10.0	14	12	12	10	10	10	10	8	8	8	8	8	6	6	6	6	6	6
1400	11.7	14	12	10	10	10	8	8	8	8	8	6	6	6	6	6	6	6	6
1600	13.3	12	12	10	10	8	8	8	8	6	6	6	6	6	6	6	6	4	4
1800	15.0	12	10	10	10	8	8	8	6	6	6	6	6	6	4	4	4	4	4
2000	16.7	12	10	10	8	8	8	6	6	6	6	6	6	4	4	4	4	4	4
2200	18.3	12	10	10	8	8	8	6	6	6	6	6	4	4	4	4	4	4	2
2400	20.0	10	10	8	8	8	6	6	6	6	6	4	4	4	4	4	4	2	2
2600	21.7	10	10	8	8	6	6	6	6	4	4	4	4	4	4	4	4	2	2
2800	23.3	10	8	8	8	6	6	6	6	4	4	4	4	4	4	4	2	2	2
3000	25.0	10	8	8	6	6	6	6	6	4	4	4	4	4	4	2	2	2	2
3500	29.2	10	8	8	6	6	6	4	4	4	4	2	2	2	2	2	2	2	2
4000	33.3	8	8	6	6	6	4	4	4	4	2	2	2	2	2	2	1	1	1
4500	37.5	8	6	6	6	4	4	4	2	2	2	2	2	2	1	1	1	1	1

* Middleton, Roger G., *Practical Electricity* (Indianapolis, Indiana: Audel and Co., a division of Howard Sams and Co., Inc., 1974).

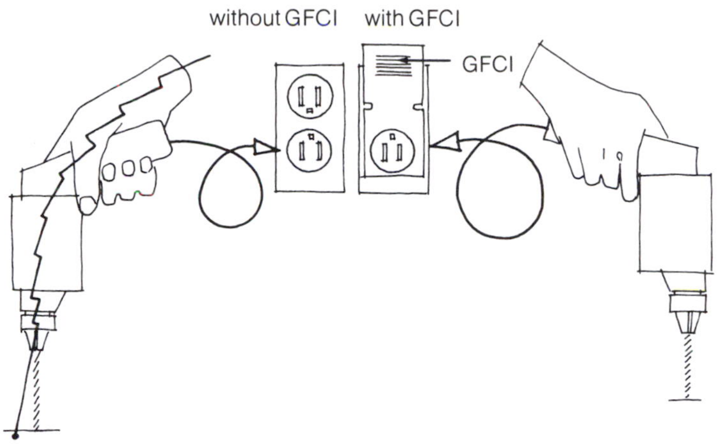

GFCIs protect people from fault currents. Be sure to check codes for grounding requirements.

maximum capacity of 20 amperes, can only be used for runs up to 30 feet on circuits that service a 2,200 watt load. No. 14 wire which is sometimes used for small lighting circuits and is rated for 15-ampere service can only be run 30 feet with a 1,400 watt load.

Cable can be purchased in standard 250-foot rolls or in smaller coils of 25, 50 and 100 feet. Unless the job requires only a few feet of cable, purchase one or two larger rolls rather than several smaller coils.

Grounding

The circuit-breaker panel in your home is grounded. This means there is a wire connecting the service panel to a rod driven into the earth. In residential areas, this ground wire is usually attached to the water-supply pipe and this in turn leads into the ground.

A grounded system reduces the effects of high voltage and lightning strikes, and reduces the danger of shock or fire, should some metal be accidentally livened.

Because of the inherently damp conditions in barns, it is also advisable to have the added protection of a *Ground Fault Circuit Interrupter* (*GFCI*). This device can be installed either in individual outlets or in the breaker panel box to protect the entire circuit. The function of a ground fault device is to shut off the circuit whenever it detects stray current that could shock you when using electrical equipment. The ground fault device will shut down the circuit even if the short circuit in the appliance is not enough to trip the circuit breaker but enough to give you a good shock.

Outdoor ground-fault protected receptacles can also be used to run heaters for water tanks during winter months.

Again, check local wiring codes to determine grounding requirements.

Planning Circuits, Lights and Outlets

When making a wiring schematic for your barn, shed or outbuilding, plan enough circuits, outlets and fixtures so switches are within easy reach, and lighting is adequate for the work you anticipate. For example, locate switches so lights may be turned on and off at two convenient locations. Also keep in mind, your wiring schematic may be used as a guide when you shop for electrical wire, fixtures and equipment. Here are some other suggestions:

•Plan on one light in the center of the alley for every two or three stalls. Install one outlet every 10 to 15 feet in the center of the feed alley. You also may need outlets for ventilating fans.

•Barns, box stalls and pens require one light outlet every 150 square feet of open pen area.

•For portable milkers, clippers and other power equipment, you will need outlets every 20 feet.

•In the milking room, there should be a light in front of every three cows, with an outlet behind every four to five cows.

•For every 100 square feet in the milkhouse, there should be a light and an outlet. Special-purpose outlets also may be needed for fans, coolers, sterilizers, heating equipment and water heaters.

•For hog and farrowing houses, one light is ample for two hog pens. Provide outlets for brooders and water warmers. Poultry-laying houses will need both dim- and bright-light circuits.

•Outlets are also needed for motor-driven feeders, and farm shops need outlets for each fixed piece of equipment.

•Special 240-volt outlets will be needed for power-driven equipment of ½ horsepower or more. Overloading and poor voltage can damage motors and other electrical equipment.

•Locate lights in feed and tack rooms, in lofts, stairways and alleyways. For horse or animal stalls, only use unbreakable light fixtures, such as those enclosed in wire housing.

•To enhance safety, use floodlights outside.

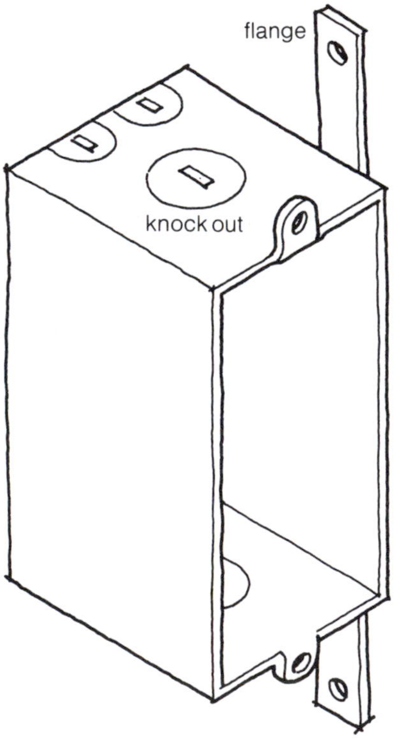

An outlet box like this may be nailed to a stud by its flange.

Installing Outlet Boxes

At switch and receptacle locations, mount rectangular outlet boxes. Buy standard 2½-inch deep boxes with side flanges for

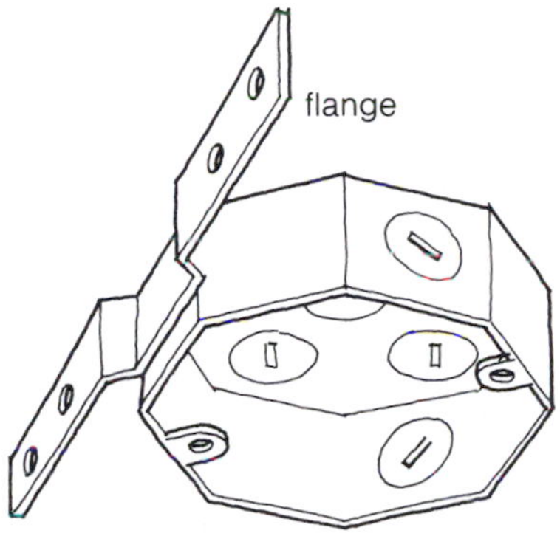

Octagonal junction boxes are used for overhead interior lights.

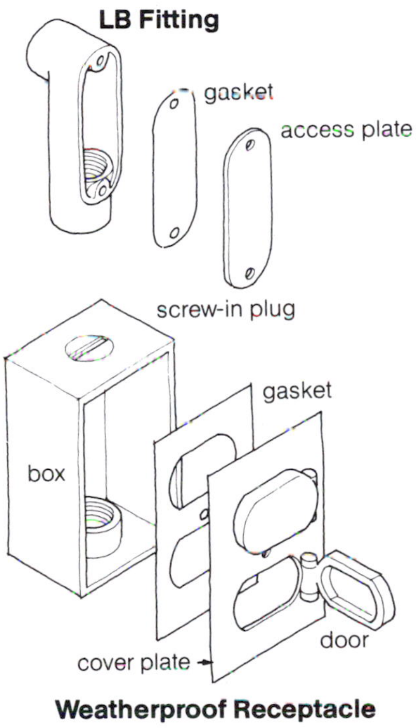

An LB fitting (L shaped with back conduit opening) protects wires passing out through a wall. A special outdoor cover plate makes the receptacle weatherproof.

nailing into studs. Mount the boxes so they will be flush with the finished wall (or studs if there will be no interior wall covering). Mount switches 4 feet from the floor and outlet boxes, 13 to 15 inches off the floor, or as required.

Unless you are using fluorescent light fixtures with built-in junction boxes, you will need to mount 4-inch octagonal junction boxes for overhead lights. You can buy these with side nailing flanges that mount directly on joists or with a hanging bar that spans between the joists and allows the box to be positioned at any point between them.

When the boxes have been mounted, run the circuit wires between them without installing switches or outlets. *Don't hook the wire up to the service panel yet.* Run the wires the most direct routes possible, but remember most codes prohibit running type NM beneath joists or over studs where it could be damaged. Therefore, you will have to drill holes through the studs and joists when you have to cross them. When drilling joists or studs, drill in the center of the wood to minimize structural weakening. If you have many wires running across a room, you can run them all together in a ceiling trough (wooden box) mounted under the joists.

When installing the wire into the junction boxes, strip about 6 inches of the plastic sheathing off the wire. Insert the wire through one of the back knockouts and pull it out the front until the plastic sheathing just enters the box. This will give you enough wire to hook up an outlet or switch. If the circuit continues from the box, another 6-inch length of unsheathed wire should be inserted in the box and the wire clamp tightened at the back of the box to hold both pieces of cable.

When running wire between boxes, be careful and do a neat job. Run the wire tightly enough to prevent sagging, but not so tight that it is stretched. Staple it every couple of feet with wire staples where it runs along joists or studs. When rounding corners and going through tight spaces, don't crimp the wire, but make easy and gradual bends so as not to damage the copper conductors or the plastic sheathing.

Wiring Lights, Switches and Receptacles

After the wiring has been run to all the boxes, *but before the circuit is hooked into the service panel,* install the switches, receptacles and lights.

In normal, two-wire circuits, there is a black wire that is "hot" and carries current to a fixture, a white wire that is neu-

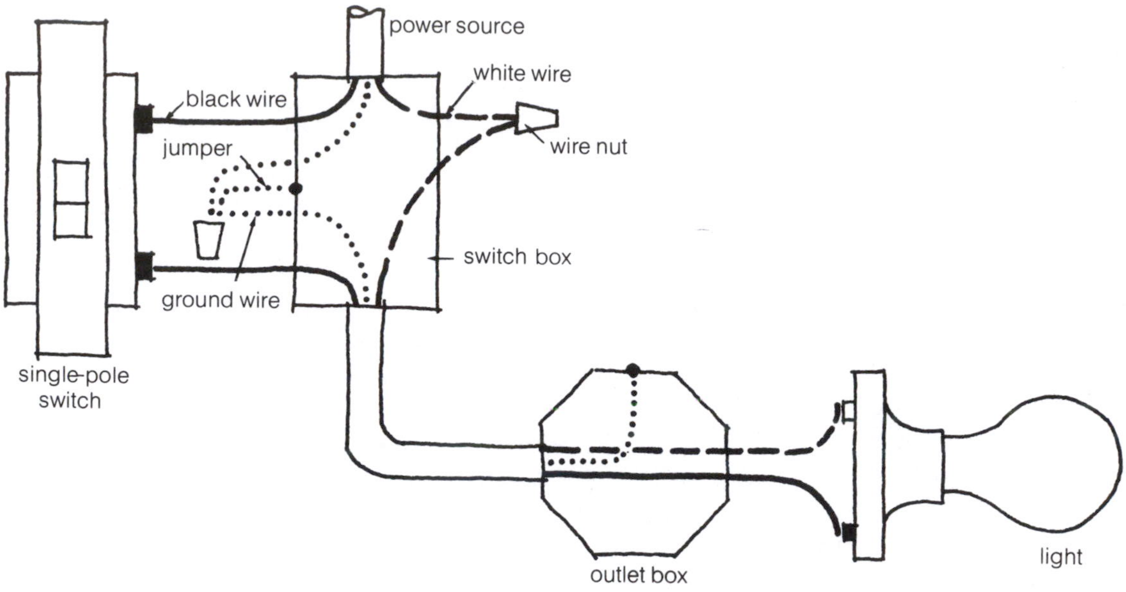

Light circuit with single switch. Note wire code: black is solid; white, dashes; and ground, dots.

tral or at ground potential and carries the current back from the fixture, and a bare ground wire that grounds all the boxes and receptacles to prevent shocks. For safety, the white wire and the ground wire should be continuous throughout the circuit. Thus, only the black wire is ever interrupted by a switch.

Several common wiring diagrams are presented to familiarize you with the flow of electricity in a circuit. For further information on more complicated circuits consult a wiring handbook or competent electrician.

Light circuit with single switch. Power is supplied to the switch box where the black wire is interrupted and attached to the switch terminal. The white wire is connected with a wire nut to the white wire running to the light fixture. The two ground wires are also connected together and joined to a jumper which grounds the switch box. This jumper should be attached to a screw on the back of the box to make a firm mechanical connection. Additional lights could be put on this switched circuit by running a new cable from the outlet box, attached to the black, white and ground wires.

Light circuit with two three-way switches. In this circuit, the light can be controlled from two locations. There are three wires *and* a ground running from the switches to the light. Three-wire cable is commonly available that has black, red, white and ground wires. In this circuit, the power comes to the switch box where the black is interrupted and tied to the "common" terminal of the three-way switch. The white neutral wire continues on uninterrupted to the light fixture. Red and black "traveler" wires are then attached to the two switched terminals on the switch and are run to the other similar terminals on the other

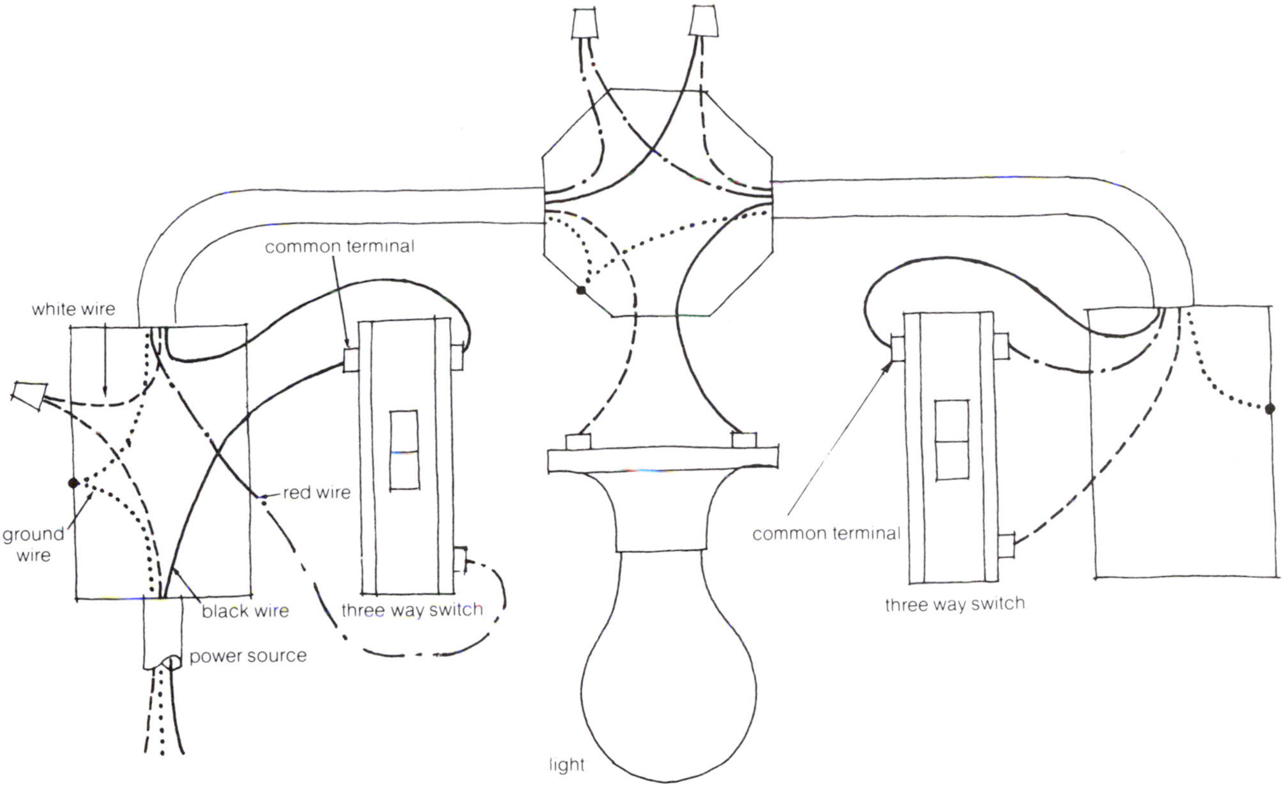

Light circuit with two three-way switches and four-strand wire.

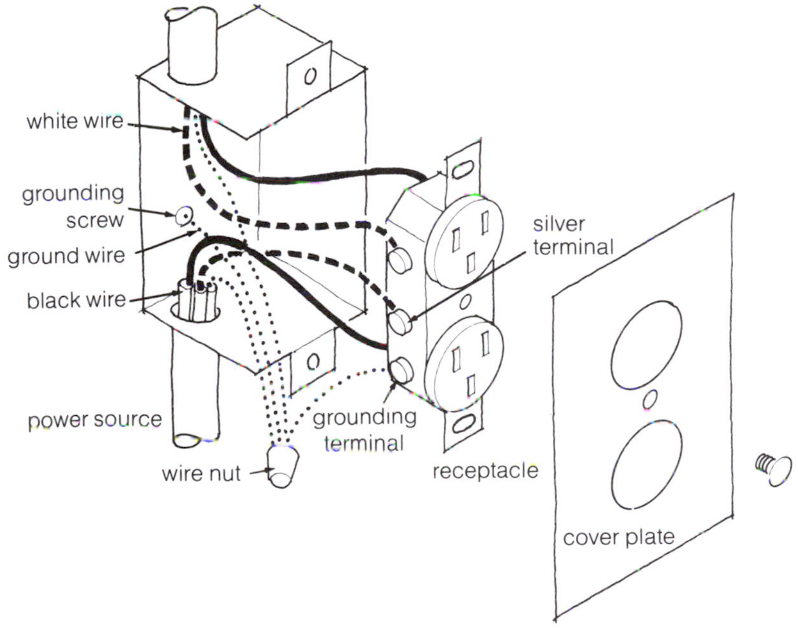

Duplex receptacle, using three-strand wire.

three-way switch. The remaining wire is then attached to the common terminal on the second switch and run back to the light.

Notice in the diagram that a black wire is attached to a white wire in the ceiling fixture box. The white wire going to the second switch is actually serving as a black "hot" wire. Whenever

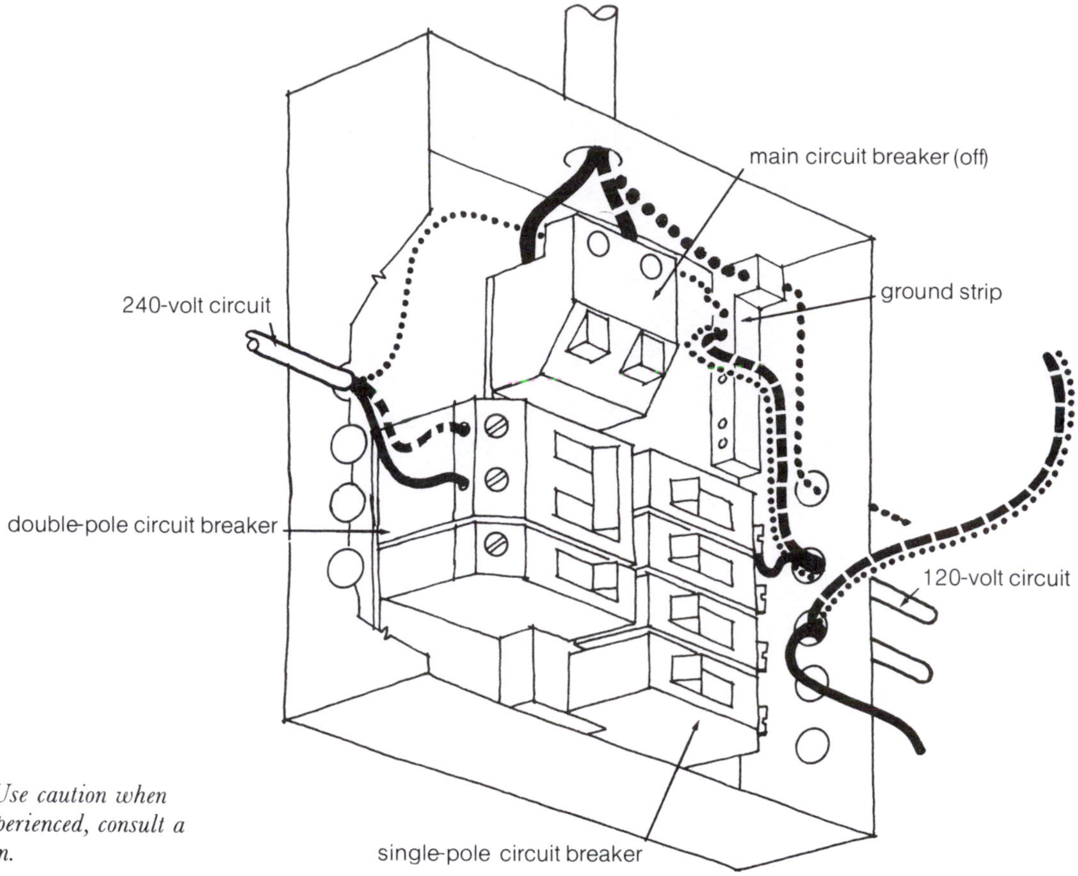

Main service panel. Use caution when working here; if inexperienced, consult a professional electrician.

this is done for switching purposes, the end of the white wire should be painted with a black magic marker to indicate that it is a black wire and, in fact, is carrying current.

Duplex receptacle. An illustration on p. 123 shows the wiring for a duplex receptacle with power coming in and exiting to another box. The two black wires are attached to the brass terminals on one side of the outlet, and the neutral white wires are attached to the silver terminals on the other side. Two short pieces of jumper wire are made up to connect the ground wire to the ground terminal on the outlet with a mechanical screw connection on the back of the box.

Connections To the Service Panel

After all the outlets, switches and lights have been installed, the final circuit connections to the service panel can be made. In some locations, *this must be done by a licensed electrician.* In others, it can be done by homeowners on their own property. If you do all the wiring yourself, always have it inspected by an electrician or building inspector *before* your new circuits are energized.

At the service panel, first *check to see that the main breaker is off*

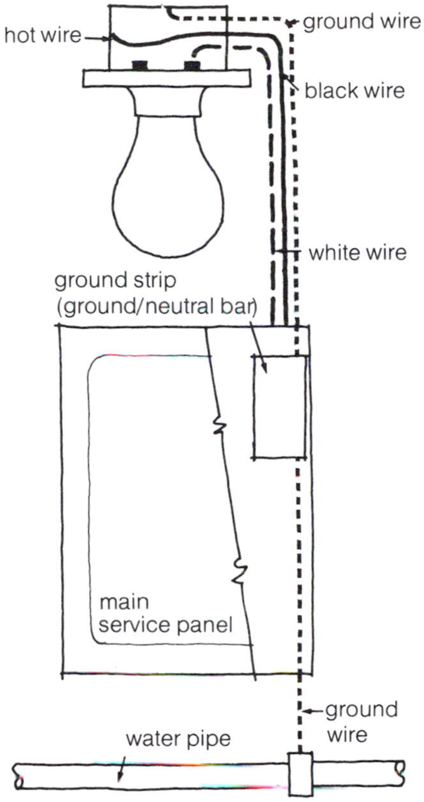

When a hot wire comes loose and touches the metal box, the ground wire removes this dangerous current, carrying it first to the neutral ground strip, then on to a water pipe that goes to the earth.

before starting to hook circuits into the box. With the main breaker off, you can work safely in the box, but remember, the entrance cable coming into the main breaker is hot and contact with it must be carefully avoided.

Bring the circuits into the box one at a time through the side knockouts that can be removed with a large screwdriver and a pair of pliers. Romex connectors should be inserted in the knockout holes to securely clamp the cable as it enters the box. Remember to leave enough wire so you can run the black lead to the circuit breaker and the white and ground leads to the ground strip. Starting from the top circuit breaker position, attach the circuits in a descending sequence with the wires run neatly inside of the box so they don't interfere with new circuit wires and breakers as they are added.

After all the circuit breakers are in place that you need, prepare the front cover of the service panel to fit over them. The front panel has knockouts for each breaker position, and using a pair of pliers you should remove the ones that correspond to the breakers you just put in. The unused breaker positions should remain covered until they are needed. Carefully fit the cover over the service panel and label all the circuits you have installed on the circuit directory on the cover door.

A final step before closing up the service box is to ground the panel and circuits. Using No. 4 solid copper wire with a metal casing like BX wire, run a ground cable from the ground strip in the box to either the building's plumbing or a copper rod driven in the earth. Using a copper rod is the surest way to ground the service panel, but if you use the plumbing, make sure the ground connection is to metal pipe that enters the building from underground. Some codes require that if you can't ground the system right where the pipe enters the building, you should drive a ground rod and attach the wire to it.

CHAPTER 8

Plumbing

Borg

WATER SYSTEMS FOR barns and outbuildings range from a garden hose run from the main house to a complete plumbing system with a well, pump, hot water and septic system. The water requirements of your livestock will determine the scope and size of your plumbing system. Most small barns need only an underground plastic pipe run from the main house to supply occasional cold water. A large cow barn, however, may require hot and cold running water, floor drains, septic and manure-removal systems.

In any plumbing system, there are two distinct sets of pipes, the water supply pipes and the waste-water drains. For rural locations, the water supply often starts at a well—either a spring box, shallow-dug well or a deep-drilled well. The well water is transferred to a storage tank by either gravity feed from an uphill spring or a pump. From the pressurized tank, water is fed through distribution lines or *risers* to various fixtures such as wash basins and sillcocks. If hot water is required, an electric, gas or oil hot water heater is installed with its own separate set of supply lines.

The drainage system starts at the fixtures and carries away waste water through pipes to a septic tank. Traps which hold standing water are always installed directly below sink and floor drains to prevent sewer gas from coming back up the drain pipe and into the building. The drain pipes must also be vented through the roof to allow these gases to escape and to prevent draining water from creating a vacuum in the pipes. At the septic tank, the wastes are partially digested by microorganisms and allowed to settle out. The waste water then travels to a leech field where it disperses into the ground.

Because of the health hazards associated with improperly installed water supplies and septic systems, plumbing work is strictly regulated by codes. These vary from state to state, especially in regard to who may do plumbing work and what materials are acceptable; but, the basic requirements are standard throughout the country. The National Plumbing Code is the basis for all state regulations and its recommendations should be followed.*

With the advent of plastic pipe and tubing for both supply and drain lines, plumbing has become greatly simplified and practical for many homeowners to do themselves. Before you start, however, take advantage of the free information and advice that is available. Contact an Extension Service agent to get

---

* See: Manas, Vincent T., and Eaton, Herbert N., editors, *National Plumbing Code Handbook* (New York: McGraw-Hill Publishing Co., 1957).

Table 8-1 Pipe Data at a Glance*

Type of Pipe	Ease of Working	Water Flow Efficiency Factor	Type of Fittings Needed	Manner Usually Stocked	Life Expectancy	Principal Uses	Remarks
Brass	Threading required or ask for pre-threaded. Cuts easily, but can't be bent. Measuring a rather difficult job.	Highly efficient because of low friction.	Screw-on connections.	12-ft. rigid lengths. Cut to size wanted.	Lasts life of building.	Generally for commercial construction.	Required in some cities where water is extremely corrosive. Often smaller diameter will suffice because of low friction coefficient.
Copper Pipe	Easier to work with than brass.	Same as brass.	Screw-on or solder connections.	12-ft. rigid lengths. Cut to size wanted.	Same as brass.	Same as brass.	
Copper Tubing, Rigid	Easier to work with than brass or hard copper because it bends readily by using a bending tool or by annealing. Measuring a job not too difficult.	Same as brass.	Solder connections.	Three wall thicknesses: K-thickest L-medium M-thinnest. 10- or 20-foot lengths.	Same as brass.	"K" is used in municipal and commercial construction. "L" is used for residential water lines. "M" is for light domestic use only: check codes before using.	

* *Homeowners Guide to Plumbing* (Milwaukee, Wisconsin: Ideals Publishing Co., 1981).

Type of Pipe	Ease of Working	Water Flow Efficiency Factor	Type of Fittings Needed	Manner Usually Stocked	Life Expectancy	Principal Uses	Remarks
Copper Tubing, Flexible	Easier than soft copper because it can be bent without a tool. Measuring jobs are easy.	Highest of all metals since there are no nipples, unions or elbows.	Solder or compression connections.	Two wall thicknesses: K-thickest L-medium 30-, 60-, or 100-foot coils (except "M").	Same as brass.	Widely used in residential installation.	Probably the most popular pipe today. Often a smaller diameter will suffice because of low friction coefficient.
Galvanized Steel (or Wrought Iron)	Has to be threaded. More difficult to cut. Measurements for jobs must be exact.	Lower than copper because nipple unions reduce water flow.	Screw-on connections.	10- or 21-foot rigid lengths. Usually cut to size wanted.	Very durable.	Generally found in older homes.	Recommended if lines are in a location subject to impact.
Plastic Pipe	Can be cut with saw or knife.	Same as copper tubing.	Insert couplings, clamps; also by cement. Threaded and compression fittings can be used. (Thread same as for metal pipe.)	Rigid, semi-rigid and flexible. Continuous lengths to 1000 ft.	Long life and it is rust and corrosion-proof.	For cold water installations. Used for well casings, septic tank lines, sprinkler systems. Check codes before installing.	Lightest of all, weighs about 1/8 of metal pipe. Does not burst in below-freezing weather.

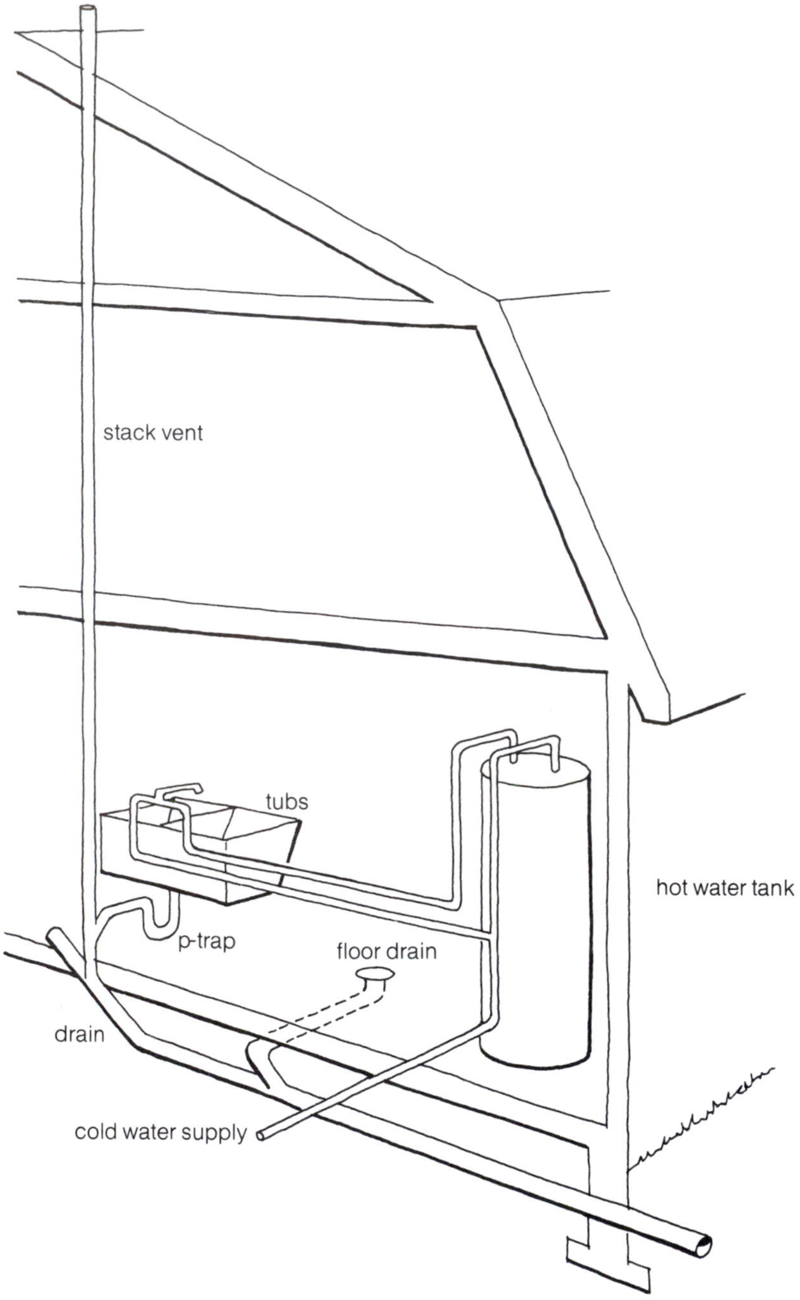

A basic plumbing diagram for a barn. Plumbing codes require venting for each plumbing fixture. Usually, hot water is only necessary for livestock operations requiring cleaning and sterilization.

recommendations on the size and type of water supply and septic system that would meet your outbuilding requirements. Also check with the local building inspector about plumbing codes and health regulations. Finally, get a copy of the National Code to guide you in your choice of pipes and layout.

Supply Pipes and Fittings

One of the hardest things about doing your own plumbing is knowing what materials and fittings are available and their proper names. A good knowledge of plumbing materials and terms will make your work easier.

Piping for water-supply lines is available in galvanized steel,

SOLDERING COPPER TUBING

Soldering copper tubing is an easy task if you observe three basic rules. The tubing must be absolutely clean, it must be dry and you must apply enough heat to the joint. Never try to join tubing that has any water in it as the water will keep the joint cold and blow tiny steam holes through the solder.

First, cut the tubing to the desired length with a tubing cutter. The cutter should be gradually clamped down to cut through the tube but not so hard that it dents it. When the cut is made, clean the outside surface of the tubing and the inside of the fitting with fine steel wool or emery cloth. Clean it just enough to brighten the metal. The tubing should also be reamed with a small hand reamer to remove any burrs left on the inside of the tubing by the cutter.

With the pipe reamed and cleaned, apply soldering flux with a small brush to the end of the tubing and the inside of the fitting. Place the fitting on the tubing and swivel several times to even out the flux.

With a propane torch, heat the fitting until the solder begins to flow around the joint. Heat a few more seconds, allowing the solder to flow into the joint and then remove the torch. Keep the solder on the joint until a line appears around the fitting indicating the joint is entirely filled. Remove the solder and let the joint cool without disturbing or touching it.

copper and plastic. Copper and plastic are the most widely used today, with plastic gaining among do-it-yourselfers. Galvanized pipe is hardly ever used except where required by code because it must be threaded with special tools to make a joint.

Copper tubing (called tubing to distinguish it from copper drainage pipe) comes in three types. Type K (thick wall) is used for underground piping, type L (regular wall) is used for high-pressure supply lines, and type M (thin wall) is used for interior, low-pressure lines only. Some state codes prohibit the use of type M altogether. Copper tubing also comes as rigid, 10-foot sections or as flexible tubing in rolls. Both types are joined together with copper "sweat" fittings, using a propane torch and lead-free solder and flux.

Some common fittings used to join copper tubing are: 90 and 45 degree elbows or ells for bends or turns; tees for joining two lines together; couplings for joining straight runs; and male or female adapters to go from a sweat fitting to a threaded one. Unions are also used to allow copper to be taken apart without unsoldering joints.

In recent years, plastic pipe has won code approval in most states for potable water supply lines. It is cheaper than copper, easier to install, lightweight and less susceptible to mineral scaling. There are two types, CPVC (chlorinated polyvinyl chloride) and PB (polybutylene). PB is the type to use if it is available

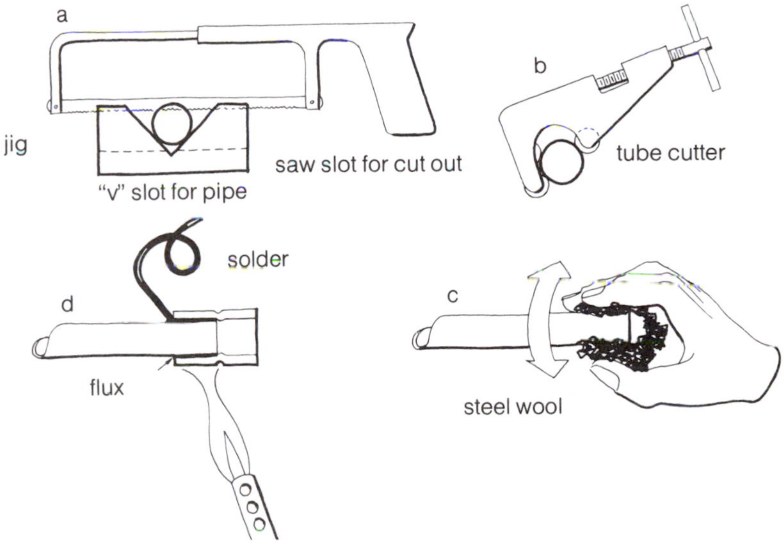

Copper tubing can be cut neatly with a hacksaw in a jig (a) or with a small tube cutter (b). Before a connection is made, the outside of the tubing and the inside of the fitting have to be cleaned with steel wool or some fine abrasive paper (c). Don't try to heat the solder wire directly (d).

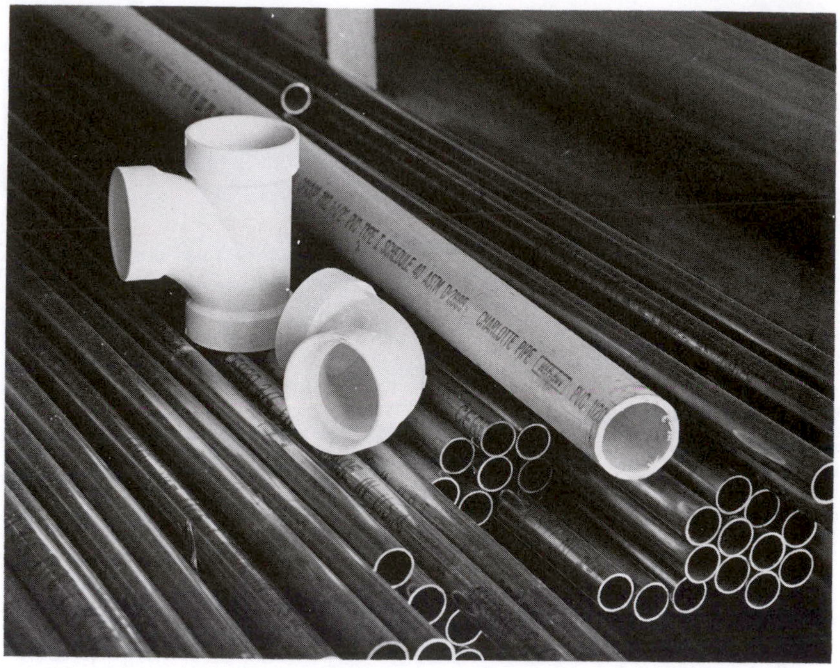

Copper tubing and white, plastic pipe are often used for small-outbuilding, water-supply lines.

because it is flexible and can be run around corners without expensive fittings. The plastic casing also provides some insulation for the water line, and it won't burst open even when frozen because it is so flexible. PB is available in rolls up to 1000 feet long and comes in two colors, gray for hot water and blue for cold. It has been approved by the Food and Drug Administration for potable water supplies.

Valves and Faucets

Valves are used to shut off water flow in a line when it is no longer needed or when a fixture must be repaired. There are two standard types, the gate valve and the globe valve.

A gate valve does not restrict the flow of water when it is fully open and therefore it is preferable for main shut offs that are only closed in emergencies. Globe valves do restrict the flow of water even when they are fully open, but have the advantage of being able to control the rate of flow from full force down to a trickle. Gate valves don't do this very well. Therefore, globe valves are used in branch lines and other places where the flow of water is to be controlled or turned on and off often. Also, globe valves are cheaper and easier to shut off.

Globe valves are available with a side drain on the non-pressurized chamber that allows you to drain shut-off pipes. These are called *stop and waste* or *stop and drain* valves. These valves must be installed in the proper direction, with the arrow on the body of the valve facing in the direction of the water flow.

For outbuildings, the only commonly-used fixtures are various faucets for controlling water to sinks, tanks and hoses. A *hose bib*

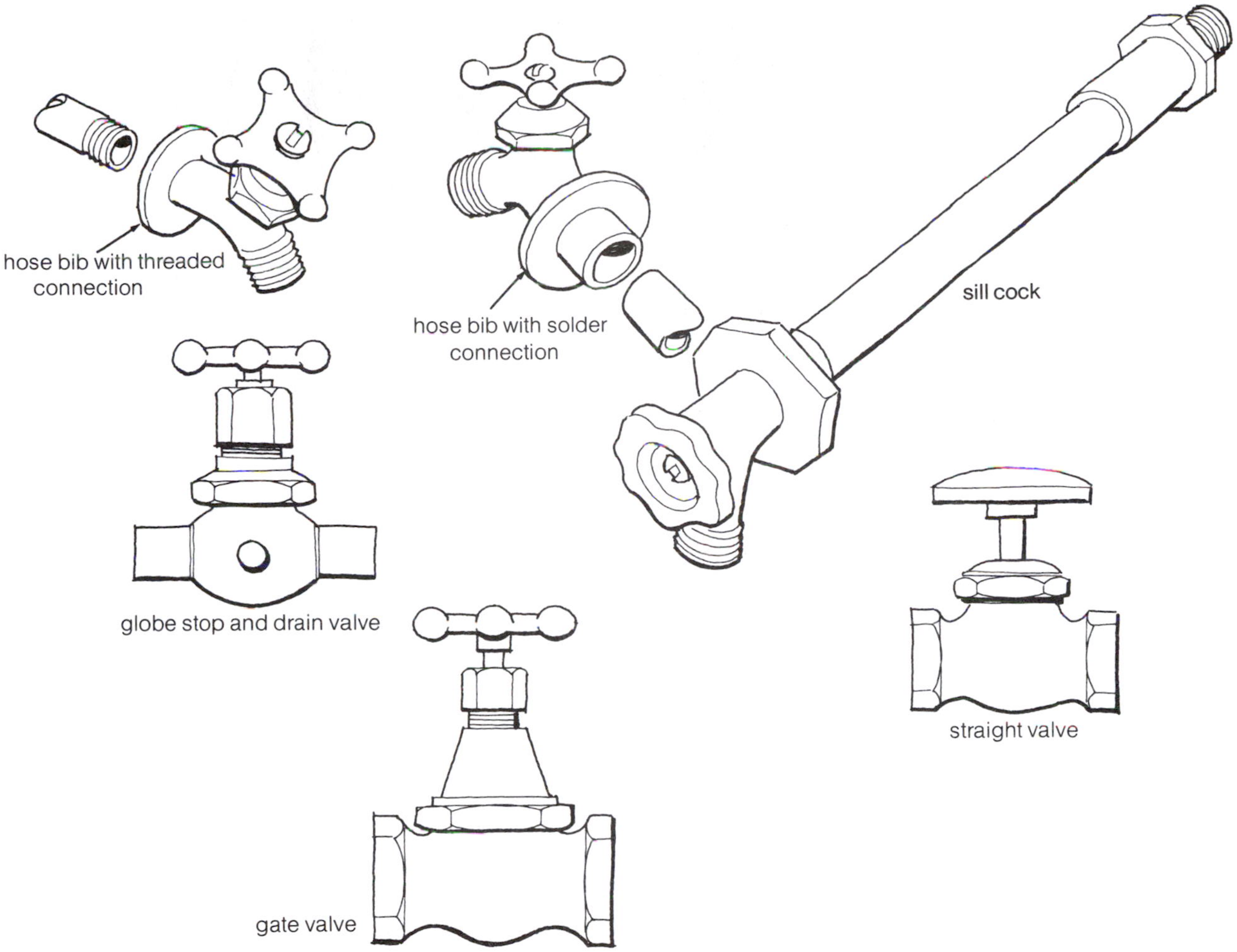

Valves and faucets suitable for small barns. A frost-proof faucet known as a sill-cock is located on the outside of the building, but its workings are on the inside where it's warmer.

is a valve with a threaded spout for drawing off water or hooking up a hose. These come with sweat or threaded connections and in different mounting styles. A frost-proof *sillcock* is a hose bib with a long extended valve. With the valve shut off located inside a warm building, water freezing is less likely.

There are two practical ways to control the flow of water to livestock tanks. You can mount a hose bib and fill the tank manually, or you can install a valve that opens and shuts when activated by a float. Float valves used in toilets to regulate the flow of water can be modified and installed in livestock tanks. There are also float valves made especially for tanks; they have protective boxes that simply mount on the side of the tank.

A laundry sink for washing and cleaning is often useful in a livestock barn or equipment shop. You can purchase old, stone laundry sinks second hand for next to nothing, or you can buy new, inexpensive fiberglass sinks. These should be equipped with a double-handle faucet (if you have hot water) and a large, moveable spout. If grease, oil or hair is likely to get in the drain, install a special drum trap that is cleaned easily by removing a top cover.

JOINING PLASTIC DRAINAGE PIPE

PVC and ABS plastic pipe are joined with fittings "welded" together with a special solvent cement. The cement and a cleaning solvent are available in individual containers wherever plumbing supplies are sold.

Plastic pipe should be cut with a hacksaw or fine-toothed, cross-cut saw. When cutting, make sure the cut is square so the end of the pipe fits evenly into the fitting. Use fine sandpaper to smooth off the burrs on the end of the pipe after cutting. Clean off grease or dirt on the pipe and fitting with pipe cleaner and a clean cloth.

Before you apply the welding cement, test fit the pipe and fitting together to make sure the fitting goes on smoothly. With a pencil, mark the orientation of the fitting on the pipe so they will line up properly when joined. The solvent will weld the plastic together in about 5 seconds, so once it is applied you don't have much time to get the fitting and pipe aligned. The orientation of the fitting doesn't matter for couplings, but for elbows and tees it is critical.

With the fitting and pipe marked, apply the solvent to the outside of the pipe and the inside of the fitting. Fit them together one-quarter of a turn out of alignment, and, as you push the fitting on, turn it to line up the pencil marks. This will evenly spread the PVC cement and make a good seal. Let stand a few seconds to cure the joint and then wipe off any excess cement.

Waste Water Pipes

Barns that have sinks and floor drains require a waste water system. This is commonly called the DWV (drain, waste, and vent) system. Cast-iron, copper and plastic are the three types of pipes used.

Because it is lightweight and easy to work with, plastic drain pipe is the best material for the homeowner to install. There are two types of plastic, PVC (polyvinyl chloride) and ABS (acrylonitrile-butadiene-styrene). Some codes permit only the use of ABS plastic. Both types are available in 10-foot lengths and range in size from 1½ to 4 inches in diameter.

Plastic pipe is cut with a regular saw or hacksaw and the fittings are welded onto the pipe with a solvent that fuses the plastic together. It is an extremely quick and easy way to assemble drainage pipe.

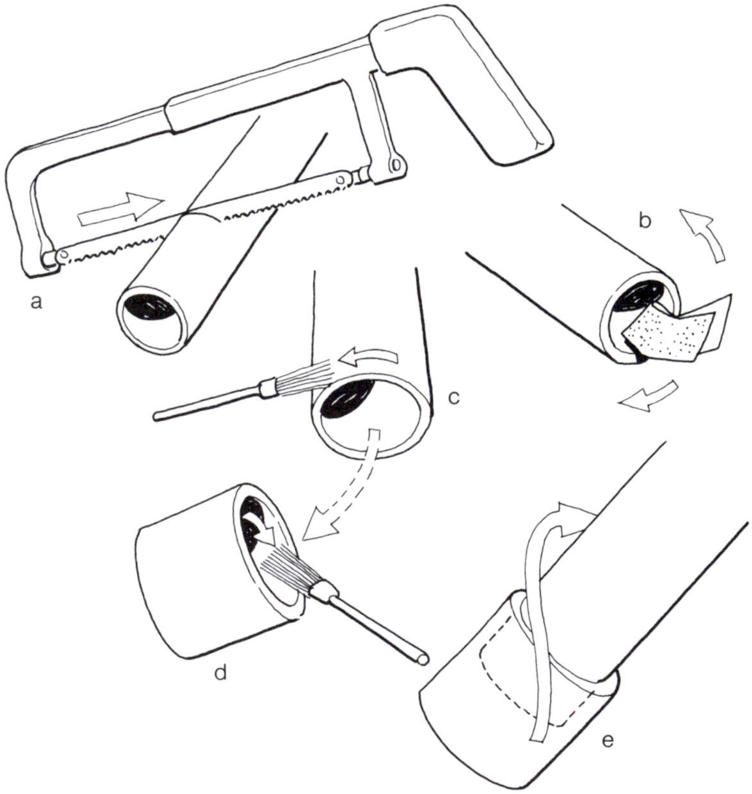

Cut the pipe to desired length (a); smooth burrs (b); test fit, then apply welding cement to pipe (c), then fitting (d); rotate fitting ¼ turn as you put it on (e).

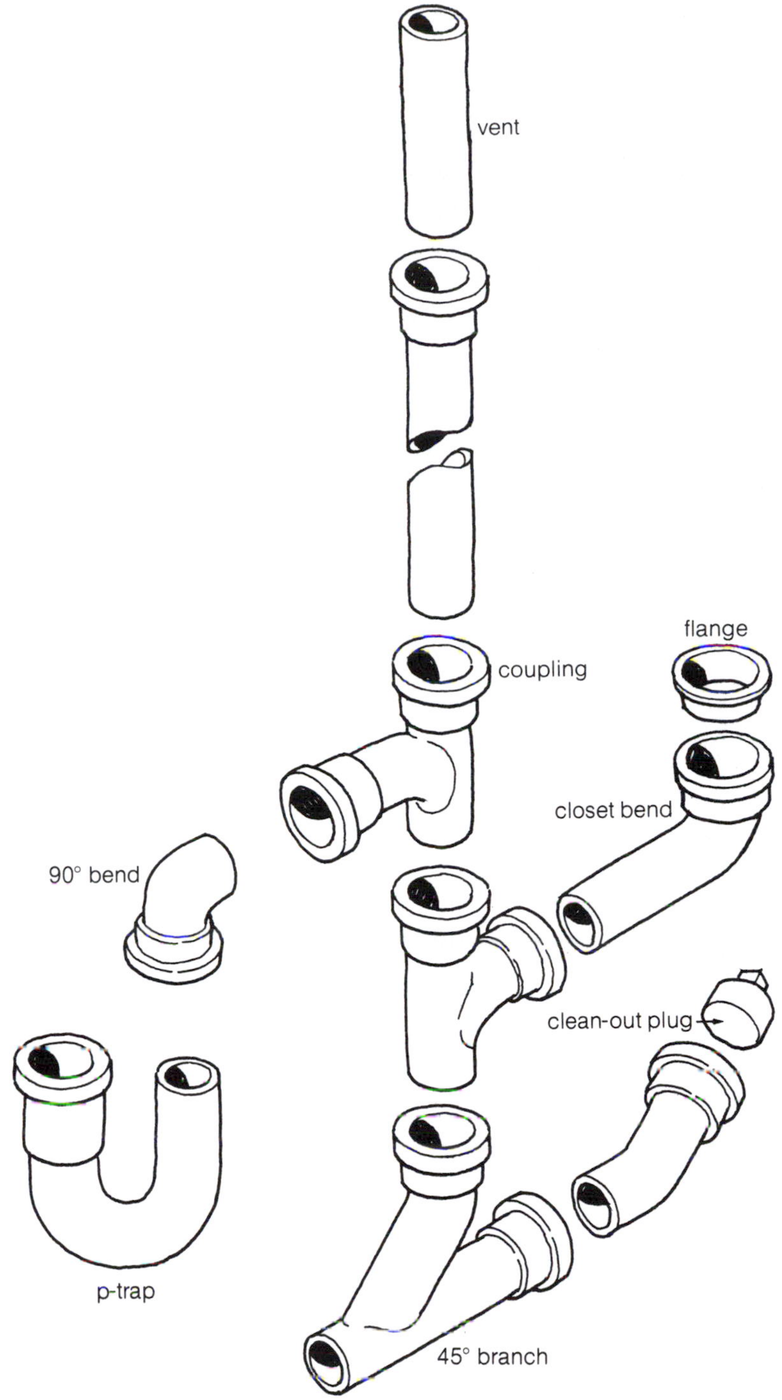

Drain, waste and vent systems include various pieces, made of cast-iron, copper and plastic.

Plastic pipe can also be run from the building to the septic tank. Special perforated pipe is made for the distribution lines in leech fields. The only places that many codes don't permit the use of plastic pipe are under concrete slabs and where the pipe penetrates foundation walls. Here, cast-iron pipe must be used, and a professional plumber must do the installation if the joints are to be leaded. For the homeowner, no-hub, cast-iron pipe is available that joins together with neoprene gaskets and clamps.

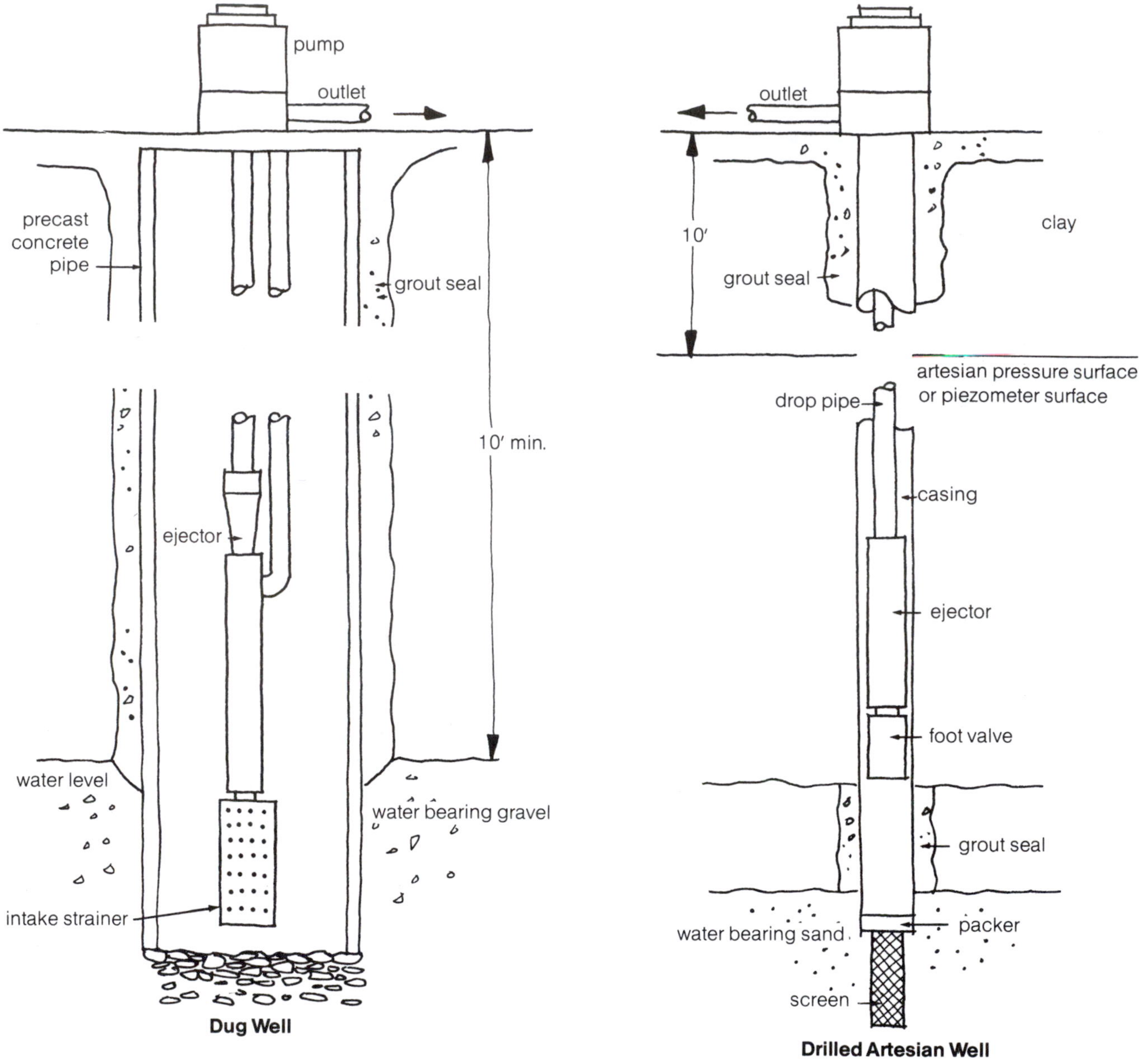

At left is a simple dug well, with a minimum depth of 10 feet and a precast concrete liner; at right is a drilled artesian well, which is dug with heavy equipment.

The Water-Supply System

Developing a new water supply is always expensive whether it be a simple spring box or an artesian well. If the main house is already served by a well, and the well can meet the additional water requirements of your outbuildings, it is best to run an underground line from the house tank to the barn. If your water requirements are greater than the well's capacity, then you must find a new source.

Finding water is not always easy. Even if a neighbor has a well 30 feet deep that supplies 100 gallons a minute, you might drill to 120 feet and find nothing. Three sources of help are available to the rural homeowner looking for water: a professional geologist can look over your land and suggest the most likely place to find water; a dowser with his divining rod can point to the location of underground water (don't laugh, more often than not it works); or you can consult neighbors about their success with wells. States often keep records of drilled wells that indicate the quality and quantity of water found in your area.

The simplest type of well is a surface spring that can be dug out and lined with concrete well casing or stone. If the spring is on an elevation above your barn, water can be syphoned off by gravity flow, eliminating the need for an expensive pump. Water from this type of well often won't meet health regulations for human consumption because of the possible surface-water contamination, and it should be tested before providing to livestock.

Three things to keep in mind when developing a spring well are: the spring box should be located above any possible source of contamination; the box should be fully encased and covered to keep out the animals and dirt; and the well must flow year-around, not just in the spring.

Normally, deep wells are drilled with heavy equipment that can bore through the earth and rock and drive the well casing into place. When the casing hits an underground water fissure, artesian pressure forces the water into the casing and up to the surface. A submersible or tank-mounted pump then lifts the water to a pressure tank in the barn. Usually, 1½-inch polyethylene pipe connects the well and tank. It must be run well below the frost line to avoid freezing.

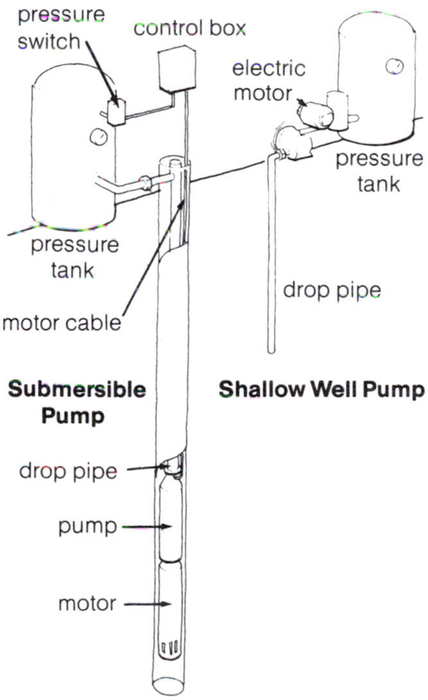

Two common types of water pumps.

When digging a new well, always ask a state or private laboratory to test the water to determine if it is fit for consumption. Shallow wells can be contaminated easily by sewer lines or spilled chemicals, and, even deep ground water, can be contaminated by nearby chemical dumps or natural radioactivity in the bedrock. Water also contains many different minerals that may not pose a health hazard, but may cause pipes to scale or deteriorate.

All but gravity-fed systems must have a pump and pressure tank to store water drawn from the well. The size and type of pump depends on the depth of the well and how many gallons per minutes (gpm) you need to supply. All pumps should have an automatic pressure switch that turns on the pump when the pressure falls below a certain point and turns it off at a pre-set high limit. Most systems are designed to operate between 30 and 40 pounds per square inch (psi) of pressure.

From the tank, the water supply lines or *risers* lead to individual fixtures. Whether you use plastic or copper tubing, keep in mind the following points when installing the lines:

•The entire system should be drainable, to protect against freezing and to make necessary repairs. Always slope the pipe at ¼ inch per foot of run and put drain valves at all low points in the system.

•Make your pipe runs as short and straight as possible. The longer the pipe run, the smaller its diameter, and the more elbows and turns it takes, the greater the loss of pressure. To compensate for pressure drop, ¾-inch pipe is commonly used for the main supply lines and ½-inch for lines to individual fixtures.

•Don't run water lines in outside walls if you can avoid it. Even pipes in insulated walls may eventually freeze causing water damage and necessitating major repairs. Install pipes where they will be warm and where they can be inspected and serviced easily.

•Always support overhead pipes with hangers or clamps set a maximum of 10 feet apart. Copper clamps that fit over different-size tubing can be nailed into the joints or studs. Galvanized and copper strap is also available in rolls to hang pipe from the ceiling at a desired height.

•Install shut-off valves on all pressure tanks, hot water heaters and at points before fixtures such as sink faucets. Then these components can be isolated for repair without having to shut off and drain the entire plumbing system.

Septic Systems

An outbuilding that has simple outdoor hose bibs for livestock water does not need a waste-water drainage system. The water simply seeps into the ground. If you have sinks and floor drains, however, regulations probably require you to dispose of waste water in an approved septic system, not just a drain that runs outside to daylight or to a drywell. This requirement is especially likely for livestock buildings where manure might mix with the waste water.

If you are required to connect barn waste-water lines to a septic system, one possibility is to tie into the existing house septic system. Unfortunately, many house systems are inadequately maintained and overburdened by high-flow fixtures such as washing machines. It is best to hire an engineer to see if your existing system can handle additional flows from the barn.

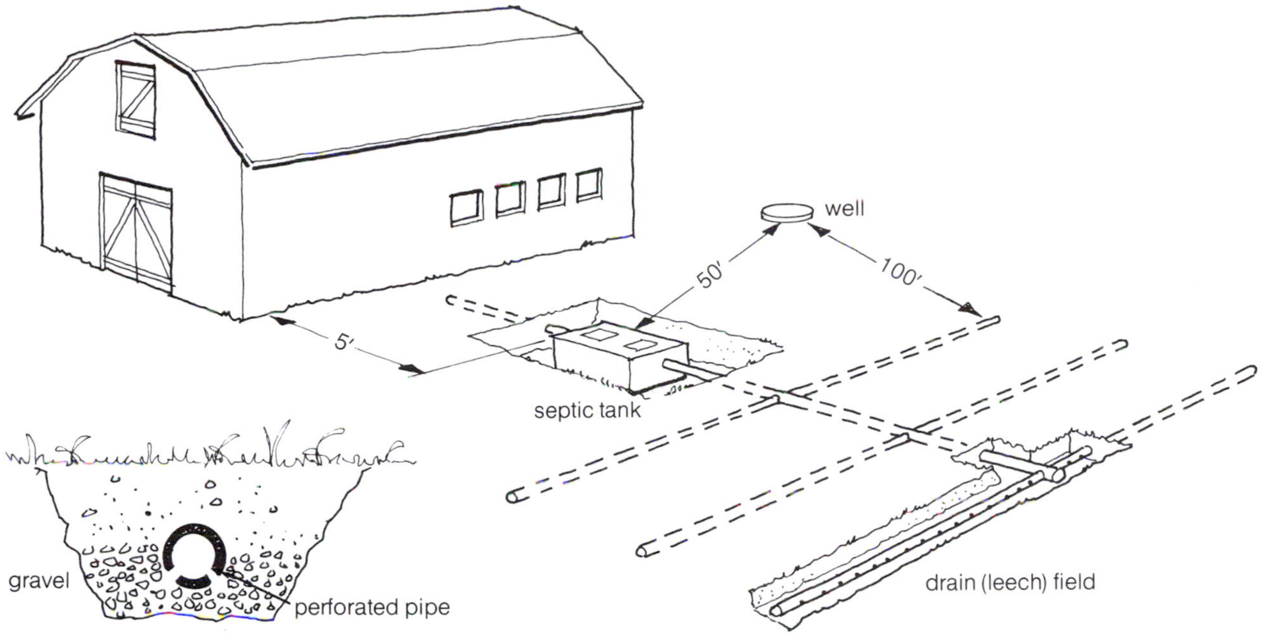

When planning a septic system, check local codes to ensure that the tank and leech field are adequate distances from the outbuilding.

If you must install a new system, most states require that a certified engineer test the soils on the septic site and approve the system design before it is put in. Percolation tests are the first step in determining the suitability of your soils to absorb waste water. With an accurate measurement of the soil percolation rate, an engineer can determine how large a septic tank and leech field you will need to handle your wastes. Design of septic systems varies with the type of soil, the terrain and state regulations.

CHAPTER 9

Barns

Borg

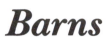

EACH OF THE following barn designs was selected with several considerations in mind, including simplicity, appearance, practicality and flexibility. Horses, cows, sheep, goats, rabbits and other animals may be housed in one or more of these small barns. In most cases, there's plenty of room for equipment as well. Using the first eight chapters as a guide, you can build one of these structures; or, with a little imagination, you can adapt a design for your own specific purposes.

The Bennett barn features a wide overhang for outside wood storage, doors shaped as ellipses and low-cost materials. The other barns are: the Pratt barn, with an attractive saltbox design; the Dickerson barn, with a flexible interior suitable for many purposes; the gambrel, with its classic roof affording extra space; the general barn, another multi-purpose building; the horse barn, with an L-shape; and two shed-style barns, suitable for cows, horses and other livestock.

Bennett Barn

Bob Bennett built a pole barn to provide housing for his rabbitry and for a tractor and other equipment moved from his crowded garage. Using the blueprints for his new home, Bob matched the roof angle, eave details and door design of the pole barn to his existing house and garage.

He and his neighbor, Dick Pratt, who was also building a new barn, got together for joint purchases of materials. They also contracted to have foundation work for both barns done at the same time. These two steps meant significant savings. Materials were purchased at steep discounts.

Most of the wood in the Bennett barn is rough-sawn, green spruce and hemlock. It was stacked properly with stickers to separate each layer, and, by the time construction began, it was reasonably well air dried. By using discounted, native lumber and making joint purchases, Bob built his barn for the incredibly low price of $3.30 per square foot (excluding labor).

Sometimes, pole-barn plans show only one 2 x 10 girt at the bottom, but Bennett had a good reason for using two. If the bottom one should rot, he said, it could be removed simply. If there was only one 2 x 10 and it rotted, a complicated job, removing the siding, would be required before the girt could be removed.

The Bennett barn measures 24 x 24 feet (excluding the overhang on the north side), and it has a gravel floor. It is framed

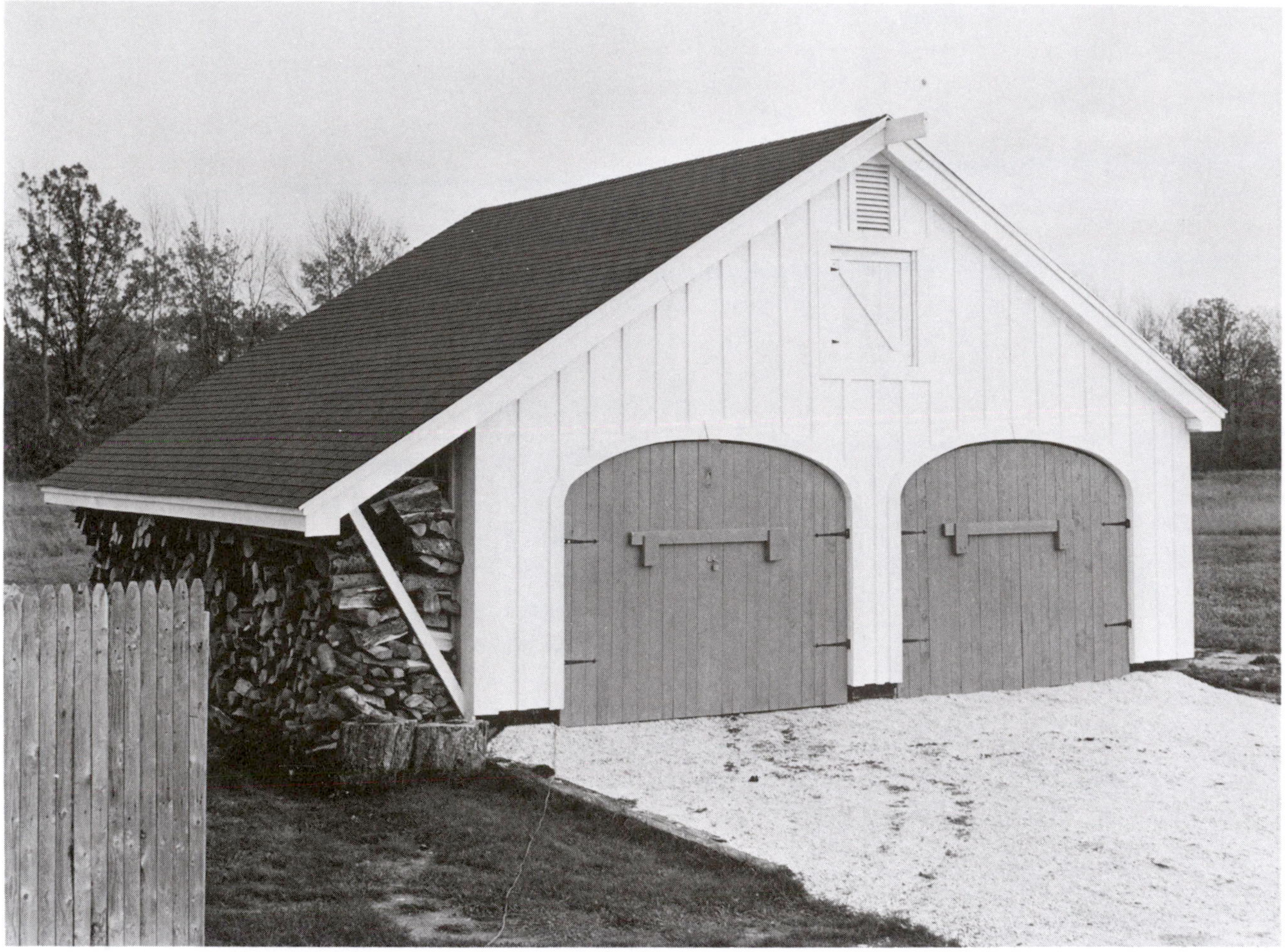

Bob Bennett built a carriage-style, combination garage/barn with a long overhang on the north side to protect his wood supply.

with 6 x 6 corner posts and 4 x 6 wall posts, spaced 8 feet on center. These posts are pressure treated and anchored to concrete footings set below the frost line. The posts are tied together with two 2 x 10 pressure-treated girts at the bottom (one partially embedded) and 2 x 4 girts spaced at 2-foot intervals above.

Where the wall posts meet the upper girts that support the rafters, braces have been added. Some are simple, one-piece knee braces; others are diagonal 2 x 4s, mounted in a V and reinforced with plywood gussets (these could not be used near windows).

Double 2 x 10 rafter girts support the ceiling joists and roof rafters. Ceiling joists, made from long 2 x 6s overlapped at the center, span the building every 2 feet o.c. Three of the joists are nailed to 2 x 6 center king posts and all of the joists are nailed to diagonal 1 x 6 ties. Collar ties (2 x 6s) are installed 4 feet o.c. The posts and ties add further structural strength. Not only do the joists help tie the outside building walls together, they also form a kind of attic-storage area for lumber, ladders and other long items. To add lateral rigidity to the joists, Bennett nailed short pieces of lumber intermittently to their top surfaces.

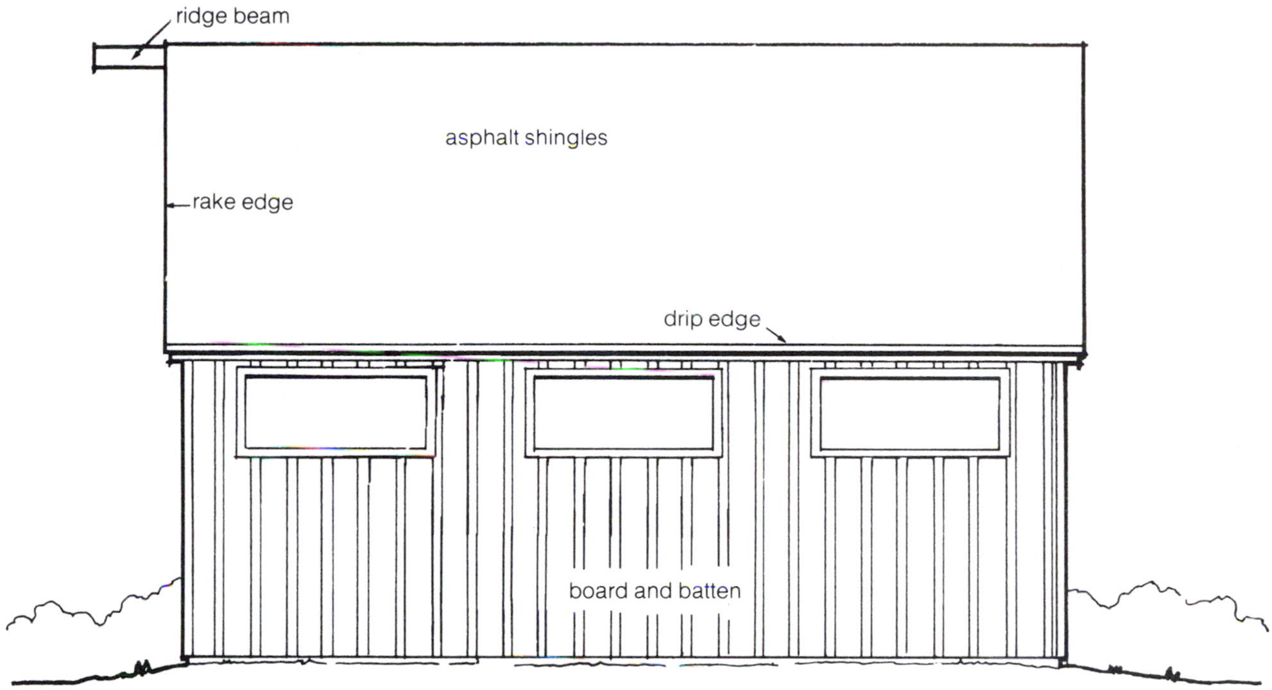

The south elevation of the Bennett barn has three windows; board and batten siding and asphalt shingles cover the barn.

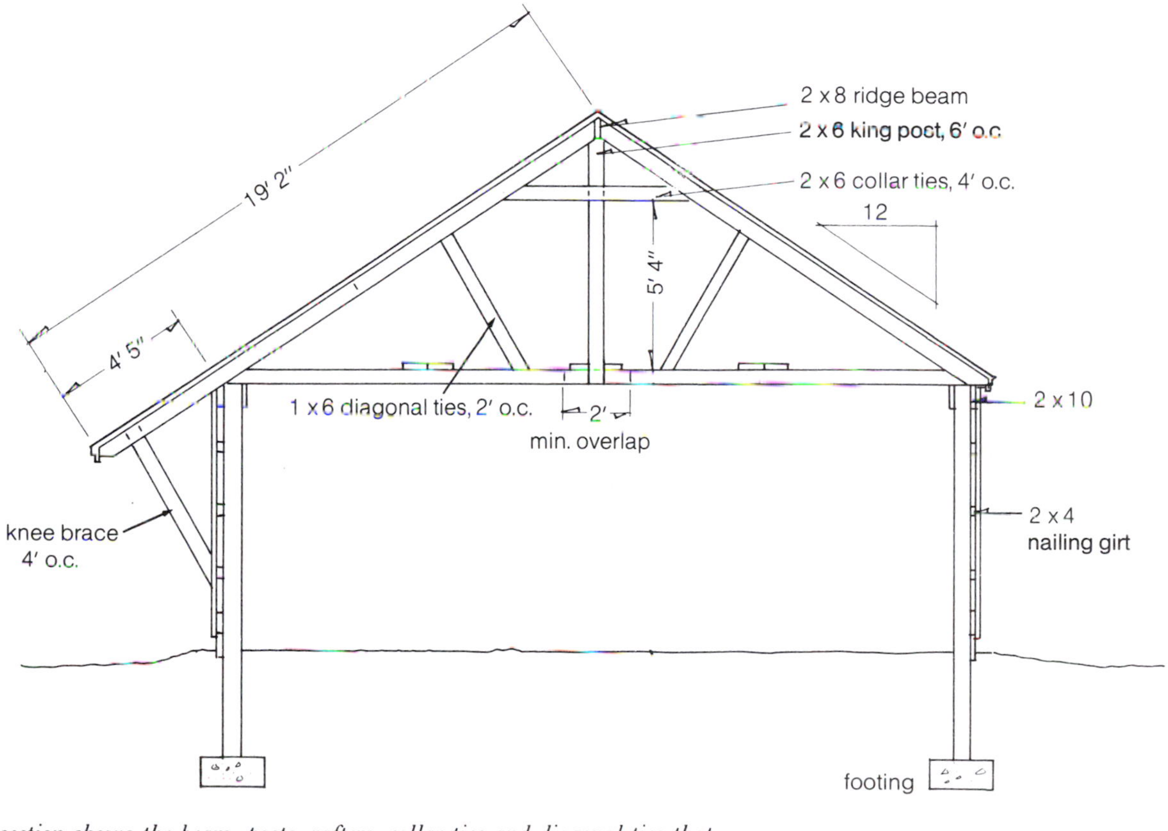

This section shows the beam, posts, rafters, collar ties and diagonal ties that support the barn roof. Footings should be on undisturbed soil, below the frost line.

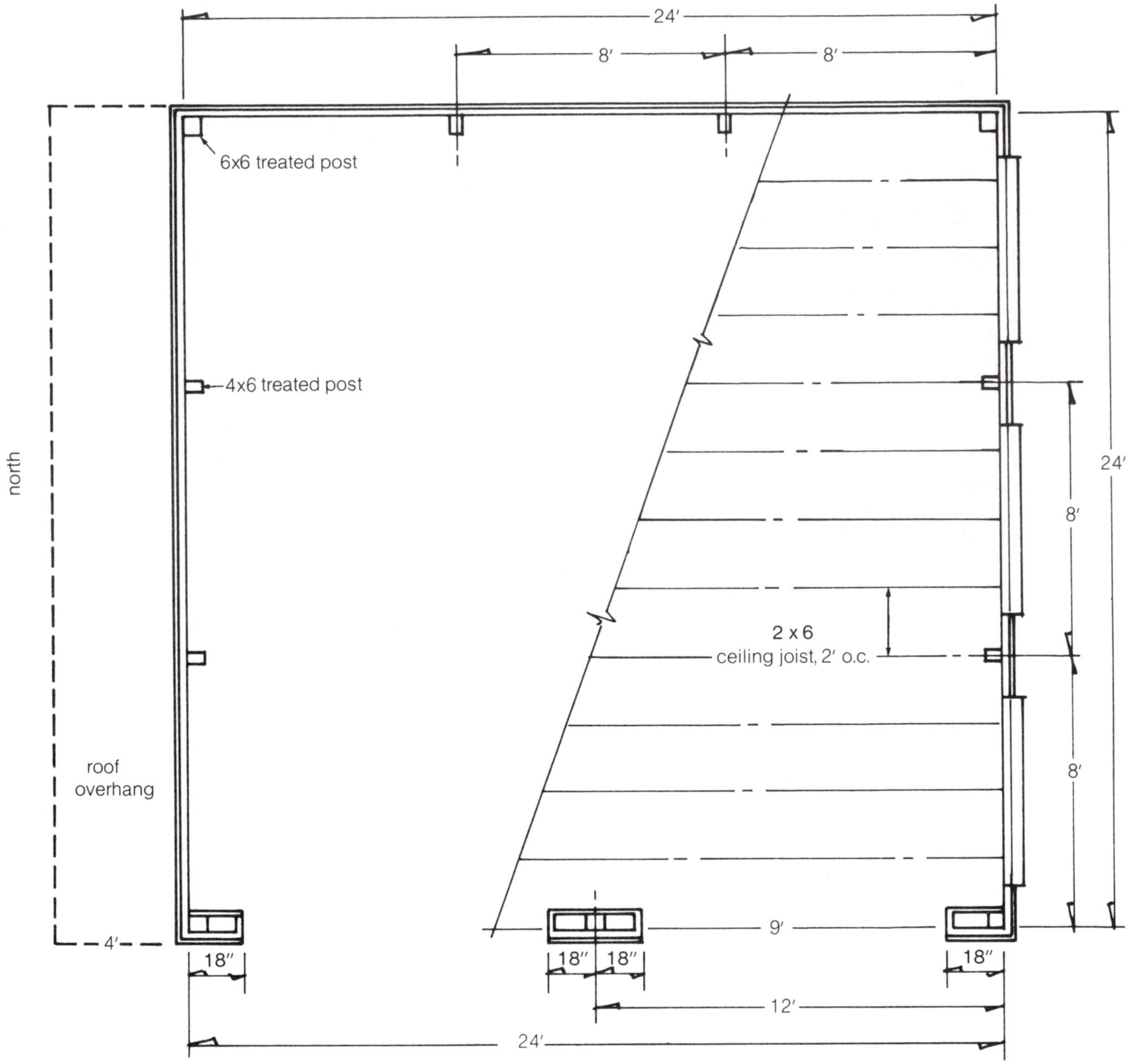

A plan view of the Bennett barn with 6 x 6 corner posts, 4 x 6 wall posts and 2 x 6 ceiling joists overlapped at the center of the span. Note the windows on south wall.

The roof has asphalt shingles on a ½-inch plywood deck supported by 2 x 8 rafters placed 2 feet o.c. On the south side, the eave overhang is only 4 inches wide, but on the front over the doors and in back, the eaves are 12 inches wide. The eaves are fully closed in with both fascia and soffit boards. On the north side of the barn, the rafters were extended to form a 4-foot overhang to protect firewood.

Rough-sawn 1 x 8 boards are used for the board and batten siding. The two front barn doors, framed in the shape of an el-

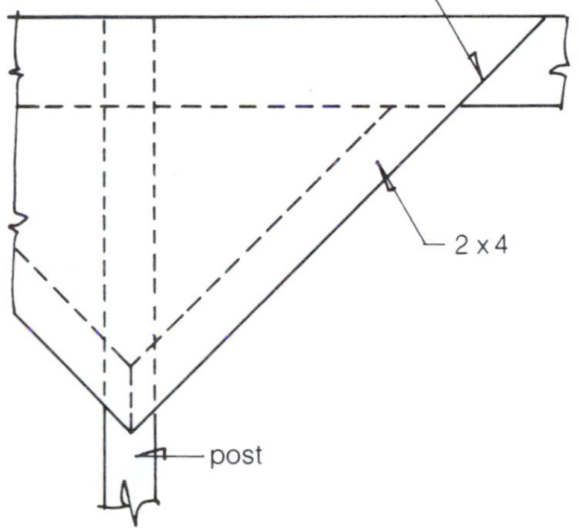

2 x 4 Brace and Plywood Gusset

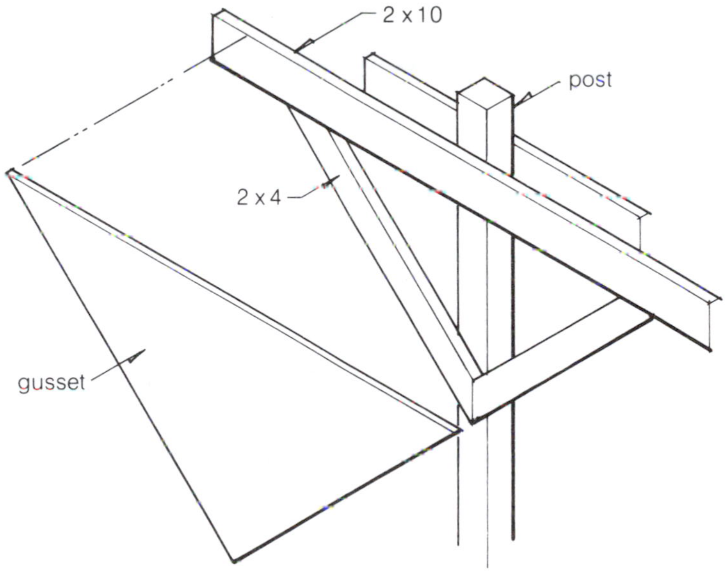

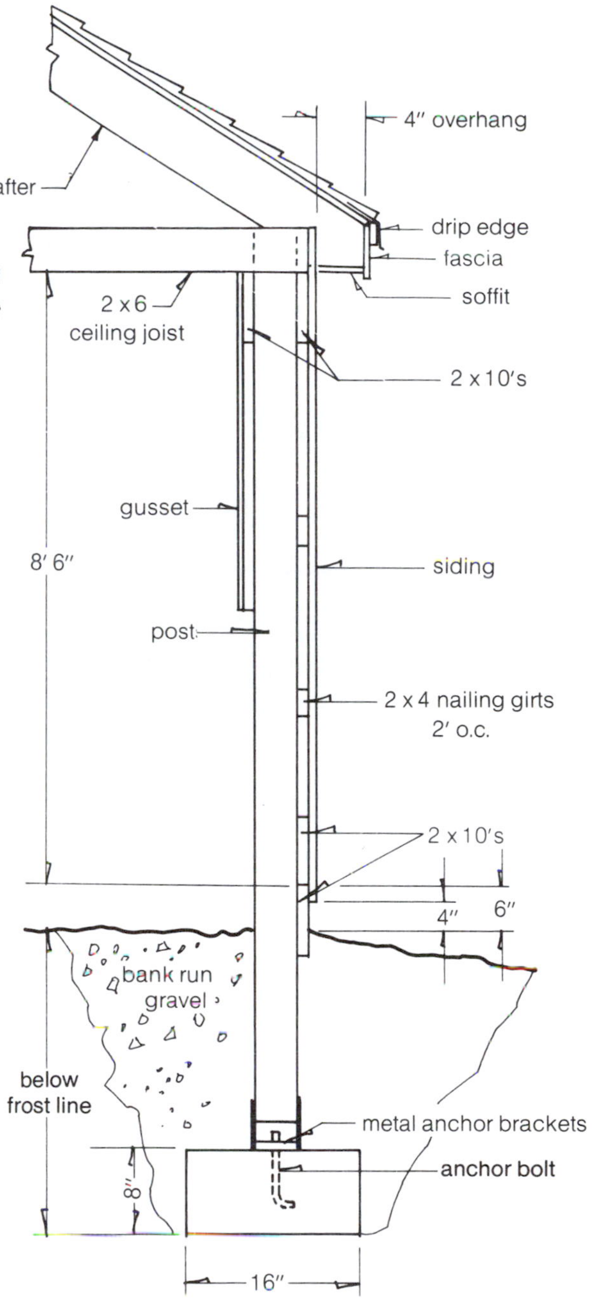

Typical Wall Section

Here's a typical wall section for the Bennett barn and a detail of bracing with a plywood gusset. Note the two 2 x 10s at ground level. These should be pressure treated or protected with a preservative.

lipse to match details on the house and garage, are made from 1 x 6 boards mounted on a Z-frame. On the south side of the barn, three 5-foot wide windows let in light and capture the sun's heat. Eventually, Bob plans to improve the passive-solar features of his barn by adding a solar greenhouse along the south wall.

The Pratt barn has a salt box design, with a long sloping roof, hayloft, entrance door and large, sliding door. The barn can be partitioned to accommodate various livestock, tools, equipment and vehicles.

Pratt Barn

The Pratt pole barn has a salt box roof to accommodate a 16- x 24-foot hayloft over the main floor area. On the east side, the long roof slopes down to a point just a little over 6 feet above ground, enclosing an area that can be used for livestock stalls. While allowing for efficient use of space and materials, the hayloft and salt-box design make the framing slightly more complicated than that in the Bennett barn.

The roof rafters are 2 x 10s spaced 2 feet o.c. The long span of the roof is supported by a pair of 2 x 10s, mounted on either side of the posts that support the east end of the hayloft, and by a 2 x 10 built-up beam. On every set of rafters, 2 x 6 collar ties lock the roof in position and keep it from spreading the outside walls.

Using a door track from an abandoned barn, Dick built a sliding door with two 5-foot wide sections that slide to either side on the outside of the barn. He installed a 36-inch wide entrance door to the left of the sliding door. Above, there's a small door for moving hay and other materials in and out of the hayloft.

The barn is supported by 6 x 6 pressure-treated posts set 8 feet o.c. on three walls and spaced to accommodate the doors and hayloft on the front (north) wall.

The hayloft floor joists, 2 x 8s spaced 2 feet o.c. and overlapped at the center, are supported by 2 x 10s and joist hangers.

1. On the north side, a double header (two 2 x 10s) for the sliding door and joist hangers support the joists.

2. At the center of the barn, another pair of 2 x 10s support the center of the joist span. (The 2 x 10s are made rigid with blocking and supported by 2 x 4 knee braces.)

3. At the west wall, these center 2 x 10s are also supported by yet another pair of 2 x 10s. This pair, running north-south, is supported by the west wall posts.

4. Joist hangers, mounted on a 2 x 10 girt, support the joists at the south wall. Notice that there are continuous *double* 2 x 10 girts on the east and west walls; only *single* 2 x 10 girts are on

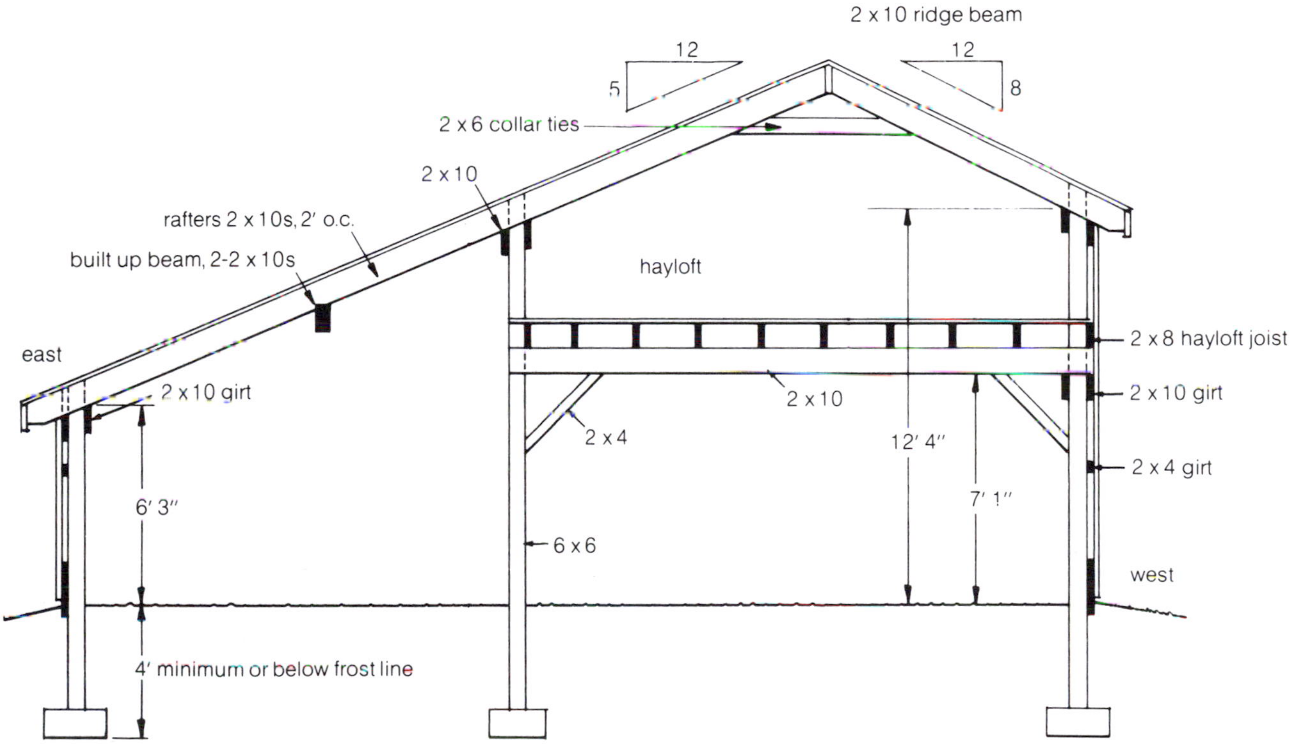

This section shows the barn's framing details. A 2 x 10 built-up beam helps support the long, sloping roof.

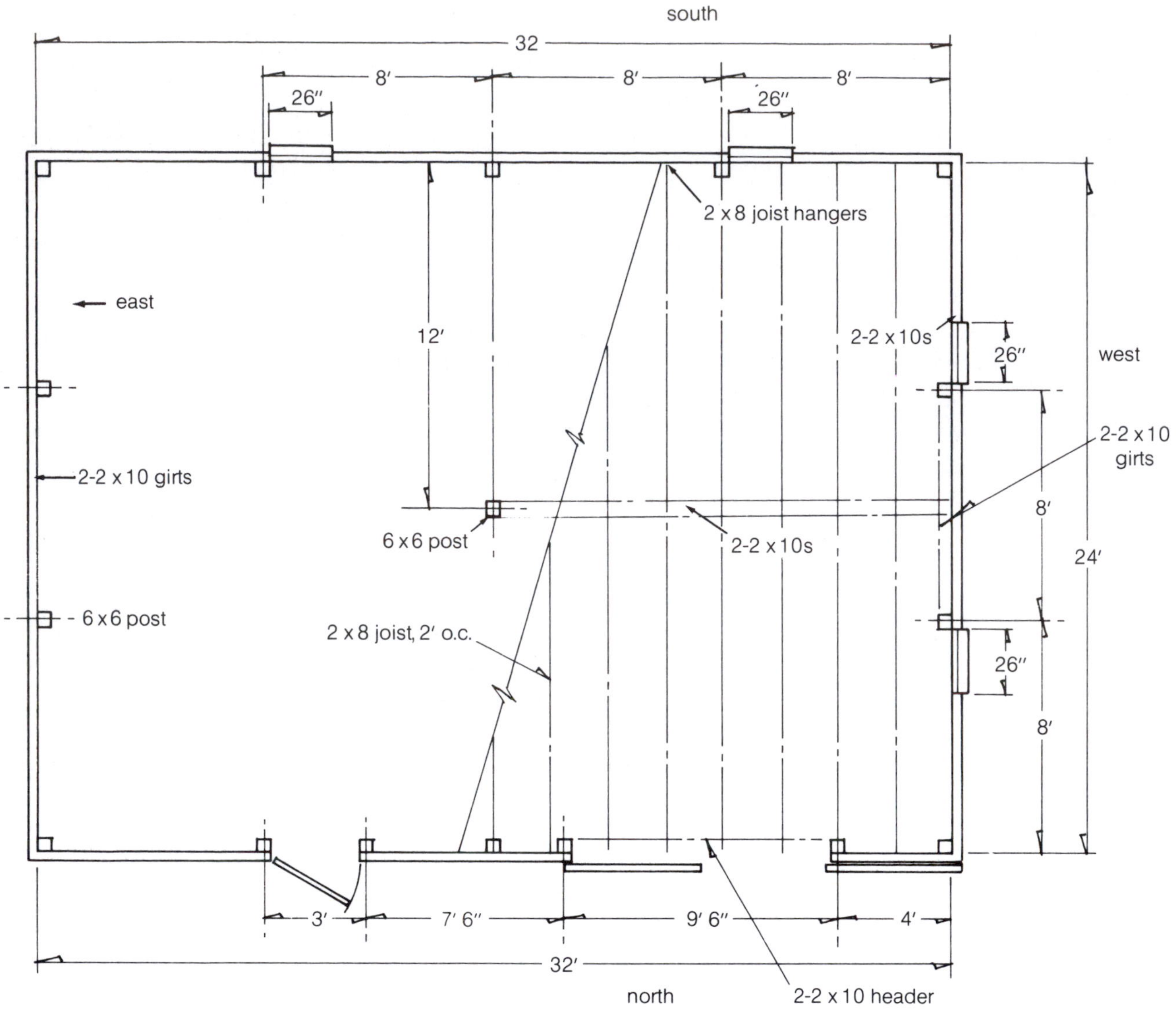

This plan view of the Pratt barn shows framing for the walls, doors and hayloft floor. Note the 2 x 8 floor joists are overlapped at the center and supported there by a pair of 2 x 10s.

the south and north walls, except for over the sliding door, where the 2 x 10s are doubled to form a header.

Wherever space permits, the wall posts are braced at their tops with 2 x 4 diagonals covered with plywood gussets. This is necessary for barns without the bracing action provided by plywood sheathing.

The board and batten siding is supported by 2 x 10 and 2 x 4 girts.

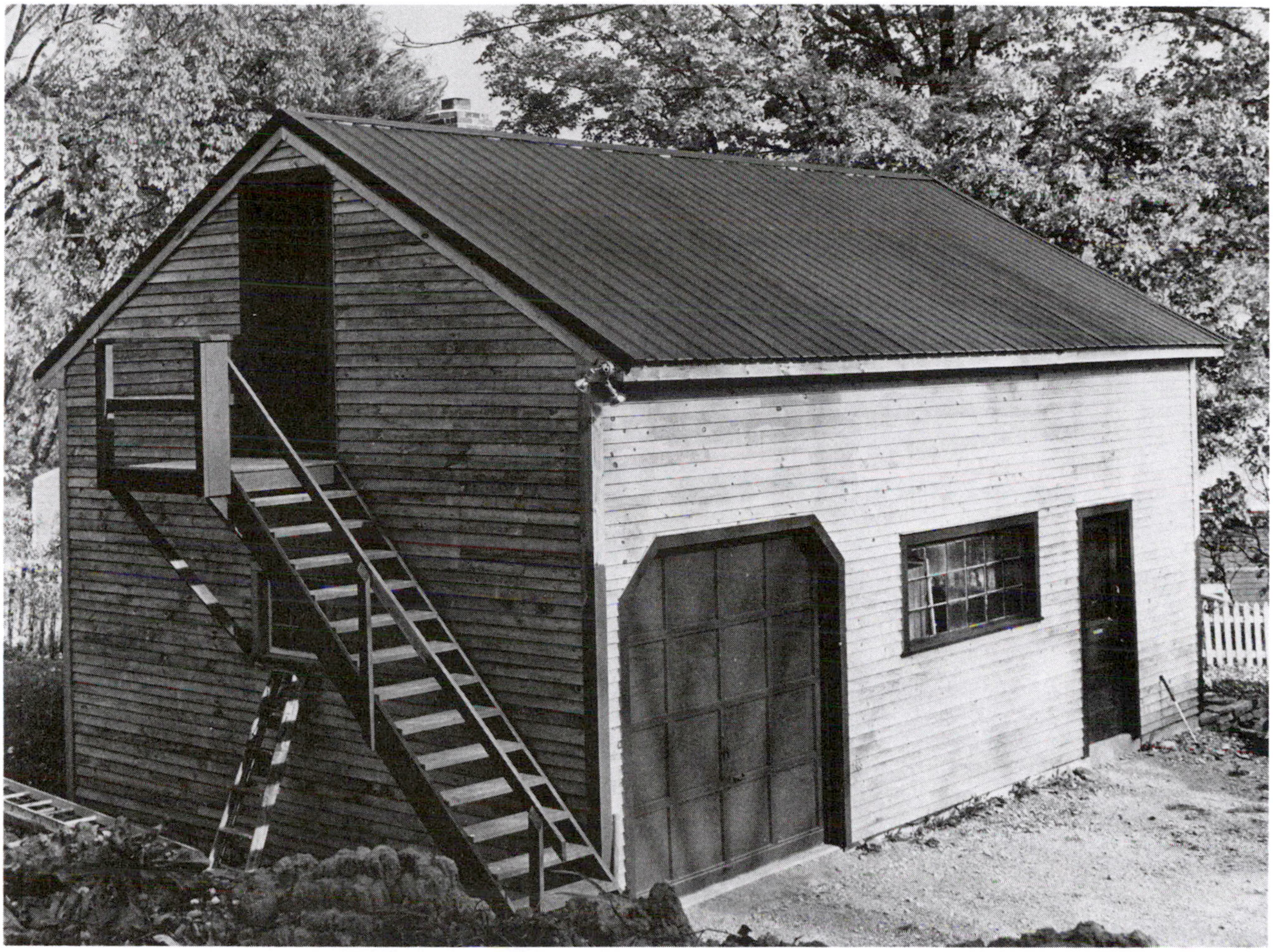

With a few modifications, the Dickerson barn can be converted easily from a garage/shop into an attractive living area or guest house.

Dickerson Barn

This 20- x 30-foot barn was built by Jim Dickerson and Lisa Boyle, who needed space to store and repair antiques, a part-time business. They also wanted to be able to convert the barn to a cottage or other uses, so they incorporated flexibility into their design.

The barn has a 10- x 20-foot garage on the west end, a 20- x 20-foot shop and a 20- x 30-foot second-floor storage area. The owners installed 100-ampere electrical service and a lined, concrete-block chimney for wood heat. With only minor improvements, the shop could be turned into a kitchen/living area and the second floor, into a loft.

The foundation is a 6-inch poured concrete slab, thickened to 1 foot beneath the walls to provide footings.

The walls are balloon framed. This means there are continuous 12-foot long 2 x 6 studs from the sill plate to the roof, placed 2 feet o.c. The 2 x 6s were selected so the walls could be insulated with 6 inches of fiberglass. If 13- or 14-foot studs could

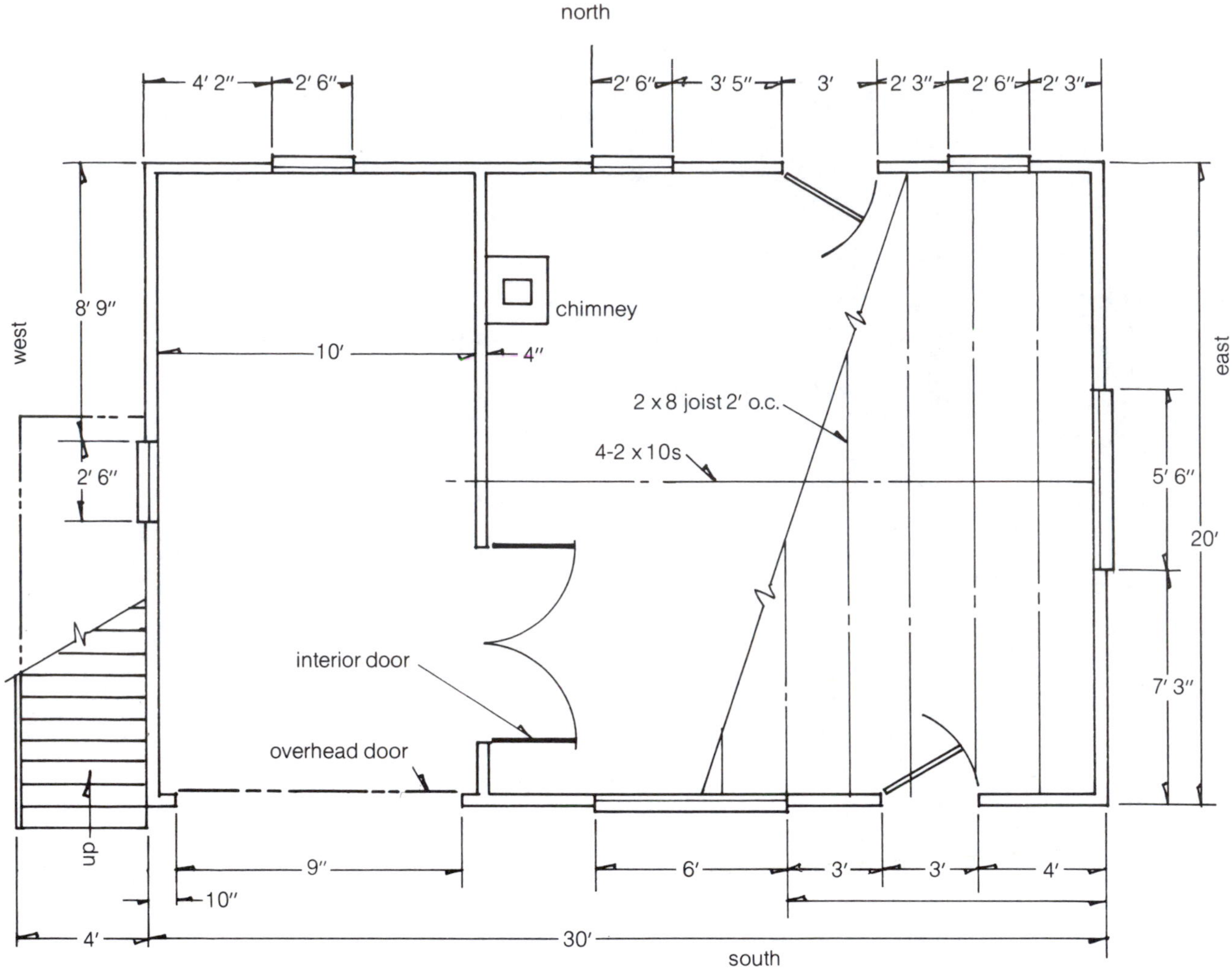

The Dickerson barn plan view, showing 2 x 8 ceiling joists supported by a built-up beam of four 2 x 10s. The overhead door at left opens to the garage, while the space at right is the shop.

be located, they would be preferable, to provide extra head room in the second floor.

The 2 x 8 joists for the second floor are supported by a built-up, 20-foot long 8 x 10 girder (four 2 x 10s). At each end, the girder rests on four 2 x 4s, nailed together and secured to the wall studs. Plywood (¾-inch) stripped from the concrete foundation forms was used for the second-floor subflooring.

The roof framing has 2 x 6 rafters and collar ties, every other set of rafters; 2 x 4 purlins support the enameled metal roof.

The barn is sheathed with ½-inch plywood and covered with No. 2 spruce bevel siding to match the main house. Dickerson plans to protect the siding with a light creosote sealer. The windows, recycled from old buildings, have various dimensions.

The shop area has an interior 6-foot wide, double swinging door, for bringing in furniture and equipment from the garage. There's also a 3-foot wide front door and a 9-foot wide garage door. While the garage door is wide enough for compact cars,

Dickerson said if he were to build another barn, he'd make it wider to better accommodate a pick-up truck.

An outside staircase permits access to the storage area above, but an interior spiral staircase or ladder could be installed easily if the barn is converted into living quarters.

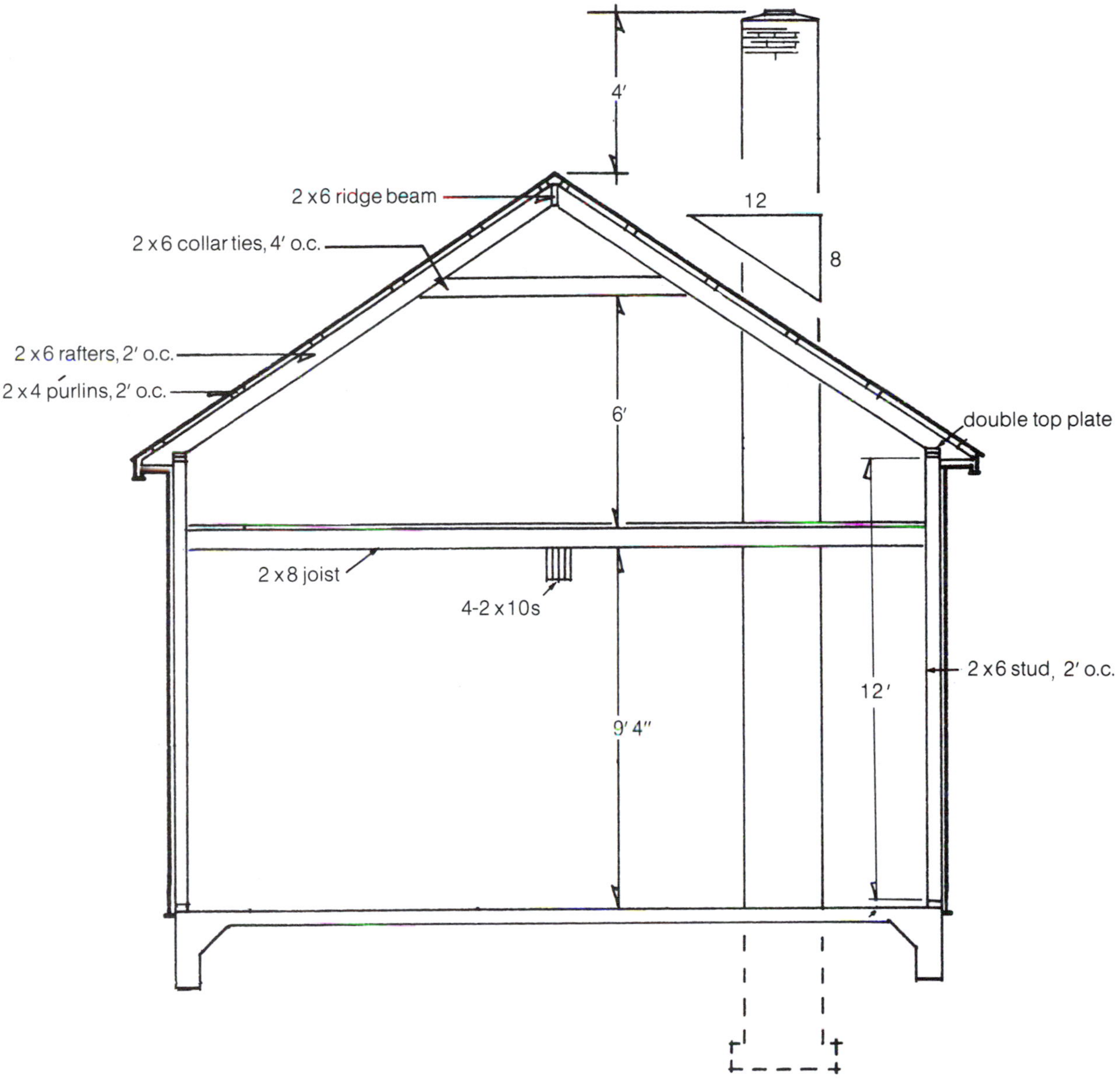

This section shows how the 2 x 8 ceiling joists are supported at the center of the span by a built-up beam of four 2 x 10s.

With its double-pitched roof, the classic gambrel barn provides extra storage space for feeds, supplies and equipment.

Gambrel Barn

This platform-framed barn measures 16 x 24 feet and has enough room for three tie stalls or two box stalls and 125 bales of hay in the attic beneath its attractive gambrel roof. The barn is built on a full foundation with concrete block walls.*

While designed primarily as a horse barn, the floor plan could be adapted easily for other animals and purposes. If you use it as a horse barn, do *not* pour concrete for the stall floors. Horses do much better on hard-packed clay.

The block-wall foundation rests on concrete footings, and the walls extend two courses or about 16 inches above the ground surface. In the plan shown, a 4-inch concrete slab is poured at ground level over compacted gravel. Be sure to add expansion joints at the edges of the slab. In cold climates, insulate the foundation and slab with 2 inches of rigid foam.

The walls are framed with 2 x 4s placed 2 feet o.c. and nailed to a 2 x 8 sill plate secured to the block wall with anchor bolts. The floor joists for the hayloft are 2 x 10s placed 2 feet o.c.; they span the 16-foot width of the barn. Tongue and groove plywood (¾ of an inch thick) forms the hayloft floor.

The gambrel roof is framed with 2 x 6s held together with plywood gussets. The eaves are formed with 2 x 4 rafter extensions nailed onto the sides of the rafters and set at a 45-degree

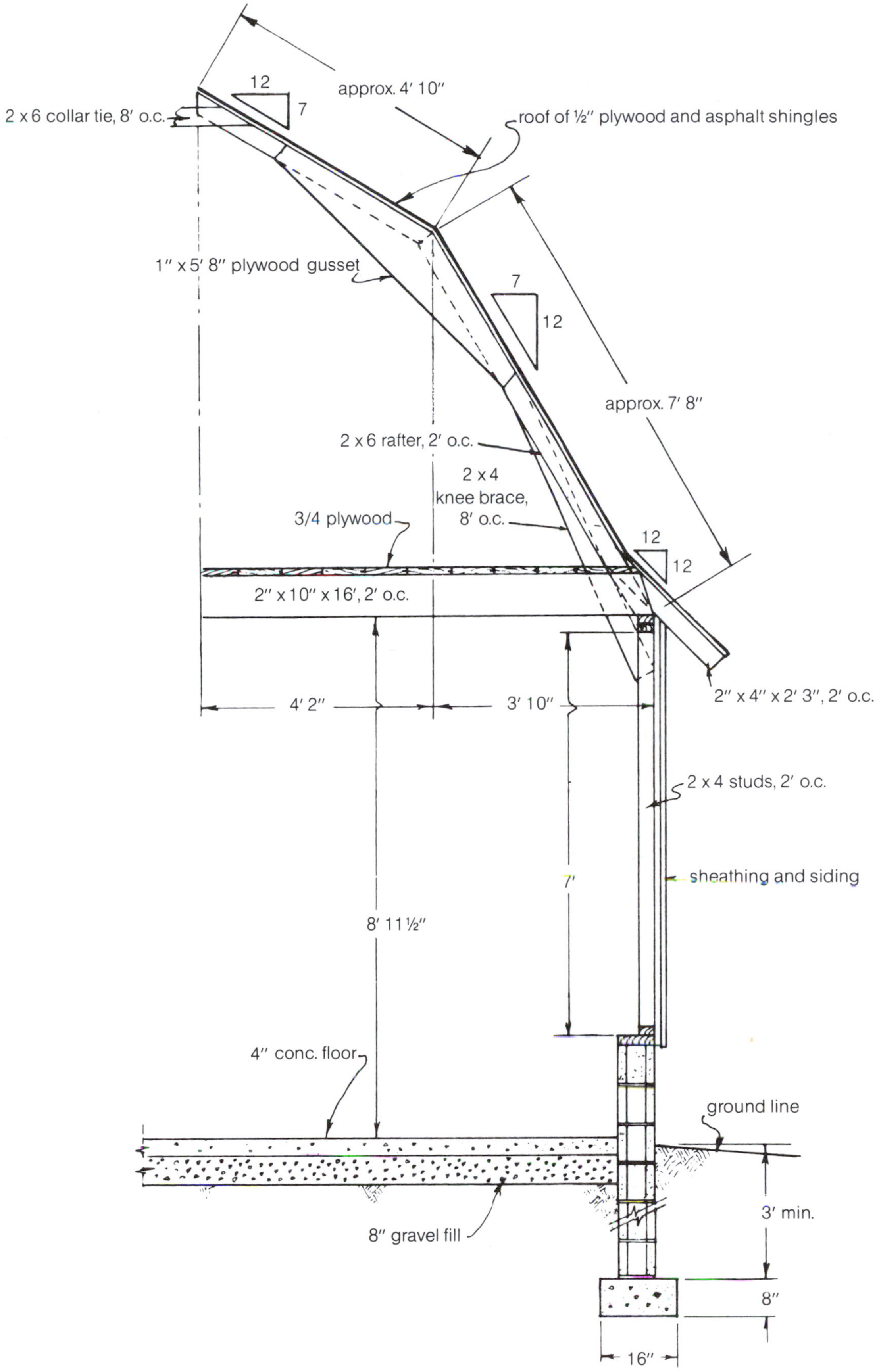

An 8-foot wide half section of the gambrel barn. Note how the roofing and rafters are strengthened by 2 x 4 knee braces, plywood gussets and 2 x 6 collar ties.

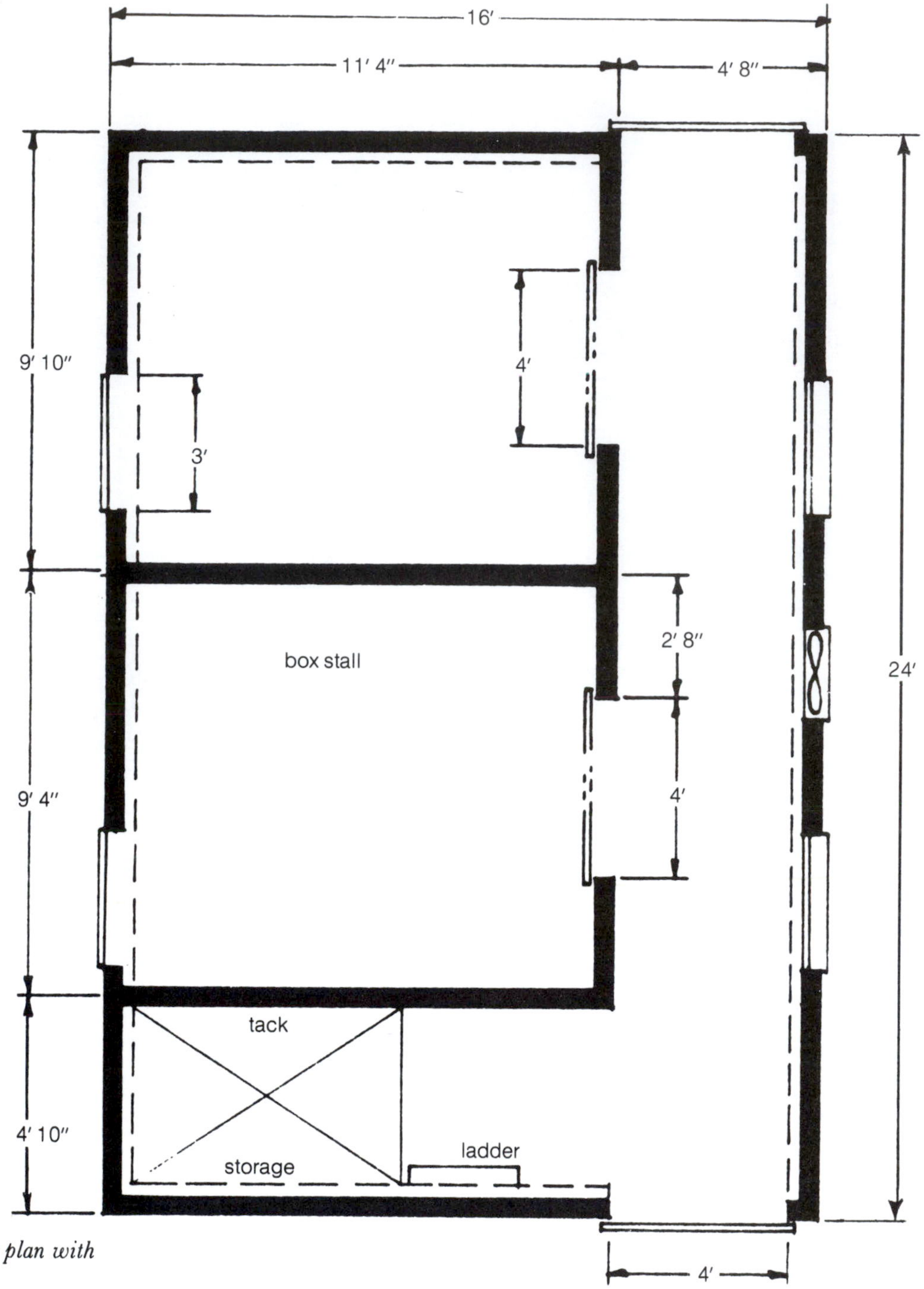

This is a gambrel barn floor plan with a box stall arrangement.

pitch; 1 x 6 lumber is used for fascia board on the end of the 2 x 4 extensions, and ⅜-inch plywood closes in the soffit underneath.

The barn may be sheathed and covered with a siding of your choice. T 1-11 plywood would serve both purposes and brace the barn as well. A 4-foot wide sliding door, made of 1 x 6s or plywood and framing, is located on the barn's gable end, as is the hayloft door. Two 3-foot wide awning windows placed on both side walls provide daylight and ventilation. A 300-cfm exhaust fan mounted on the side wall will provide enough extra air flow for summer ventilation.

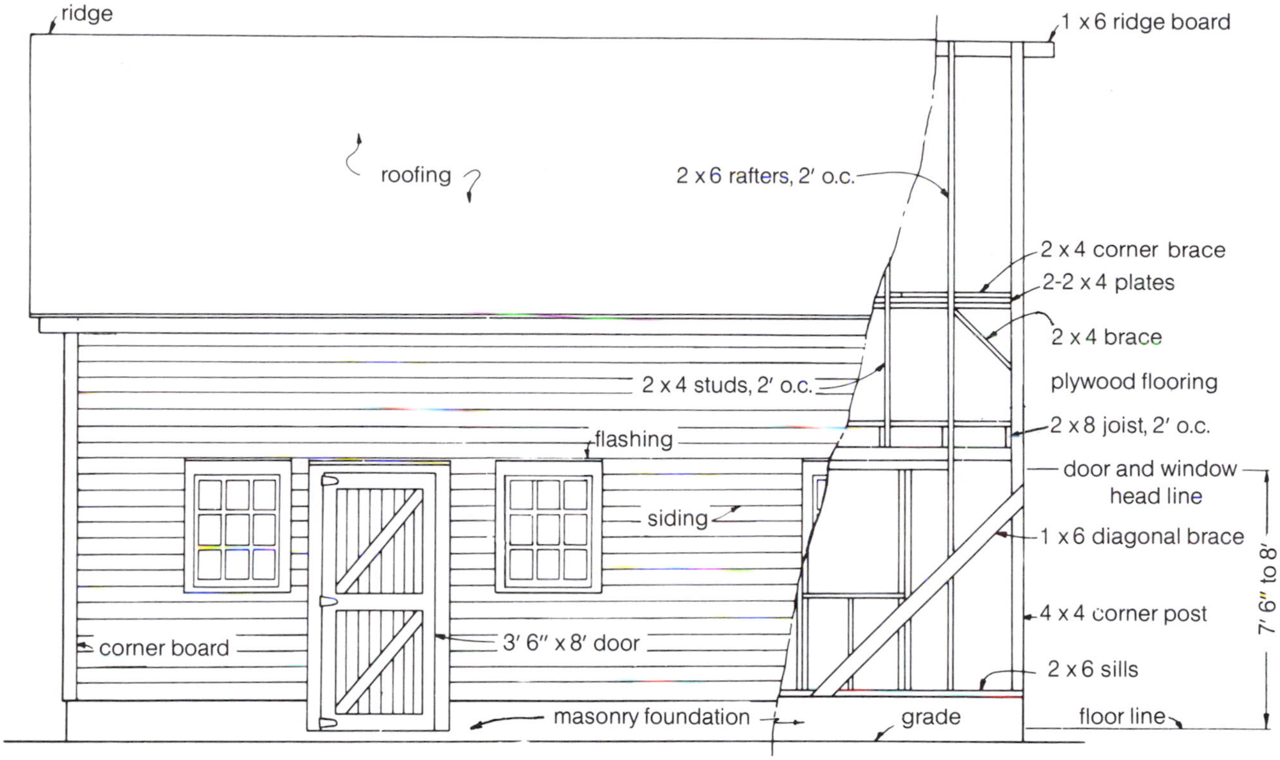

This general barn has a traditional exterior, with bevel siding and small-paned windows. The 1 x 6 diagonal braces are notched into the corner posts and studs.

General Barn

This 24- x 30-foot barn features cow stalls, a calf pen, three box stalls for horses and a second-floor, 720-square-foot hayloft. The floor plan may be rearranged in one of several configurations to suit your purposes.

The barn is built on a poured concrete foundation with a 4-inch concrete slab under the calf pen, cow stalls and feed room. The center driveway and box stalls have a clay floor. A 6-inch concrete wall supports the interior edge of the concrete slab flooring along the driveway.

The barn walls are balloon framed with 4 x 4 corner posts and 12-foot 2 x 4 studs set 2 feet o.c. Diagonal 1 x 6s are notched into the studs to brace the walls. Tongue-and-groove drop siding is nailed directly over the studs. The hayloft floor is ¾-inch plywood supported by 2 x 8 joists spaced 2 feet o.c. A 1 x 6 ribbon plate notched into the studs supports the joists.

The roof is framed with 2 x 6 rafters 2 feet o.c. with 2 x 6 collar ties at the top. Plywood (½-inch) is used for the roof decking under the asphalt shingles. Screened louvers are placed under the eaves on both ends of the barn to ventilate the hayloft.

On each end of the barn, there are two 4-foot wide, double-sliding doors so vehicles can be driven into and directly through the barn without backing up. However, this 8-foot wide opening

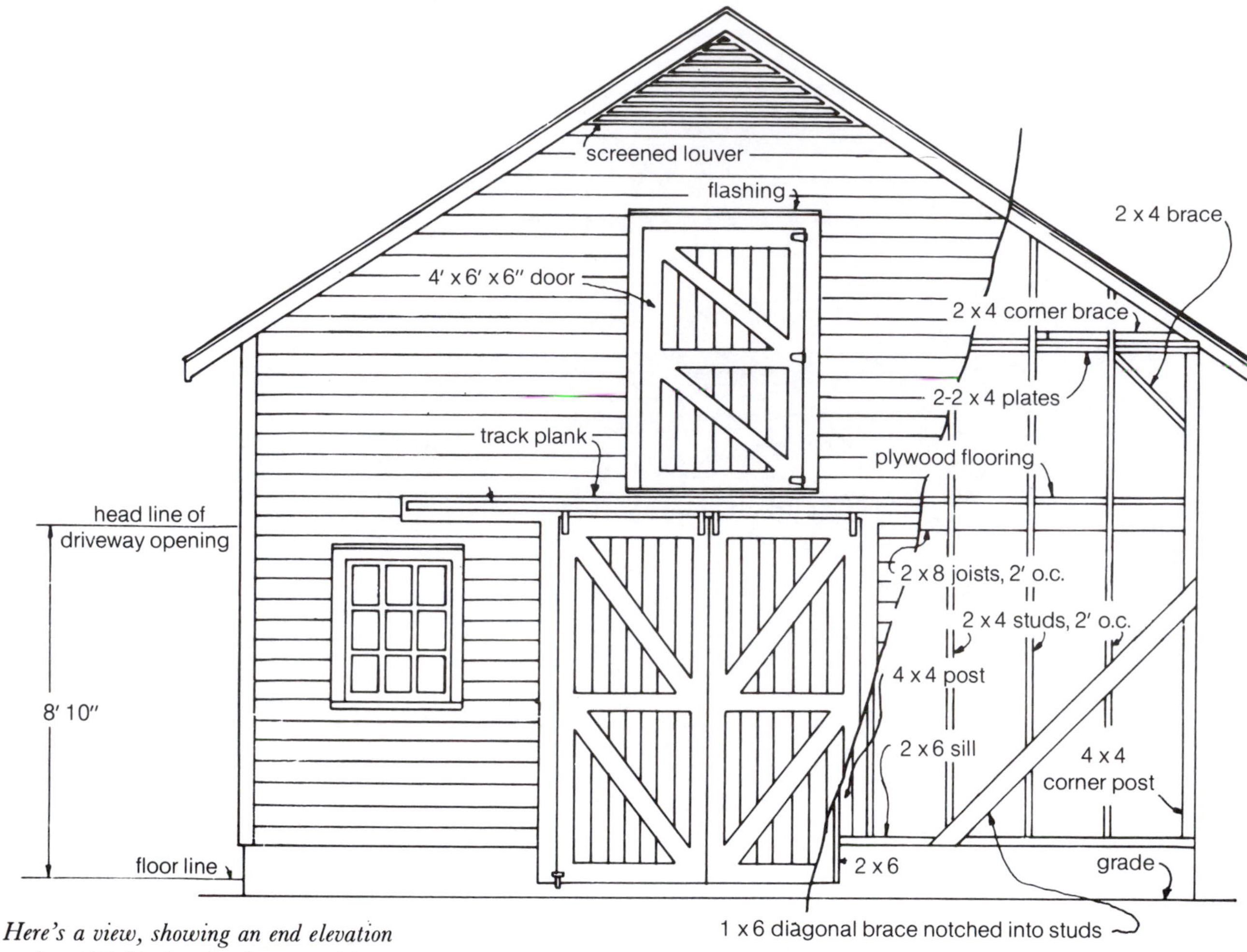

Here's a view, showing an end elevation and partial framing for the general barn. The window is optional. The door slides open, guided by a track at the top and roller guide at the bottom.

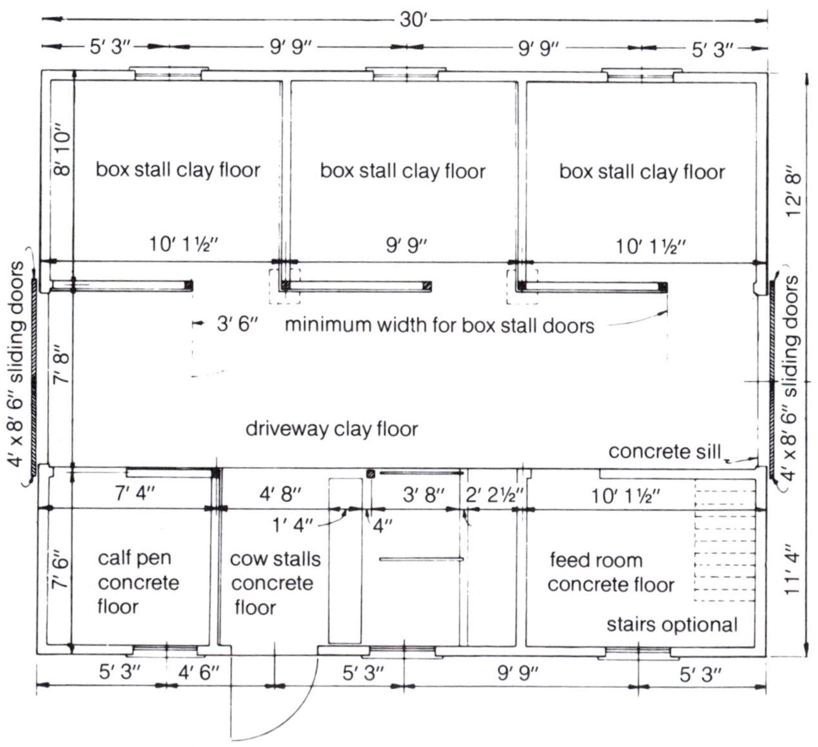

The floor plan may be arranged this way, or altered as necessary to achieve the optimum configuration for your livestock and equipment.

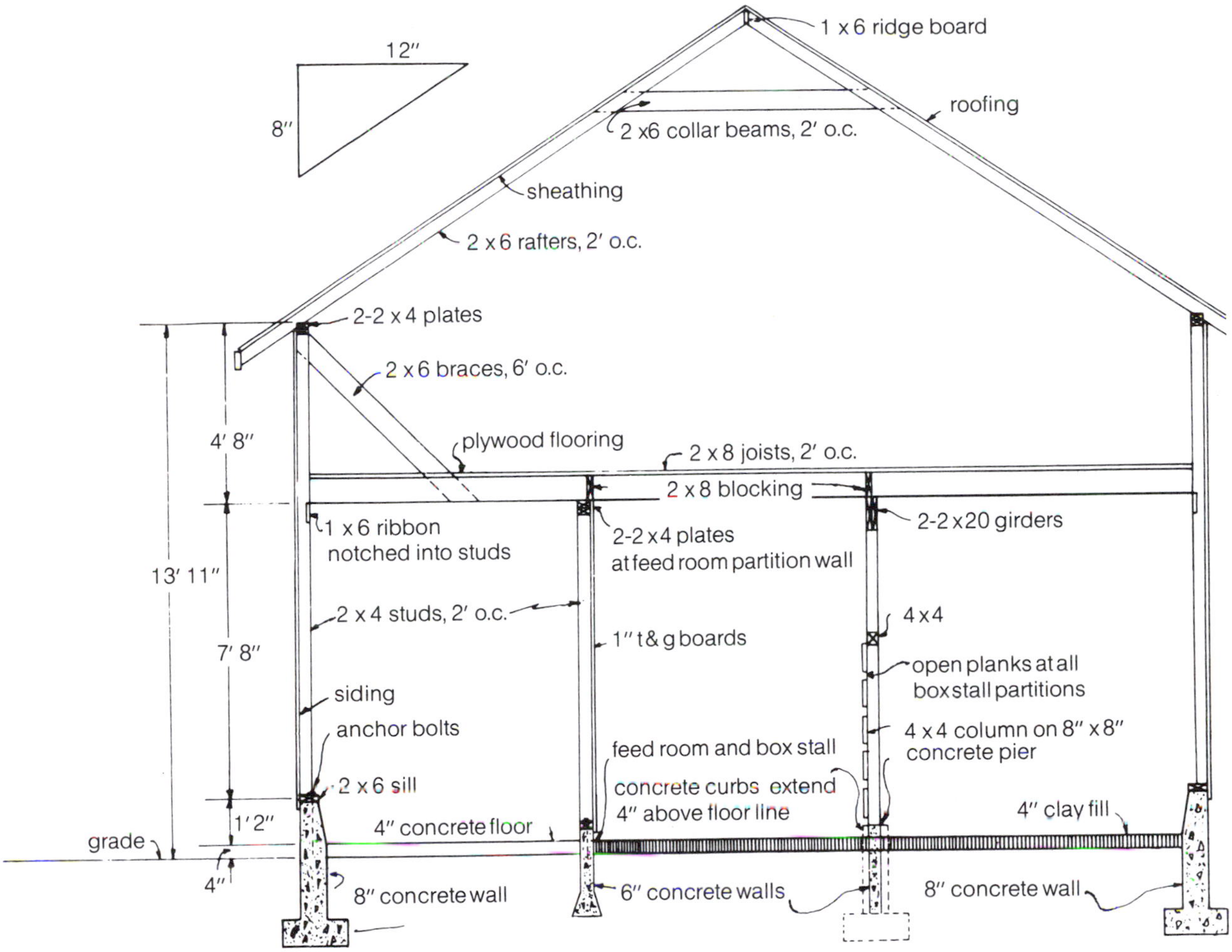

This section from the general barn shows the flooring and framing. Adjust the walls and footings to fit existing soil conditions.

is tight for a pick-up truck and it leaves little room for maneuvering around the vehicle once it's inside. If you plan to drive vehicles into the barn, increase the width by at least a foot and adjust other dimensions as necessary.

The windows on the side of the barn with cow stalls are barn sash. The windows are hinged on the bottom so the tops open in for ventilation. Over the horse stalls, the windows should be smaller (two lights high) and placed as high as possible on the wall to prevent the horses from breaking them accidentally.

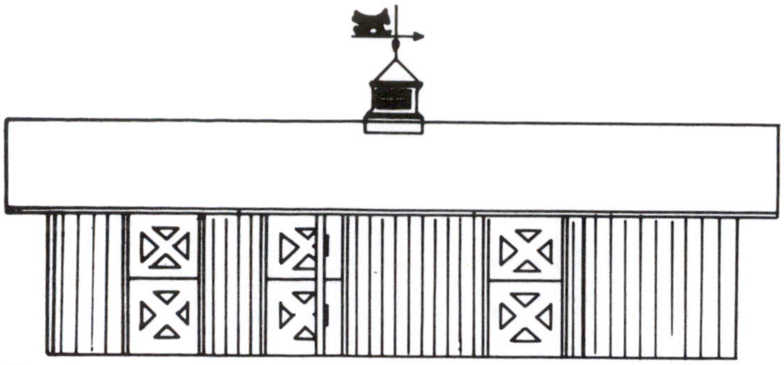

Front

End

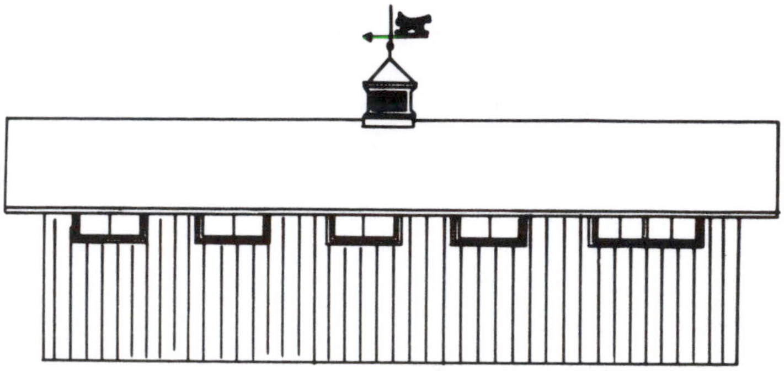

Back

This barn has a covered way, in front of the Dutch doors, for saddling and grooming horses (see front elevation).

Horse Barn

This two-stall horse barn has a simple, yet flexible design—one that could be expanded easily if more stalls are required. In addition to the two box stalls, there are tack and feed rooms. A covered breezeway in front provides protection from the sun and rain while saddling and grooming the horses.*

The barn is built with 16- and 18-foot long poles, set on concrete pads. Posts set on concrete piers could be substituted. The barn has an L-shaped foundation, with concrete slabs for the tack and feed rooms and clay floors for the stalls.

The walls are tied together with 2 x 10 splash boards and 2 x 6 girts. Also, 2 x 8 and 2 x 10 girts support the 14-foot long 2 x 6 rafters. According to the plan, the rafters are spaced 2 feet o.c.; in northern climates with heavy snow loads, reduce the spacing to 18 inches o.c.

Two different roofing options are presented. You can use 2 x 4 purlins, spaced 2 feet o.c. and aluminum or galvanized-steel roofing. Another option would be ½-inch plywood, covered by shingles.

Asphalt roll roofing could also be used, and would be more durable and water tight in harsh climates. To keep the rafters from spreading, use double plywood gussets at the ridge.

* The Grambrel, General and Horse Barn plans were developed by the United States Department of Agriculture Cooperative Extension Service. Copies of these plans are available through your state university Extension Service office.

Dutch doors, 4 foot wide, are the entrances to the two stalls and feed room. The tack room has a smaller 30-inch Dutch door. Eight windows provide light and ventilation, including two 36-inch wide windows in each of the stalls. These are needed for ventilation in warm climates; for cold climates, use only one window to reduce winter heat loss. Ventilation is also achieved through two gable-end louvers and a cupola. Cupolas may be home built or purchased pre-assembled.

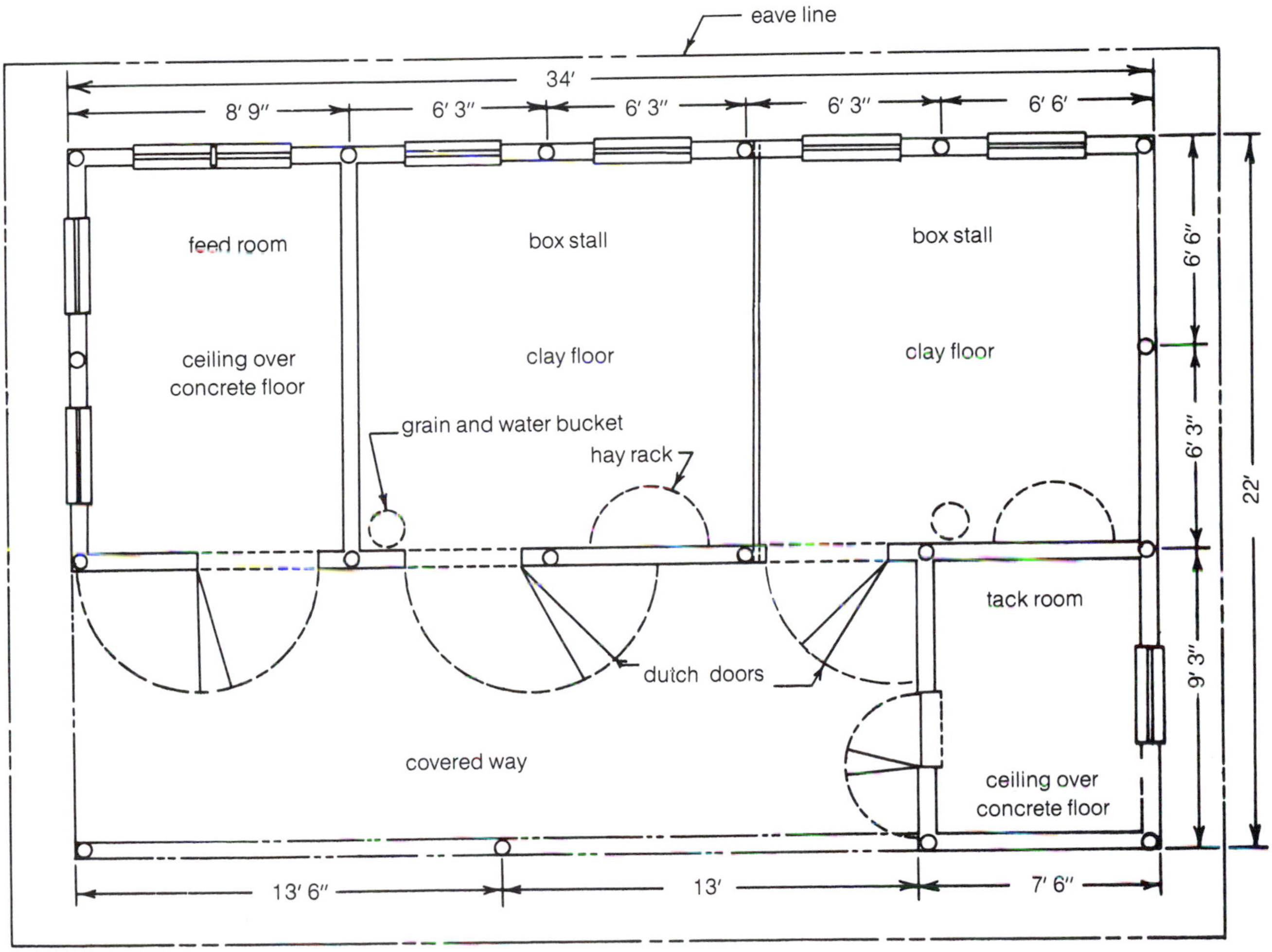

The floor plan shows the covered way, as well as the feed and tack rooms and two box stalls. There's a door to each room.

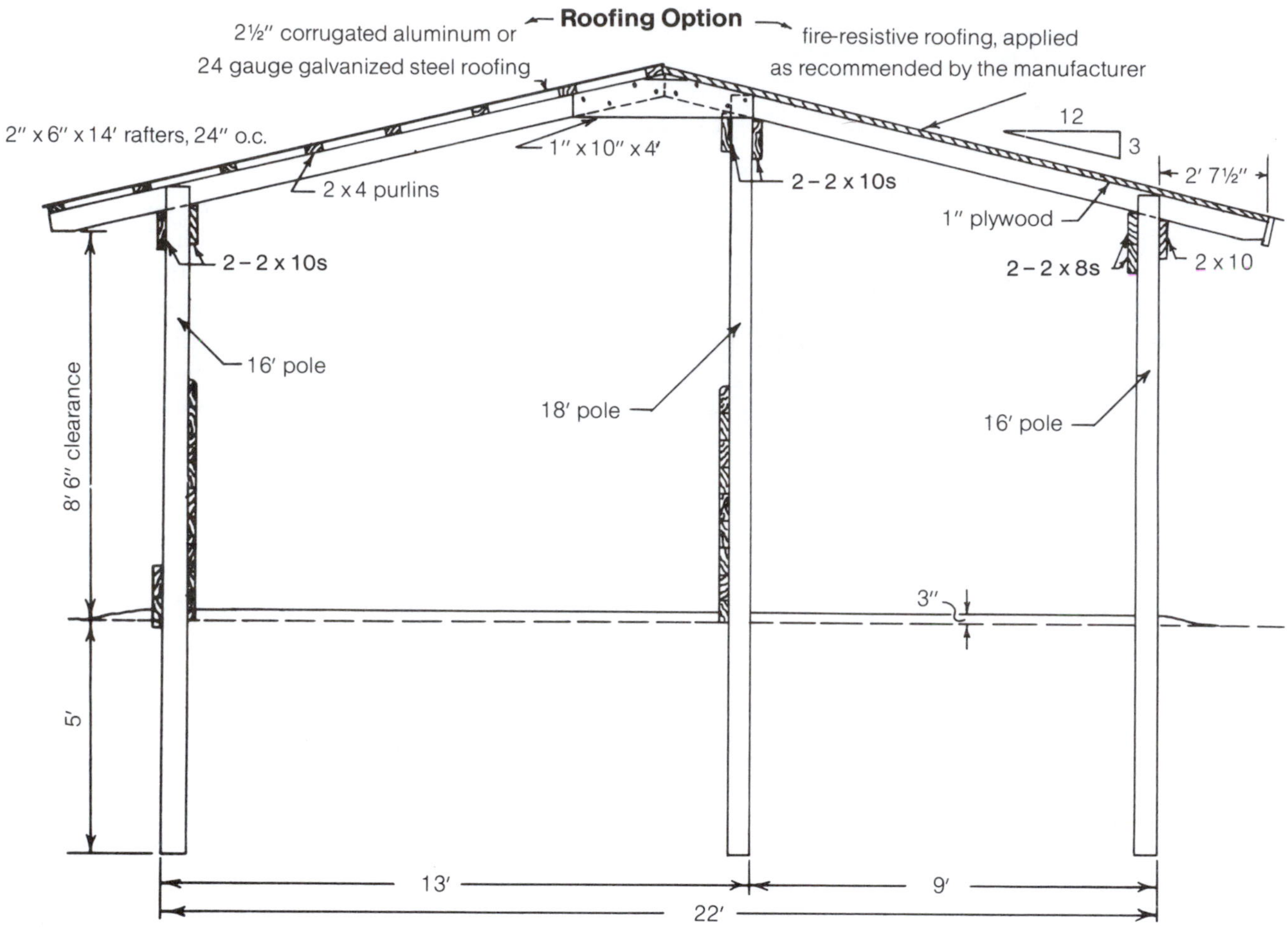

This horse-barn cross section shows two roofing options: at left, there are rafters and purlins, along with metal roofing; and, at right, there are also rafters, along with a fire-resistant roofing of your choice.

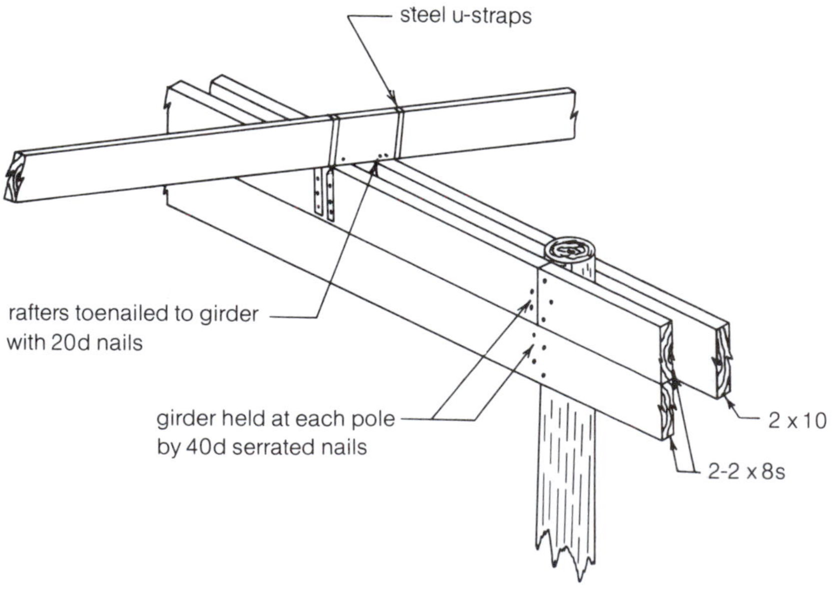

Rafters may be secured to the upper girders this way.

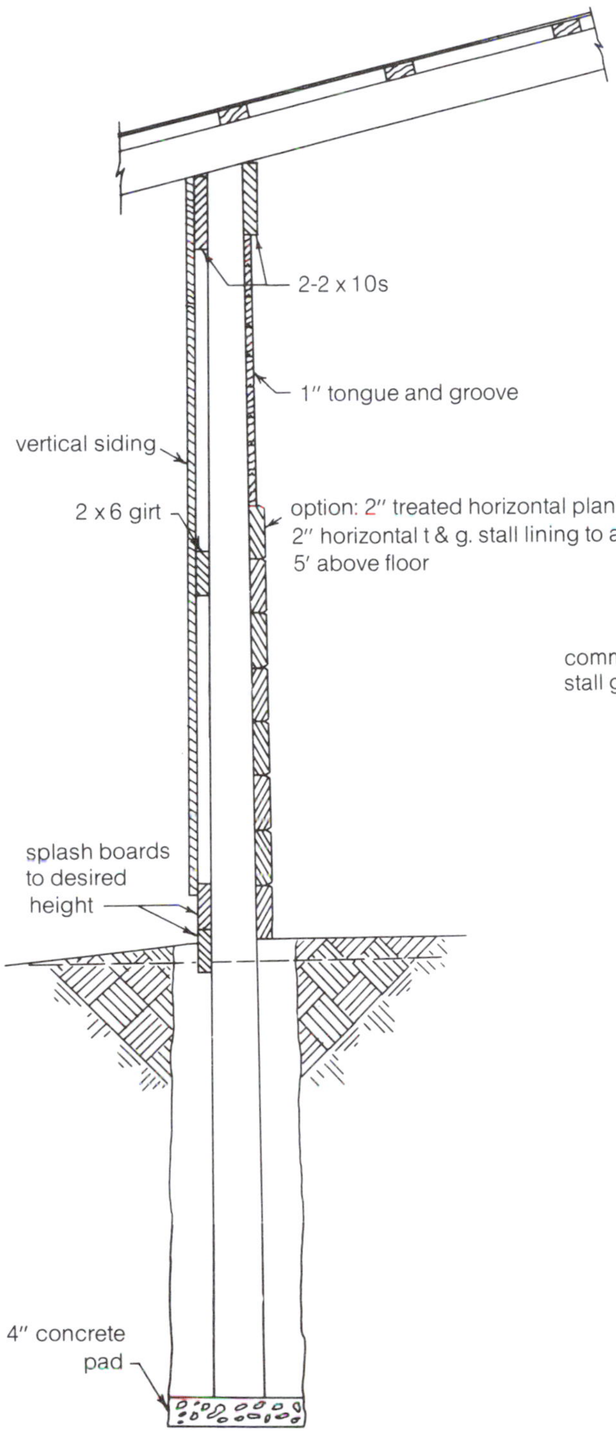

This wall detail for the horse barn shows splash boards and poles at or below the earth surface. These should be pressure treated or treated with preservatives.

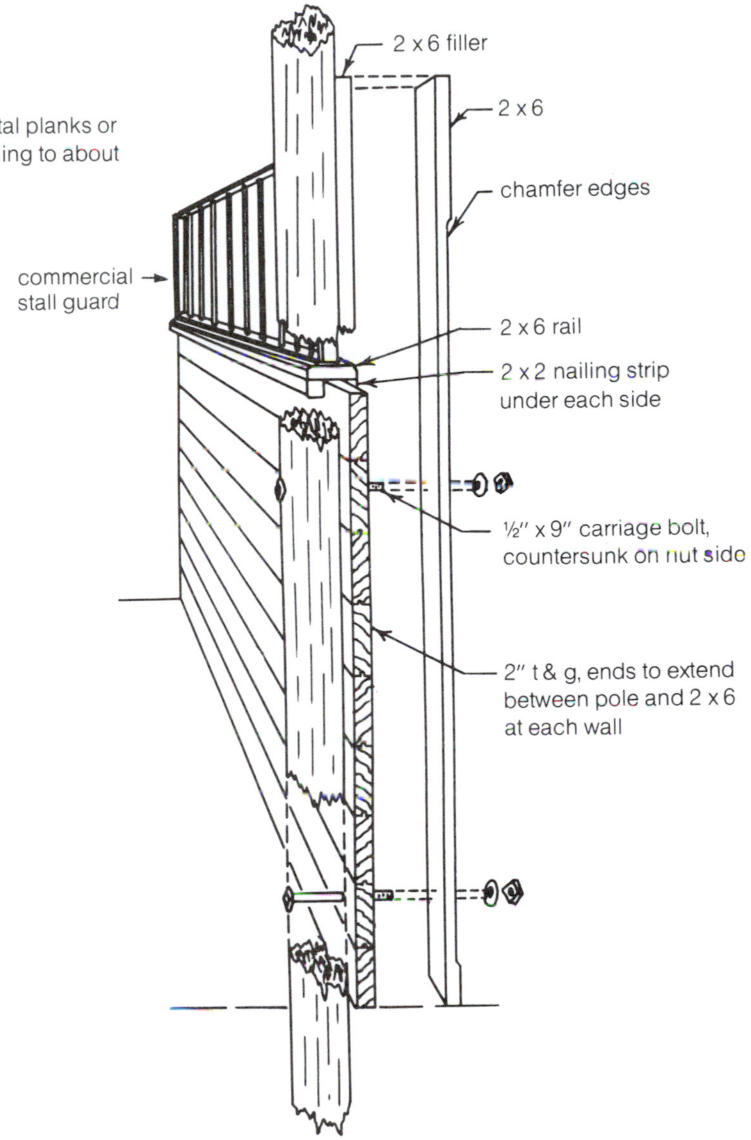

Here's a cutaway section, showing details of a stall partition, suitable for pole barns.

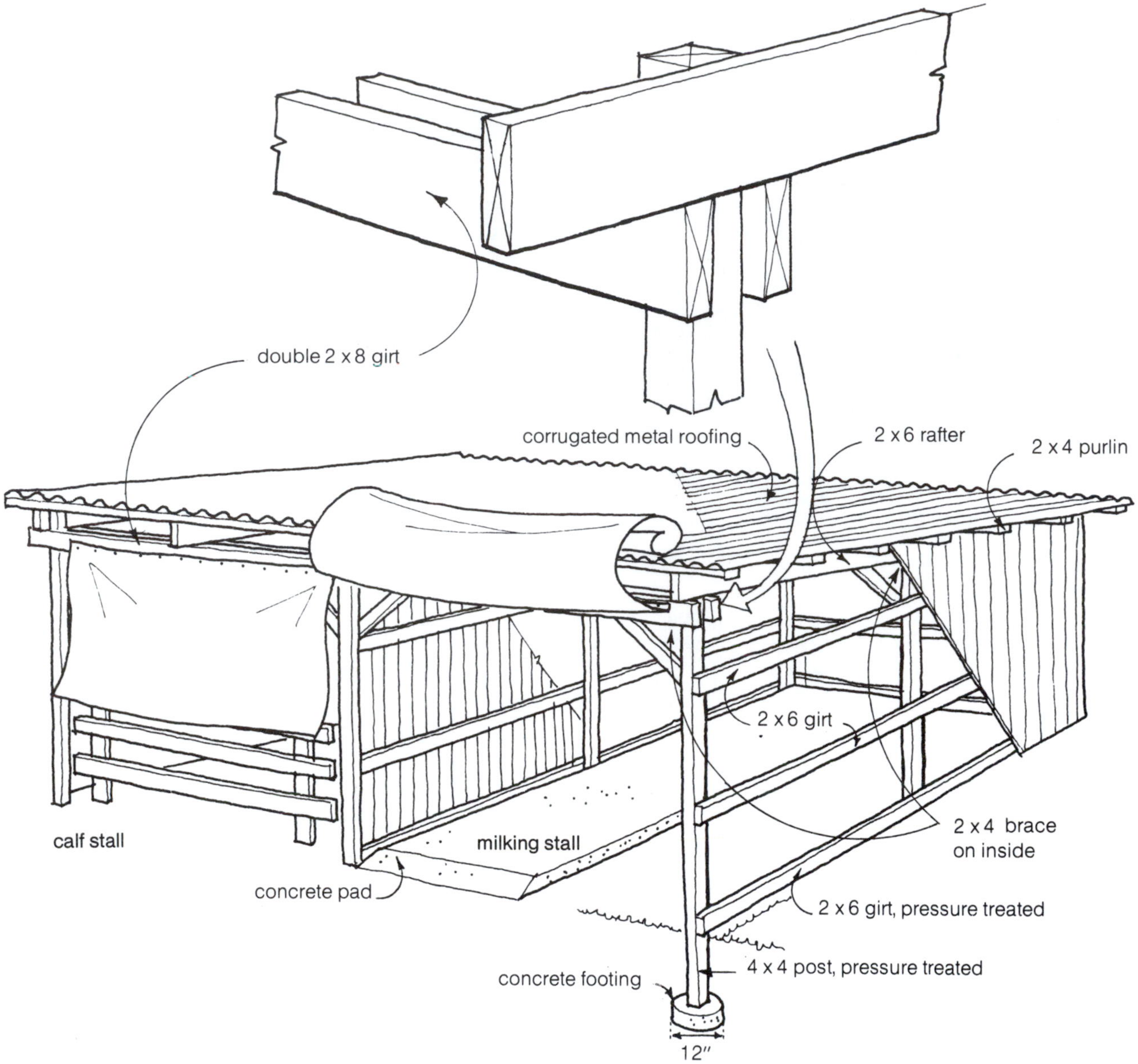

The small family cow barn needn't be a fancy outbuilding. You can use old tarp or carpet to slow winds and drafts from the front.

Family Cow Barn

Small family cow barns should provide a place for housing and milking a cow, as well as a shelter and pen area for her annual calf. The pole-barn design shown here is based on a small barn built by my neighbor, Elmas Scott. His barn serves these purposes nicely.

The barn faces south and old carpet pieces tacked to the front provide economical temperature and draft control. During the day, the carpets are thrown back to allow sun into the barn; at night, the carpets are dropped down to keep out drafts and help retain heat in the barn. The barn also has a solid dividing

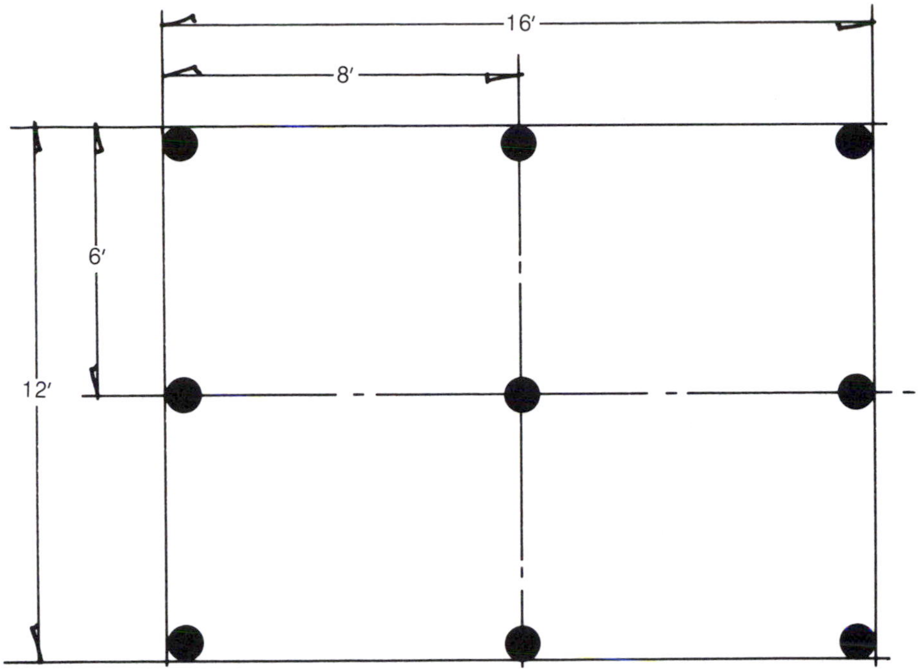

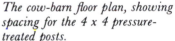

The cow-barn floor plan, showing spacing for the 4 x 4 pressure-treated posts.

wall down the middle to separate the calf and cow when necessary. If you wish to let the calf nurse after you're through milking, you can provide a small door between the milk stall and the calf stall. When you're through milking, you can let the calf in to strip the cow of the rest of the milk.

The barn measures 16 feet wide by 12 feet deep and is 8 feet high in the front; the roof slopes down to 6 feet in the back. Nine pressure-treated posts set on concrete footings support the walls and roof. In the milk stall, there is a concrete pad, 4 inches thick and 30 inches wide, for the cow to stand on. This solid pad is easier to clean than a dirt floor and it elevates the cow for easier milking. The concrete pad could be extended across the entire back of the milk stall to provide a dry, solid platform for feed boxes and other items.

The 2 x 6 girts on the sides of the walls secure the posts and carry the board and batten siding. Rough-sawn, 1 x 8 boards are used for the interior wall and exterior siding of this barn, though plywood or metal siding could be substituted.

The roof is supported by double 2 x 8 girts attached to the top of the posts—front, middle and back. On top of these, sit 2 x 6 rafters spaced every 4 feet o.c. For northern climates, space the rafters 2 feet o.c.

The barn is covered with galvanized-metal roofing, nailed to 2 x 4 purlins running across the rafters every 2 feet o.c. The ends of the rafters and the purlins should overhang the walls by 12 inches to form a protective eave. If desired, attach 1 x 8 boards to the ends of the rafters and purlins to form a continuous fascia board around the building.

To give the barn lateral stability, brace all the posts to the top girts; 2 x 4s will do the job nicely.

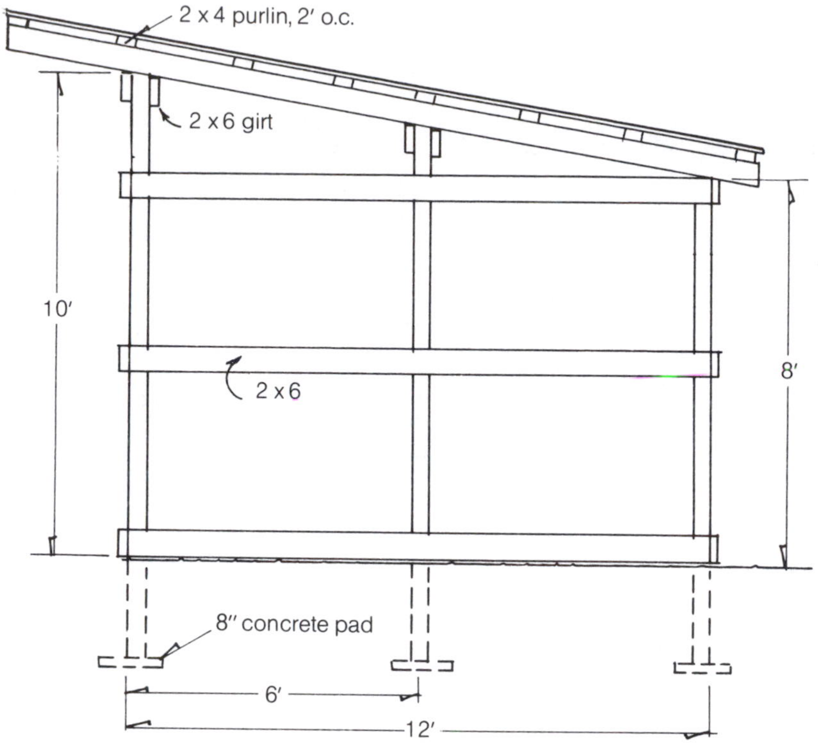

Simple pole-barn techniques are used to frame this shelter. The posts are 4 x 4s, pressure treated, while the wall girts and rafters are 2 x 6s. The footings are 8-inch diameter concrete pads.

Two-Horse Barn

You can build a shed-style barn for two horses using simple pole building techniques. The finished barn measures 24 feet wide by 12 feet deep and is divided into two 12-foot wide stalls. Each stall has a 4-foot wide Dutch door.

Seventeen 4 x 4 pressure-treated posts support the walls and roof. These are set on concrete footings 4 feet apart on the front and back and 6 feet apart on the sides. The posts in front are 10 feet high, sloping down to 8 feet in the back. A post in the center of the barn helps carry the dividing wall of 2 x 6 planks. The siding is ½-inch plywood which also serves to brace the posts, thereby eliminating the need for plywood gussets.

The roof is framed with double 2 x 6 rafter girts which carry 2 x 6 rafters spaced 4 feet o.c. On top of the rafters, 2 x 4 purlins 2 feet o.c., carry the metal roofing. Both the rafters and purlins overhang the siding to form the eaves; 1 x 8 fascia boards can be attached to the end of the rafters and purlins to finish the eaves.

You can make the Dutch doors from either 1 x 8 boards held together with a Z-frame on the back or from pieces of ¾-inch plywood. Attach the doors with strap hinges to 1 x 6s that form a door casing. Wooden turn latches can be screwed into the door casing to keep the doors closed. Windows of any size may be framed into the sides of the building.

To design a 12- x 12-foot, single-horse pole barn, simply "cut" the two-horse design in half.

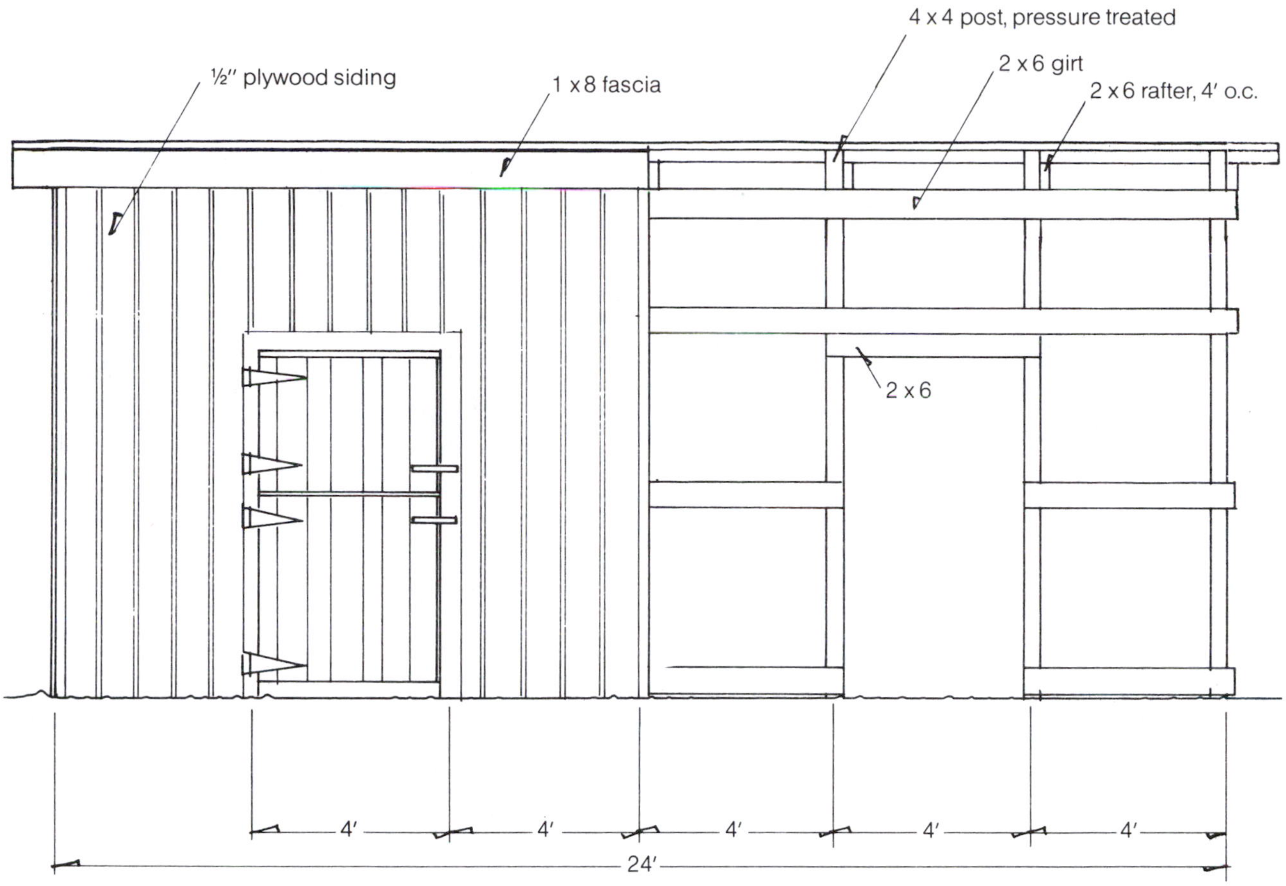

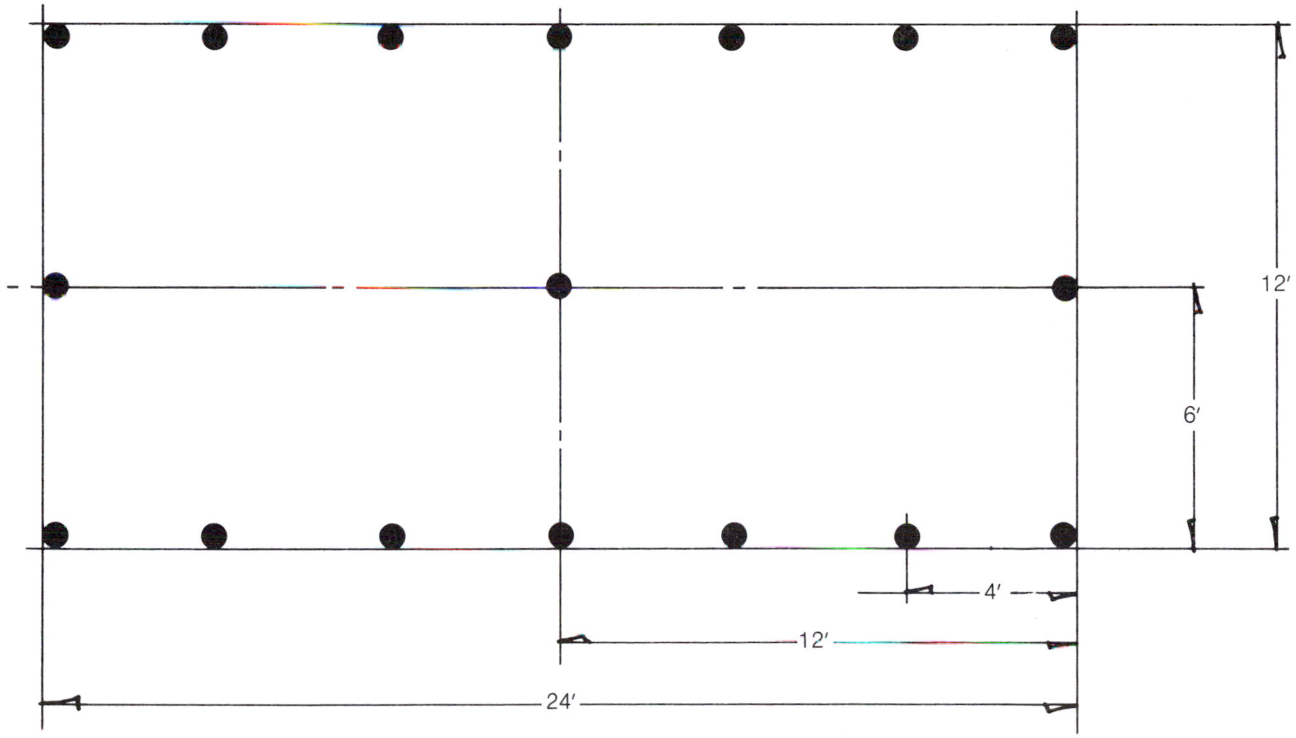

Posts or poles for the two-horse barn may be positioned as shown here. Also, the plan may be divided in half to make a one-stall shelter.

CHAPTER 10

Root Cellar and Storm Shelter

Bubel

HAVING A ROOT cellar in your back yard is like having your own private supermarket of fresh fruit and vegetables. Any time you need potatoes for dinner or an apple for a snack, just open the door and walk inside. Anything you need is right there on the shelves. But, a root cellar is more than a convenience. In addition to providing the security of having a year's supply of fruits and vegetables on hand, a root cellar is a sanctuary. In an emergency, it could shelter your family for a few hours, a few days or even a few weeks if necessary.*

A good root cellar should provide cool, above-freezing temperatures and good circulation of moderately humid air. The combination root cellar and storm shelter described here has these features. For example, the dirt floor takes advantage of the naturally cool, even temperature of the earth. It also cuts costs and provides needed humidity.

The reinforced concrete-block walls are durable and sturdy; they are reinforced with rebar inserted in some of the middle cores. In extreme climates, the remaining cores can be filled with loose insulation to keep temperatures more even. The walls are topped by a wood frame made of 2 x lumber. The roof is protected by sheathing, roll roofing and plastic film, then covered with 2 feet of soil. The cellar is vented for good air circulation, and a drainage system protects the interior from water seepage.

The ideal location for a root cellar is on a hillside or slope facing away from prevailing winds. If the cellar is so located, you may find you won't need a stairway. Even without a slope, choose the highest ground you have for the best drainage possible. Avoid any low-lying areas, or your cellar may fill up with water in the spring.

Materials

The materials needed to build this concrete-block root cellar cost about $1,200, a large sum, but a bargain if you amortize the cost over the 20 or more years the cellar could be used.

If you are looking for ways to trim the costs, it can be done by making the most of any materials you may have or can buy cheaply in your area. Remember, in a root cellar, you are not interested in appearances. The only thing that counts here is size and strength.

---

* This chapter has been adapted from Hobson, Phyllis, *Build Your Own Underground Root Cellar* (Charlotte, Vt.: Garden Way Publishing Co., 1981).

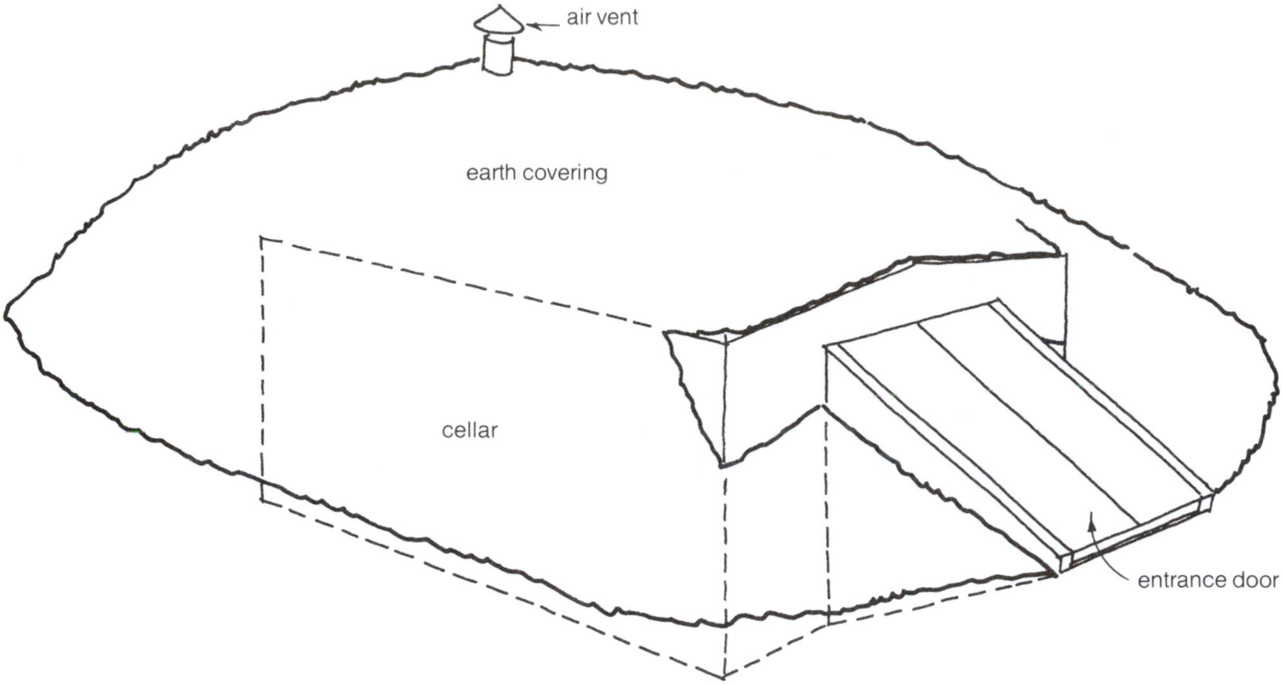

The root cellar is covered with earth. The entrance door and stairway may be omitted if the cellar is built into a slope.

All lumber should be pressure treated or treated with a wood preservative, such as copper naphthenate sold under various brand names. Wood should be soaked in the liquid, if possible, or generously painted with the solution, according to the manufacturer's directions. Be prepared to store all materials under cover to keep them as dry as possible before and during construction.

Building the Cellar

Excavating

For your 12-foot by 8-foot cellar, dig a hole 16 feet long by 12 feet wide by 4 feet deep. Dig an entrance ramp 4 feet wide that slopes from the ground level to the bottom of the hole. The floor should be as level as possible.

Marking Off

Beginning 36 inches from the entrance end, pound four stakes into the ground to mark off an area 8 feet wide by 12 feet long. This marks the *outside* of the cellar wall. Check to make sure the corners are square by measuring diagonally from corner stake to corner stake. Complete the wall and footing layout, using Chapter 3 as a guide. The footings are 16 inches wide and the

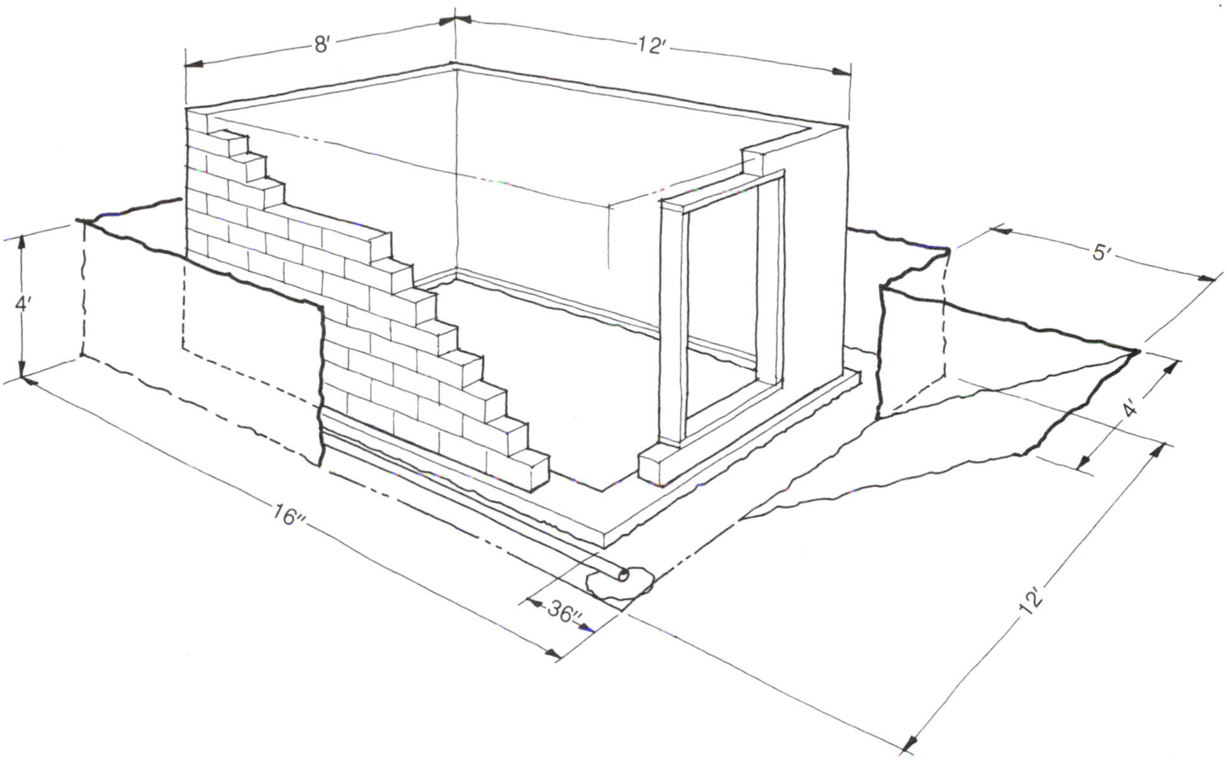

Actual dimensions for the cellar, foundation hole and sloping entrance ramp. The walls are 12 courses of blocks.

blocks are laid so their outside edges fall on the outside dimensions of the wall.

Digging the Footing

Dig a trench 16 inches wide and 8 inches deep all the way around the hole between the footing lines. Keep the corners square and the bottom level.

When the trench is dug, scan the length of the string lines from each corner to make sure the trench is straight. Install footing form boards. Insure that they are straight and level.

Pouring the Footing

When the stakes and form boards are all in place, dampen the trench with a garden hose or a few buckets of water; then fill the trench with wet concrete. Level it just to the tops of the boards using a wooden float.

One way to do this is to take a 2 x 4 and place it on top of the footing forms. Then move the board in a lateral sawing motion as you move it slowly down the footings. This will create a smooth footing surface for the concrete blocks.

Give the wet concrete at least one day to harden before you start laying the blocks. Spray the surface with water once a day to keep it damp. In cold or very hot weather, keep it covered with straw, canvas or old blankets.

Laying the First Course

To make sure the blocks are going to fit the footing, make a trial run by setting the first course in place without mortar (one row of blocks all the way around). Use the batter boards' string and plumb bob to re-establish the outside wall lines. Once this has been done, dry lay the blocks, starting from one corner.

Place the blocks on the concrete footing, and slip a 2-inch by 8-inch strip of ⅜-inch plywood between each two blocks to allow for the mortar. Lay nine blocks lengthwise down the two side walls. Then place five blocks between them on the two end walls. Blocks should always be placed with the large core side down. Adjust, if necessary, to make the blocks fit, then mark the placement on the footing with chalk. Remove the blocks.

Starting at one corner, spread a layer of mortar along the chalk line from the corner down both walls. The mortar should be 1 inch thick by 8 inches wide by the length of two blocks.

Set a corner block on the mortar and tap it into place with the trowel handle until the block is level and plumb, and the mortar is ⅜ inch thick. Scrape off excess mortar under the block on both sides.

Spread mortar ½ inch thick on the end of a stretcher block, and set it in place end to end with the first block in the bed of mortar. Again tap the block into place and scrape off all excess mortar. This mortar can be used again.

Repeat with another block placed at right angles to the first along the other wall.

Continue laying blocks all the way around the footing, spreading mortar on the footing and on the end of each block and checking to be sure each block is level before setting the next. Use corner blocks on each corner. Spread mortar on both ends of the last block in the course and carefully ease it in. It may take a couple of tries.

As the mortar between the blocks and between the blocks and the footing begins to stiffen, it will pull away slightly from the blocks. To correct this and make the walls more waterproof, run a mason's jointer or a metal or wooden rod along the joints to force the mortar in the cracks. This operation, called tooling, will leave a concave impression between blocks.

Tool all joints and wipe off any excess mortar.

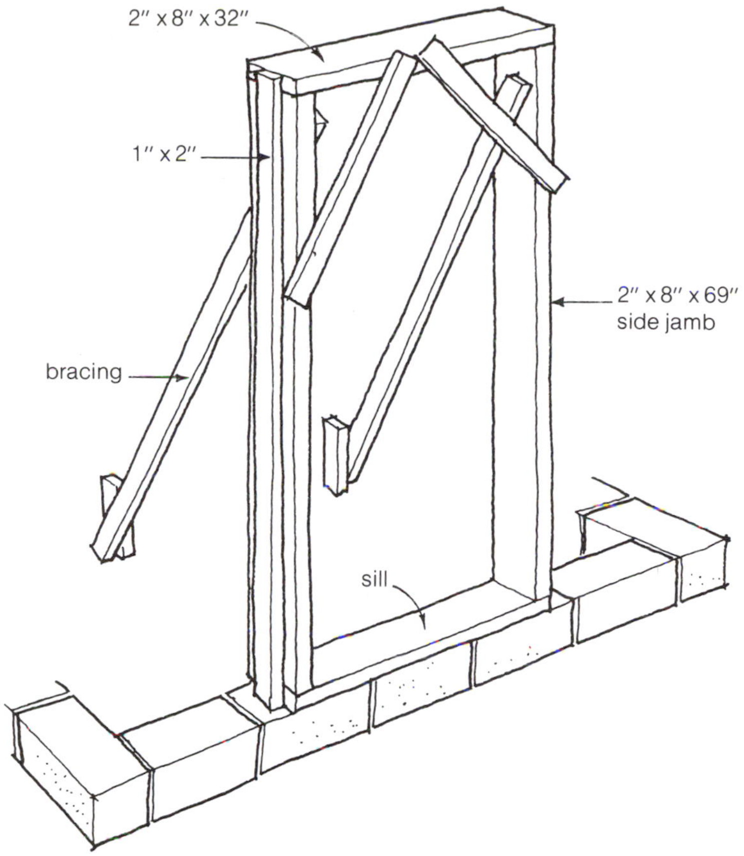

The door frame is built on the first course of blocks. Note the 1 x 2s added to the side jambs.

Laying the Drain Pipe

With the first course of blocks in place, pour 2 inches of gravel around the outside perimeter. Then, lay a line of 4-inch perforated plastic drainage pipe on top. Be sure the holes on the pipe face down on the gravel. Slope the pipe 2 inches from back to front. Join sections according to manufacturer's directions, using 90-degree elbows at the corners. Make sure the pipes are not clogged with dirt.

At each front end of the pipe, dig a hole about 3 feet in diameter (or as large as possible) and 3 feet deep. Fill both holes with gravel, and lay the pipe to the center of each. Cover pipes with 12 inches of crushed stone or gravel.

If your land is wet, and if the ground slopes away from the root cellar, it is a good idea to run the drain pipe all the way to an outside surface discharge.

Marking the Doorway

Mark off a 32-inch rough opening in the center of the entrance wall. The doorway should be opposite the entrance ramp and should be positioned as evenly with the blocks as possible to avoid the time-consuming job of cutting blocks to fit. The door is set on top of one course of blocks to discourage mice, chipmunks and water from entering the cellar.

To make the door frame, nail together the 2 x 8 lumber as shown on page 171, with a 1 x 2 centered on and nailed into the side jambs. The 1 x 2s will be mortared into the grooves of the concrete blocks. Brace the door frame temporarily with scrap lumber. The sill should be flush with the outside edge of the concrete blocks.

Building the Walls

Beginning at a back corner, spread mortar on the top edges of the first course of blocks, two blocks long in each direction. Build the corner three blocks high. Repeat on all four corners, using corner blocks where needed. As you build the walls, place ½-inch rebar in the cores of the blocks, approximately 32 inches o.c. Secure with concrete, slipped into the cores. If it's not possible to insert a single, long piece of rebar, use smaller pieces and overlap them by 6 inches.

Hook a mason's line around the ends of two corner blocks and stretch a taut line from corner to corner. Using this line as a guide, fill in the second course of blocks along one wall, then move the line to the next wall. Continue building up the walls, first on the corners, then filling in the course, all the way around. As you finish each course, move the mason's line up to the next corner block. Tool all joints as they begin to stiffen. Build the entrance wall around the door frame, using blocks cut to fit. Continue until the walls are 11 blocks high all the way around.

Over the doorway, insert a lintel. To make the lintel, lay six lintel blocks end to end so a channel is formed. Place reinforcing bars in the channel and fill with concrete. Let dry. Turn the lintel over and gently set in place. Adjust until the mortar joint is the same thickness as the other horizontal joints.

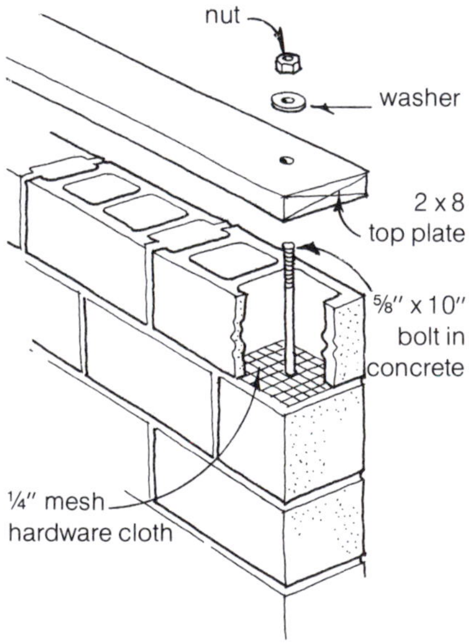

This cutaway view of the top course shows the placement of the hardware cloth, anchor bolt and top plate.

Insulating and Topping the Walls

Before laying the top (12th) course of blocks, insulate the entrance wall by filling the cores of the blocks with sawdust or commercial insulation pellets. In a northern climate, you may want to insulate all walls.

Cut ¼-inch mesh hardware cloth into 6-inch-wide strips, and lay the strips over the tops of the eleventh course of blocks on all four walls. The strips should be cut in pieces as long as possible.

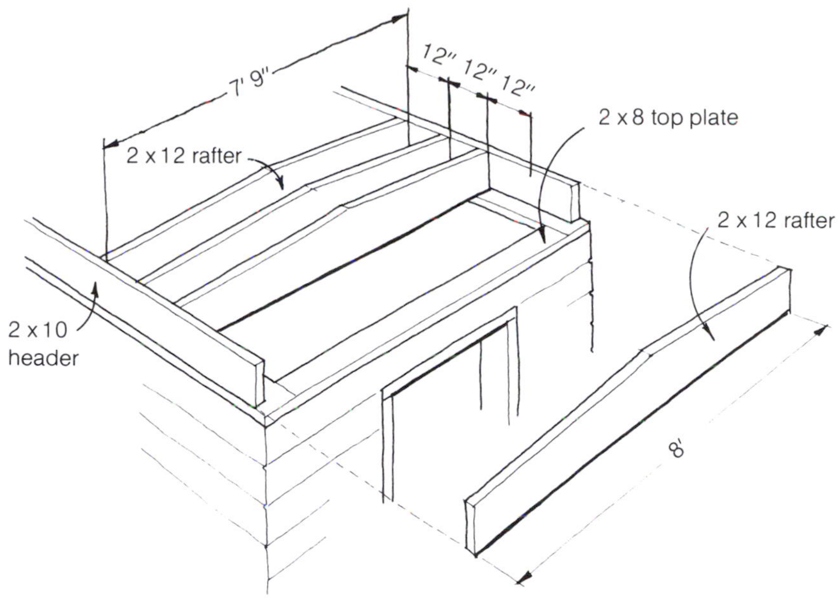

Frame the cellar roof with 2 x 10 headers and 2 x 12 rafters. Use a saw or plane to taper the rafters so they slope down from the middle to the edge, to meet the tops of the headers.

Now proceed with the final course of blocks, spreading the mortar on the hardware cloth as necessary.

When the final course is in place, mix a batch of concrete, and fill the cores of the top course of blocks, smoothing off the top. As you fill the cores, imbed a ⅝-inch by 10-inch anchor bolt in the concrete every other block. The nut end should protrude from the concrete about 2 inches.

Allow the mortar and concrete fill to cure at least three to five days. If the weather is very cold or very hot, keep it covered, but do not wet down.

Roofing the Cellar

Attach 2 x 8 boards to the top of the block walls by drilling holes in the boards for the bolts you embedded in the concrete. The 2 x 8s may be pieced together, if necessary. Fasten the top plate securely with washers and nuts.

Taper both ends of the 13 rafter boards (2 x 12s) with a saw or plane. Set aside.

The 2 x 10 headers should be set on edge on the 12-foot top plate you bolted to the top of the block walls and be toenailed in place 1½ inches from each end.

Set the two end rafters (2 x 12s; 8 feet, ⅜ of an inch long) at each end of the headers. Nail to the top plate and to the ends of the headers, creating an open box on the block walls.

One at a time, set 11 rafters in place across the top of the building, between the two headers, every 12 inches the length of the building. Toenail the rafters to the top plate, and nail through the headers into the ends of the rafters on each end.

When all the rafters are nailed in place, cover the top with ¾- or 1-inch exterior grade plywood, and nail in place. Fill any holes around the edge of the roof with batt insulation, stapled in place. Then cover the entire roof with overlapping roll roof-

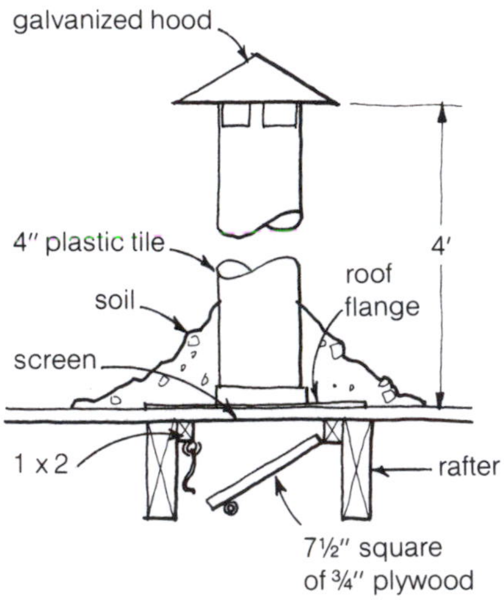

Side view of the roof air vent assembly. Pile soil around vent to hold it in place.

ing. Begin the layers at each side, overlapping each layer 2 inches and extending it at least 1 foot over on all sides. Nail in place.

Installing the Air Vent

Between the second and third rafters from the rear wall, drill a small hole in the roof. Using a saber saw or a keyhole saw, and starting from the hole you made, cut a 6-inch hole in the roof, centered between the rafters.

Cut an 8-inch square of screen. Staple this to the underside of the hole to keep out insects.

Cut two pieces of 1 x 2 to measure 7½ inches long. Nail these to the rafters on both sides of the hole you just cut. Cut a piece of ¾-inch plywood to measure 7½ inches square. Attach the plywood piece to one side of the 1 x 2s with a ¾-inch leaf hinge. Close the plywood hatch door against the other piece of 1 x 2 with a simple hook and eye. Weatherstrip around the edges of the plywood.

On top of the roof, screw the plastic flange to the plywood, centering it over the hole. Knock out the center, if necessary, by tapping it with a hammer. Then slip a 4-foot length of plastic sewer pipe into the flange. Caulk around the bottom of the vent and install the galvanized hood on top.

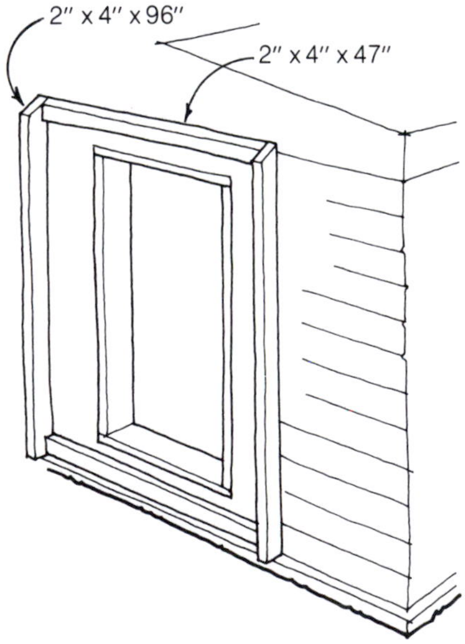

This second door frame, made of pressure-treated 2 x 4s, supports the sides of the hatchway. The hatchway and hatchway doors are added later.

Waterproofing the Exterior

Cover the roof with roofing tar. Pile soil around vent pipe to hold it in place. Over the rest of the roof, place 1½ feet to 2 feet of baled straw "slices." Top this with two to five layers of 6-mil plastic sheeting. Fold the plastic over the edges on all sides well below the top plate. The soil covering will hold it down securely.

Cover the exterior of the block walls with a ½-inch coating of cement plaster (1 part cement, 3 parts sand). Then coat with foundation coating.

Caulk well along the bottom edge between the first layer of blocks and the footing with commercial caulking.

Building the Door Frame and Stairway

Using pressure-treated 2 x 4s, build a door frame around the door opening. Attach the frame to the block wall, using concrete nails. The door will be installed later (see p. 176).

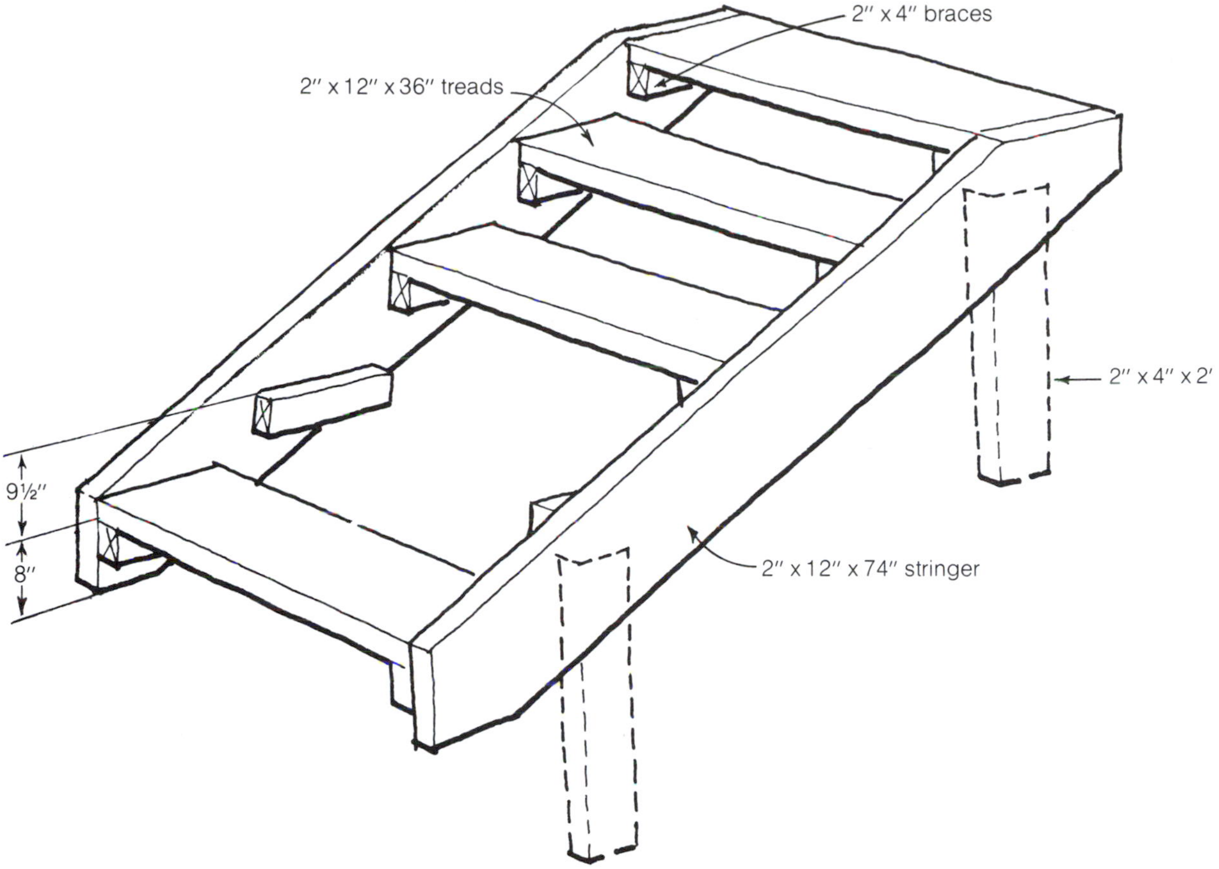

The stairway is set in place on the entrance slope, supported by 2 x 4s that may be set in concrete. The 74-inch long stringers are trimmed to fit the slope.

Stairway. Dig four holes, 6 inches in diameter and 3 feet deep, at the top and bottom of the stairway slope.

To make the stringers, cut two pieces of 2 x 12s so each is 74 inches long (before trimming to fit the slope). Then trim the ends so they fit the slope of the entrance ramp.

Securely nail ten 12-inch pieces of 2 x 4s to form braces on both stringers for treads. The first braces should be nailed so that the top edges of the braces are 8 inches from the bottom edges of the stringers. The other braces should be nailed 9½ inches apart, measuring from the top of one brace to the top of the next brace.

Make five treads from 2 x 12s to measure 36 inches, and nail them in place on the top edge of the braces.

Set the stairway in place on the slope. Nail 2-foot-long pieces of 2 x 4s to the inside of the stringers so they extend into the holes you dug. Place a few rocks in the bottom of each hole. Then fill the holes with wet concrete mixture to hold the stairway in place.

Building the Hatchway and Doors

Using pressure-treated 2 x 4s, build a framework for the stairway wall and hatchway. Cover the exterior sides of the frame-

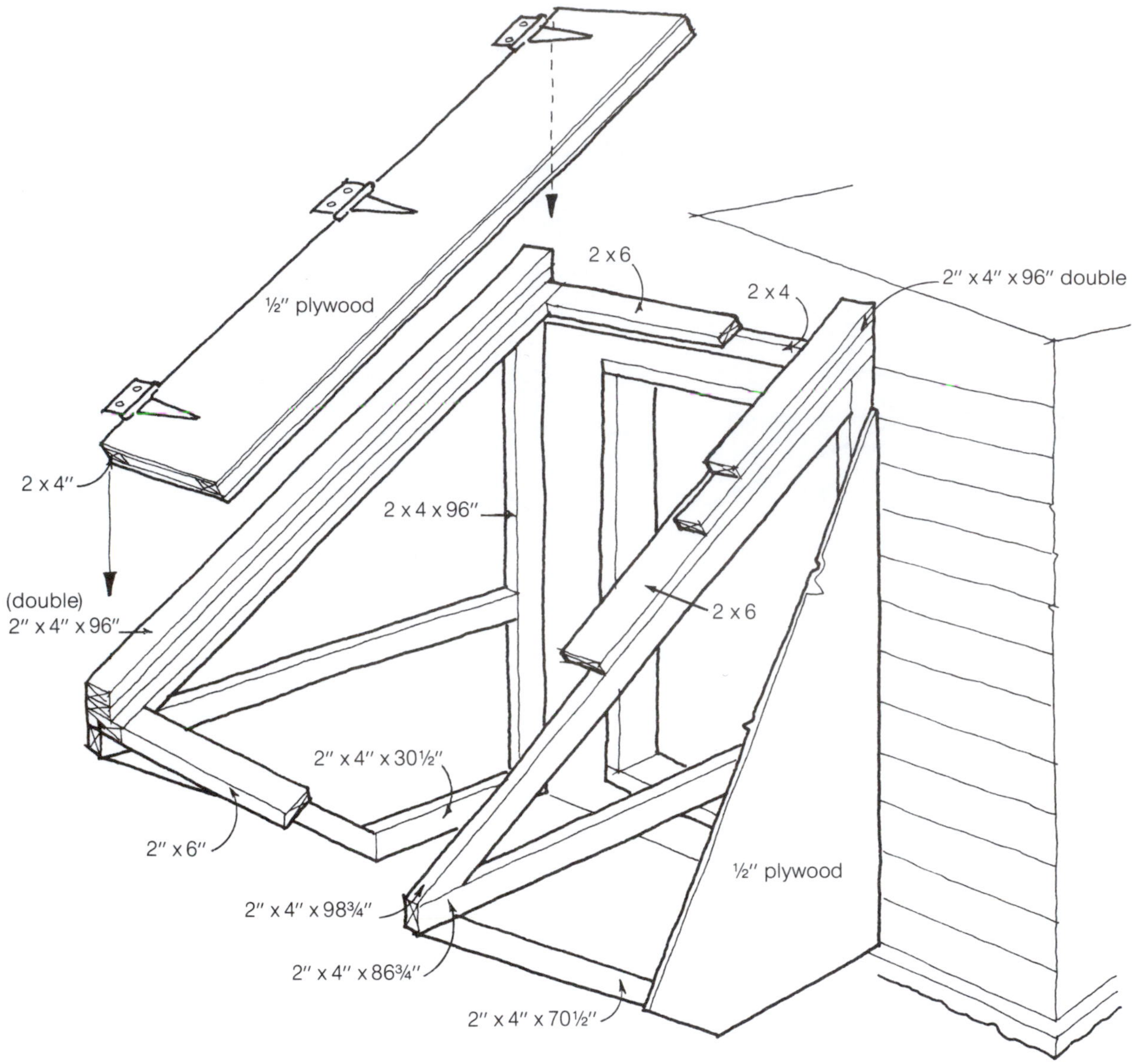

Basically, the hatchway consists of 2 x 4 support framework; 2 x 6 facing; double 2 x 4s on either side; framing for the hatchway doors; ½-inch exterior grade plywood covering; and hatchway doors (one shown).

work with ½-inch exterior-grade plywood that has been painted with wood preservative.

Face the top of the stairway wall frame with four 2 x 6s laid flat on the 2 x 4 supports. Attach double 2 x 4s (3 inches thick), to the outside edges (each 21½ inches) of the 2 x 6 frame. The two hatchway doors will fit between these 2 x 4s and be hinged to them.

Hatchway doors. Cut four pieces of ½-inch plywood, each measuring 21½ x 96 inches. Nail 2 x 4 pieces along the outside edge and across the middle of two of the plywood pieces. Fill the center cores with insulation and nail the other two pieces of plywood on top.

Nail a 1 x 6 board the length of the right hand door with the board extending 2 inches over the edge. This board will overlap the opening and shut out drafts when the doors are closed.

Lay the two doors on the 2 x 6 frame and attach with three 6-inch T hinges on each door. Attach large handles at a convenient position for opening.

Interior door. Nail 2 x 4 pieces along the outside edge and across the middle of a piece of ½-inch plywood. The dimensions of the plywood should be 28⅞ x 68⅞. Fill the center cores with insulation and nail a second plywood sheet on top.

Before installing the interior door, insulate between the 2 x 8s and the concrete blocks. Stuff pieces of batt insulation into any gaps between the blocks and the frame.

Nail 2 x 4 lumber to the edge of the 2 x 8 door frame, all the way around, on the outside of the wall. If necessary, shim to form a snug fit.

Attach the interior door to the 2 x 4 frame with screws, using three 6-inch T hinges. Install a latch or hook and eye on the outside to hold the door closed. Nail a 1 x 2 stop to the side jamb located on the inside of the root cellar along the edge of the door.

Attach an easy-to-grasp handle (metal or homemade) at a convenient position.

Covering the Cellar

Using the soil you removed when excavating, fill in around the three sides of the block walls, tamping it down well with your feet. Pile the dirt up to the roof, then cover the roof, being careful not to disturb the air vent or the roofing.

Cover the roof with 2 feet of soil, gradually tapering it out to the sides so there is at least 1½ feet covering all three sides. Pile the soil carefully around the hatchway door frames, packing it well.

You may need to bring in additional soil, but earth is the cheapest insulator for your cellar, so don't skimp on this step. When the cellar is well covered, seed with a good ground cover to make the area more attractive and hold the soil in heavy rains.

CHAPTER 11

Carports and Sheds

Borg

AMONG THE MORE versatile structures you can add to your homestead are attached shelters or nearby sheds. An attached shelter might serve as a carport, boatport, storage area or, if left open and uncovered, a patio. In this chapter, I also describe a woodshed, a free-standing tool and equipment shelter and, finally, a smaller, portable storage building.

Three of the structures are shown with aluminum siding and roofing. But, other materials, including plywood, steel, fiberglass and native lumber, could be substituted. Futhermore, you can adapt each plan to suit your particular space requirements. If you decide to modify these plans, remember to select dimensions that enable you to use standard-size, "off-the-shelf" lumber and materials. This saves money and reduces waste.

Attached Carport

The attached carport or multi-purpose shelter (p. 180) should have a minimum roof pitch of 1 inch to 12 inches (1:12). You may want to increase this pitch to at least 3:12 if you live in a region with heavy snows. Dimensions might be adjusted as well, if your space requirements demand this. Here are the steps to follow when building the carport:

1. Locate five footings 6-feet o.c.; excavate; and pour footings and concrete piers so their top surfaces are level. Set anchors or anchor bolts in concrete.
2. Attach and plumb 4 x 4 posts; brace temporarily with 2 x 4s. Attach 2 x 8 rafter support.
3. Level and nail 2 x 6 header to house or existing structure; where possible, nail into studs approximately 32 inches o.c. Mark header for rafters; 2 feet o.c. is suggested spacing.
4. Mark and cut first rafter to desired length. Notch rafter, 1½ inch, where it crosses rafter support. Place rafter in position to check fit; trim if necessary; if fit is satisfactory, use as pattern for remaining rafters.
5. Secure two end rafters, using framing anchors at header plate and nails at 2 x 8 rafter support. Fasten remaining rafters. Remove temporary 2 x 4s.
6. Nail 2 x 4 purlins flat on rafters with two 16d nails at each rafter. Purlins should be 24 inches o.c.
7. Nail 2 x 4 horizontal blocking, 24 inches o.c., between 4 x 4 posts. Add 2 x 4 diagonal bracing.
8. Paint or stain framing.

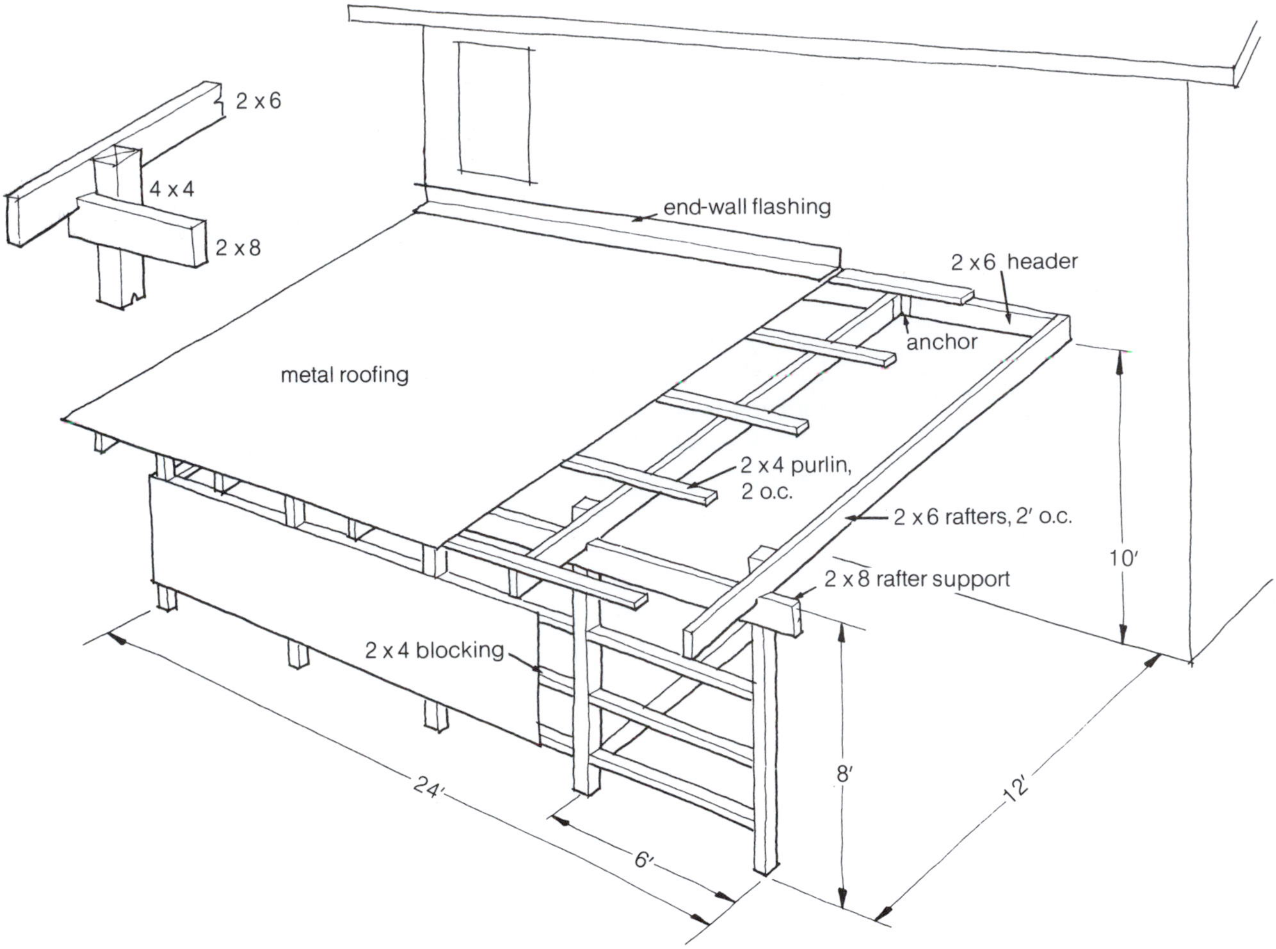

The attached carport may be built as shown, or dimensions may be varied to suit your vehicles and storage requirements.

9. Apply covering or sheathing. If covering with aluminum, use weatherproof nails with washers. Predrill nail holes to avoid shock marks around nails. To avoid possible leaks, nail through the tops of the metal corrugations. Paint or finish as required.

Tool Shelter

Construct the tool shelter using many of the steps followed for the attached carport. Both may be built over a gravel, dirt or concrete floor. Condensation is sometimes a problem in buildings with metal siding and roofing, so a space has been left between the side panels and roof for ventilation.

Here's a sequence of steps for building this tool shelter:

1. Prepare four concrete-pad footings for corner posts. Locate, plumb and set four pressure-treated 4 x 4 corner posts. Locate, plumb and set a fifth 4 x 4 for the right side of the door. The remaining three uprights may be either 4 x 4s or 2 x 6s.

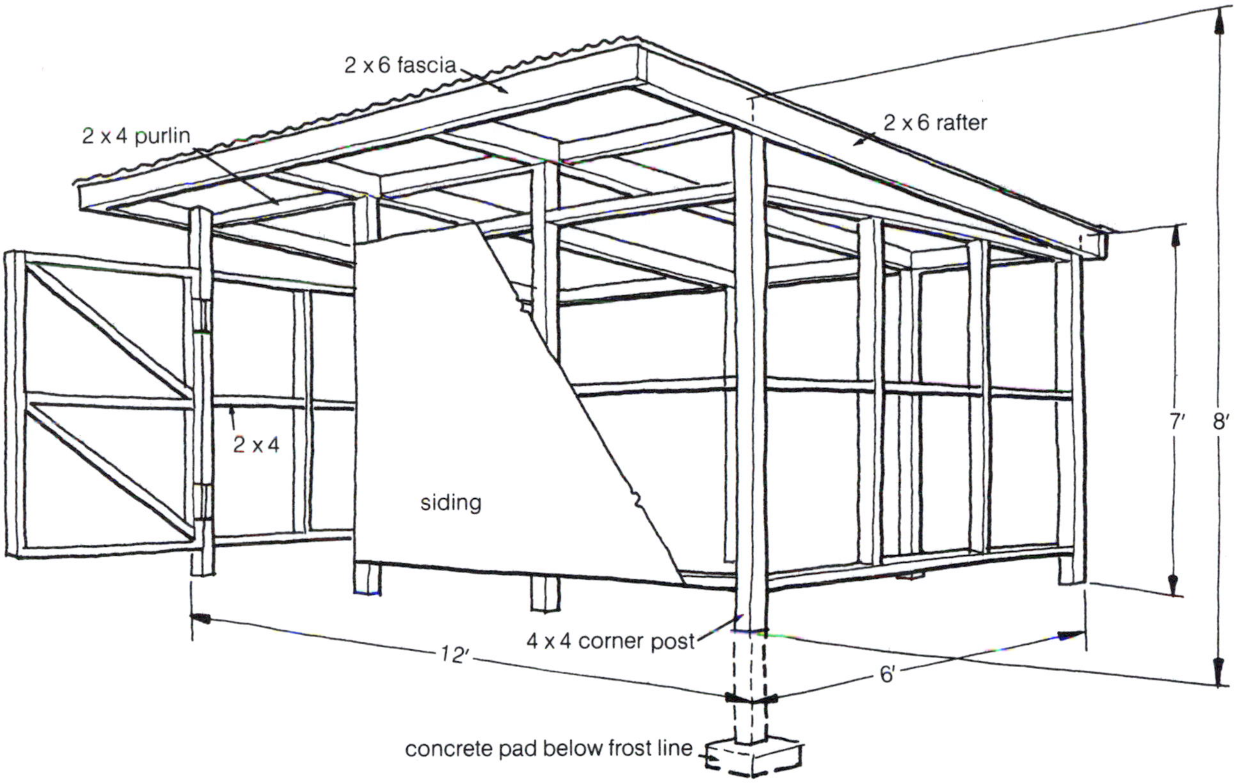

The tool shelter has openings at the base and at the top of the siding. These could be closed in if a tighter building is required.

2. Establish a level line 6 or 8 inches above ground level on uprights. Use line for locating and attaching 2 x 4 cross members.
3. Nail additional cross members as shown.
4. Cut and attach four 2 x 6 rafters.
5. Nail 2 x 4 purlins between rafters, 24 inches o.c.
6. Nail 2 x 6 fascia at end of rafters on front and back of structure.
7. Make door from 2 x 4s, then hang as shown.
8. Paint or stain frame.
9. Apply sheathing or covering.

Woodshed

The old-fashioned woodshed (p. 182) features an 8- x 8-foot enclosed area that can be used for storing lawn mowers, chain saws and other garden and yard tools.

The roof of the center portion extends out 8 feet and provides a sort of porch which can be used for storing other items out of the weather or as a dry place for splitting kindling. The side sheds are for wood storage.

The original shed was constructed using platform framing on a rock foundation; however, a more practical method today is to

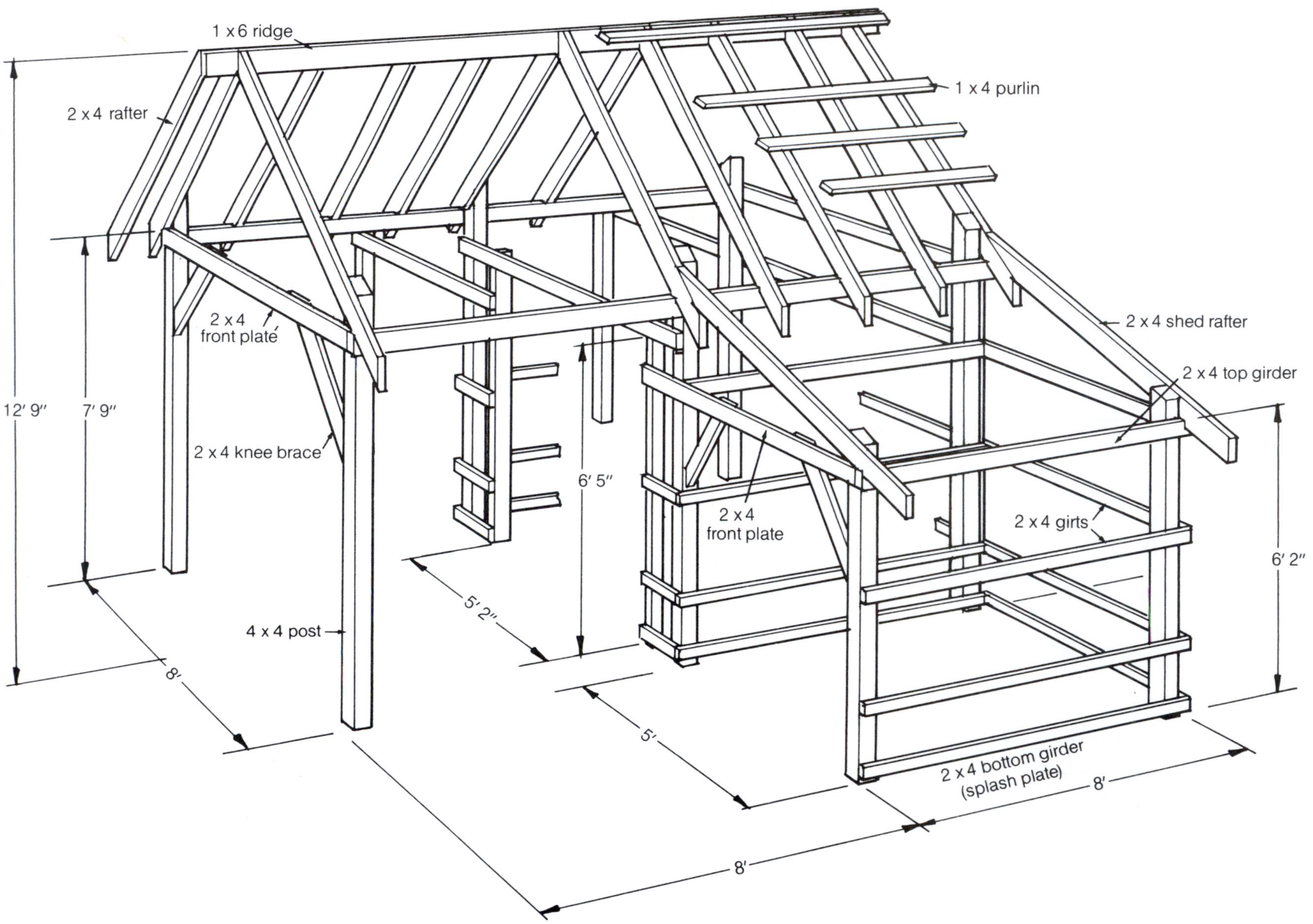

The woodshed has an 8-foot porch on the front, a good place for splitting firewood. Attached sheds on either side of the main structure are for wood storage.

use pole-barn framing methods. The exterior is covered with rough-sawn siding, or native materials over which 1 x 2 battens are added. The roof is shingled with wooden shingles although other materials may be substituted.

Building the Woodshed

The first step is to lay out the shed and locate the positions for the 4 x 4 posts. Then dig the holes, pour concrete piers, and erect, plumb and brace the posts.

Nail the bottom girders or splash plates in place all around the sides and front and back making sure they are level in all directions. Nail the top girders in place; then nail the back nailing girts. After the back girts are fastened, nail the side girts on both the inside and the outside walls. Then nail the front plates across the front of the sheds and the extended porch.

Frame in the door in the center section using 2 x 4s and then add the blocking on the side posts for nailing the siding to the front of the posts.

Install the knee bracing on all corners. The knee bracing on the front is nailed into the surface of the posts, then nailed to the back of the top girder

Cut the upper 2 x 4 rafters and the 1 x 6 ridge board to length. Install the ridge board and end rafters, allowing the ridge board to extend 8 inches past the outside edge of the shed. The end rafters are nailed to the outside edges of the posts and allowed to sit on the top girders. Then install upper rafters at the center posts and follow with the rest of the rafters spacing them about 24 inches o.c.

After the rafters have been installed, cut off the tops of the posts flush with the top edges of the rafters.

Cut the shed rafters. The end shed rafters must be cut to fit flush down on the upper rafters. The remaining rafters can be merely extended past and nailed to the sides of the upper rafters.

Nail 1 x 4 collar ties (not shown) to the upper rafters.

You may use siding and sheathing of your choice. One option would be to cover the main shed, and leave the sides of the two smaller attached sheds open, for better air drying of wood.

For the roof, purlins spaced 24 inches o.c. and metal roofing is one alternative; shingling is another. If you use wood shingles, apply a sheathing of plywood or 1 x 4s, spaced about 2 inches apart.

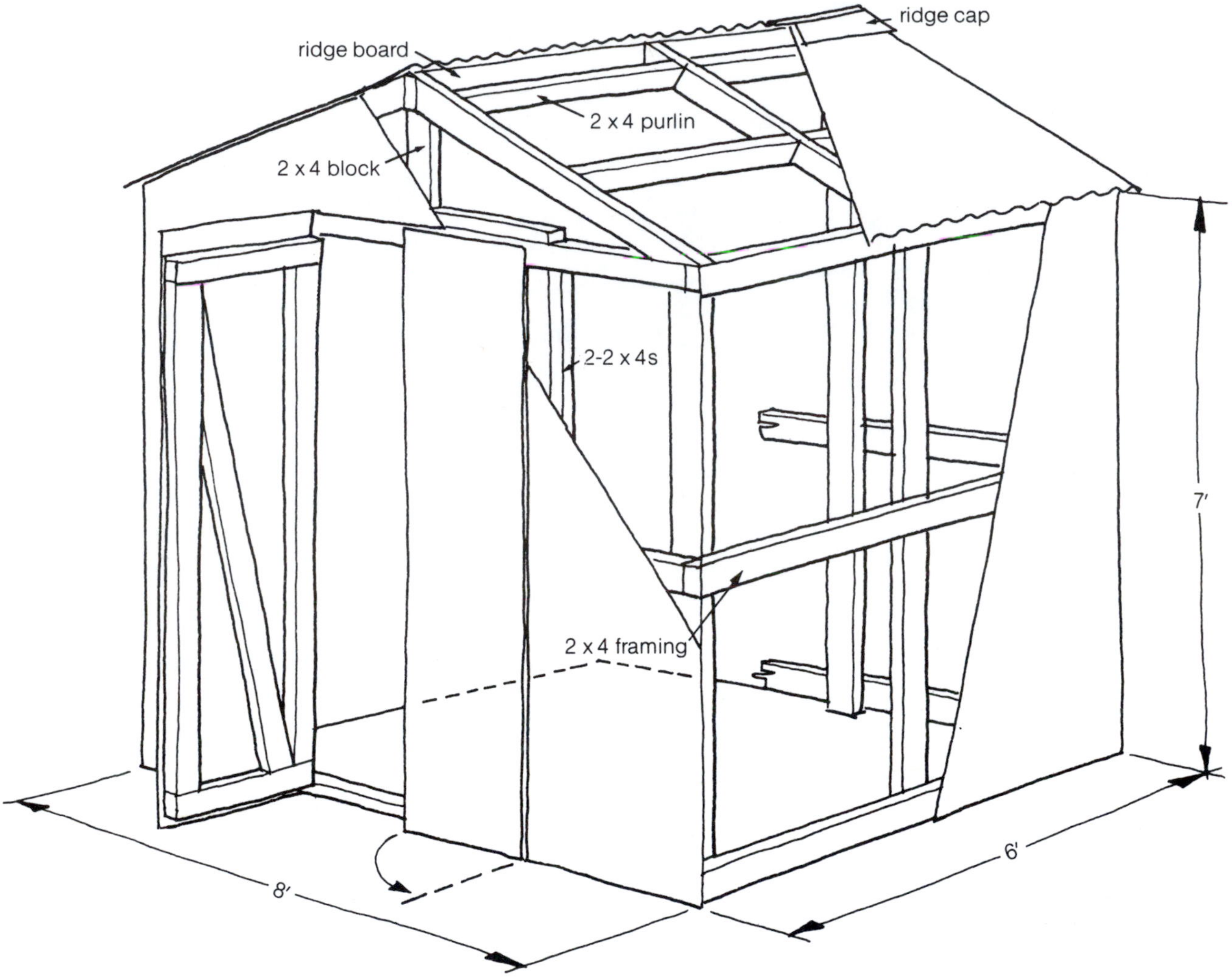

This small storage building has 2 x 4 framing throughout. If you wish to add stability, use 4 x 4s for corner posts.

Cut 2 x 4s for lookouts and nail them between the end rafters and fly rafters.

Make doors using a Z-fame and 1 x 12s. The best option here is two doors for the main shed, designed so they swing out into the porch area.

Storage Building

The portable, free-standing storage building has 2 x 4 framing, but treated and embedded 4 x 4 posts might be used at the corners if you want to add more stability and anchor the building permanently. Here's how to build this structure as shown:

1. Begin by framing the walls with 2 x 4s. These are doubled

at the door to provide enough surface for the hinges. And in the back wall, the 2 x 4 studs extend up to the rafters and ridge board.

2. Fasten the four walls together, by nailing horizontal 2 x 4 girts at base, midpoint and top.

3. Fasten ridge board to top of block in front and to extended 2 x 4 stud in rear wall.

4. Cut rafter ends at required angles and nail to ridge board and top wall pieces. Install purlins.

5. Frame doors, using two 2 x 4s to accommodate hinges.

6. Paint or stain framing.

7. Apply panels or covering.

8. Hang doors and install door hardware. Add door stop at bottom by toenailing 2 x 4 to door frame; face 4-inch side to ground.

CHAPTER 12

Poultry Housing

Grant Heilman Photography

A POULTRY HOUSE or pen should provide a clean, dry, comfortable environment for the birds throughout the year. It must be tight enough to prevent drafts on the birds and help maintain a uniform temperature economically. Poultry facilities should have floors that can be easily cleaned and disinfected. Most poultry pens or buildings should provide for ventilation to control moisture, remove gases and dissipate heat. In northern climates, the houses should have insulation in both the side walls and ceilings.

Piped-in water is desirable but may freeze in severely cold climates, especially in small-flock situations where body heat is not sufficient to keep the pen temperatures above freezing. It is possible to correct this particular problem by wrapping the pipes with electric heat tapes.

Houses should have artificial light to provide the right lighting for both layers and growing stock. The poultry house should be strong enough to withstand high winds and snow-loads.

Poultry House

The 12- x 10-foot poultry house (p. 188, 189) is ideal for a small flock.* It has room for feeders, nests and a feed-storage area. The house may be fitted with fiberglass windows that tip in from the top. Or, you can cover the front openings with wire. Then, in the winter, staple polyethylene over the openings to conserve heat. This latter alternative is suitable for warm climates, where good summer air circulation is needed. Notice that wire screening is also suggested for the rear opening.

Floor

To protect the poultry from rodents, construct the house on a wooden floor made of 10-foot long 2 x 6 joists and plywood and mounted on level concrete blocks. For more stability, substitute concrete piers for blocks. Secure the floor joists to the blocks or piers with anchor bolts. Two 12-foot long 2 x 6s are needed for the front and rear joist headers.

---

* Plan adapted from the United States Department of Agriculture Cooperative Extension Service. For a copy, contact your local state university Extension office.

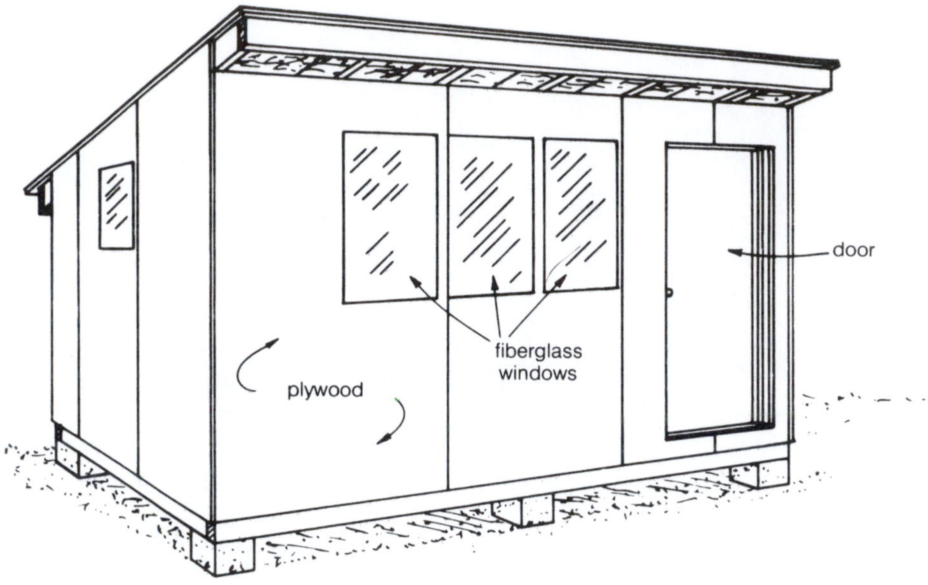

Here's a perspective of the 12- x 10-foot poultry house. To reduce costs, the fiberglass windows could be replaced by old, recycled windows.

Framing

Platform-frame construction techniques are used throughout. Begin with the front and back walls, using standard layout and construction methods, as described in Chapter 4. All studs are 2 x 4s. After assembling and erecting the front and back walls, brace them temporarily, then install the end rafters. This will tie the front and back walls together, and provide upper nailers for the side-wall studs. Now complete the side walls, spacing the studs 2 feet o.c.

Complete the roof with 2 x 6 rafters, spaced 2 feet o.c. These 12-foot long rafters provide an overhang for the front and back of the house. You can finish the roof with plywood sheathing and roll roofing. Another suitable alternative would be purlins and metal roofing. The rafters are covered, front and back, with 1 x 6 fascia boards.

Inside, the poultry house has a simple wall partition, separating the feed and nesting area from the storage area. The roosts and nests should be at least 2 feet above the floor. Space the roosts 12 to 15 inches apart.

The siding used for this poultry house is ½-inch exterior-grade plywood.

If you use fiberglass panels or windows that tip in at the top, be sure to adjust them according to the weather. When the house tends to be stuffy and the ammonia fumes are strong, the house needs more ventilation. The house should never be closed tight, even on cold nights. It is always well to leave at least some of the windows slightly ajar.

Lighting

The use of artificial light in the poultry house is not to give the hens more time to eat. It is the stimulation of the light itself

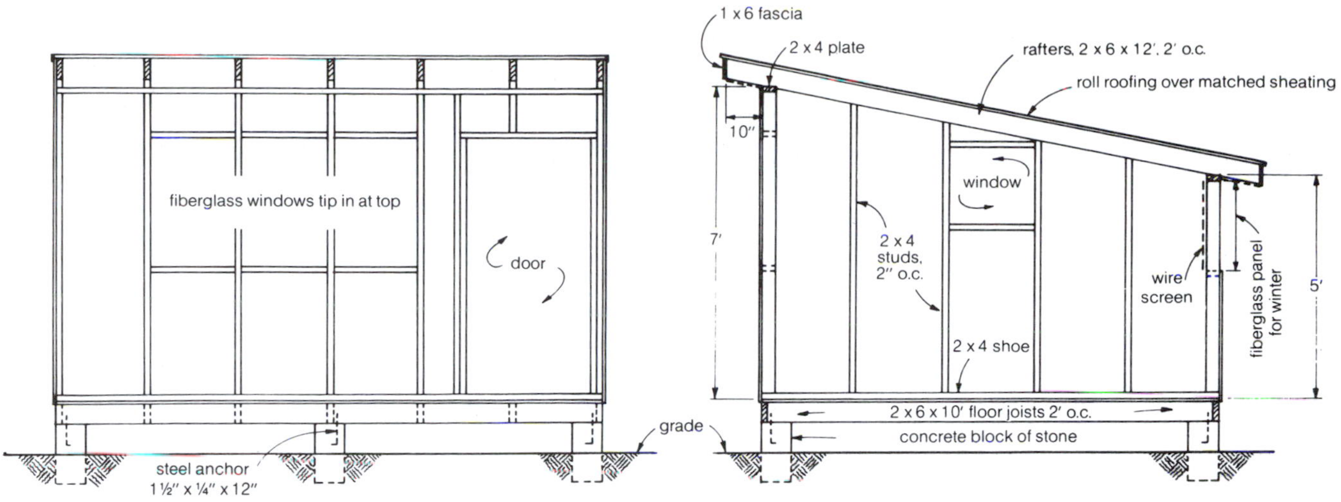

To frame the poultry house, use standard platform construction techniques.

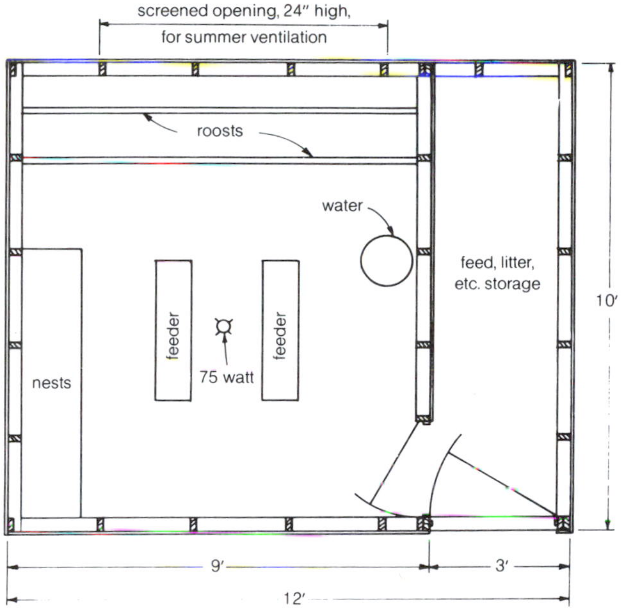

Note positions of roosts, nests, feeders, waterers and storage space.

that makes them lay more eggs. The light stimulates the pituitary gland through the eye. This gland, in turn, secretes hormones that stimulate the ovary of the hen to lay eggs.

Laying birds need about 14 hours of light daily. In the northern part of the United States, darkness exceeds the amount of daylight during many of the fall and winter months. Beginning on August 15th, the birds need to receive extra light to give them a 14 to 15 hour day. During the months of November, December and January, when the days are shortest, they will need five hours of supplemental light. The lights may be turned on in the morning or in the evening. A combination of both morning and evening lights may be used, as long as the birds have at least 14 hours of light each day. With small flocks it is a task to turn the lights off each night, or to get up early enough to turn them on in the morning. The best way to solve this problem is to use a time clock which automatically turns the lights off and on at the desired time. This generally gives satisfactory results.

As soon as the natural day length reaches 14 hours again, the lights can be discontinued.

The lights should be ordinary 40 to 60 watt incandescent bulbs. To get the most efficient use of the light wattage, a reflector can be used. This will frequently permit the use of lower wattage bulbs. One light fixture should be installed for each 200 square feet of floor area or less. A distance of 10 feet between lights generally gives good distribution of light for the birds. A rough rule of thumb for light intensity is 1 foot-candle at the feeder level. One bulb watt per 4 square feet of floor space will usually provide 1 foot-candle if the bulb is 7 to 8 feet from the litter and has a reflector.

Range Shelter

Range rearing is a popular way to raise poultry in warm climates or during warm seasons, when tight housing is not required. Good range-reared birds usually have excellent color and beautiful feathering, and they are inclined to be big-framed

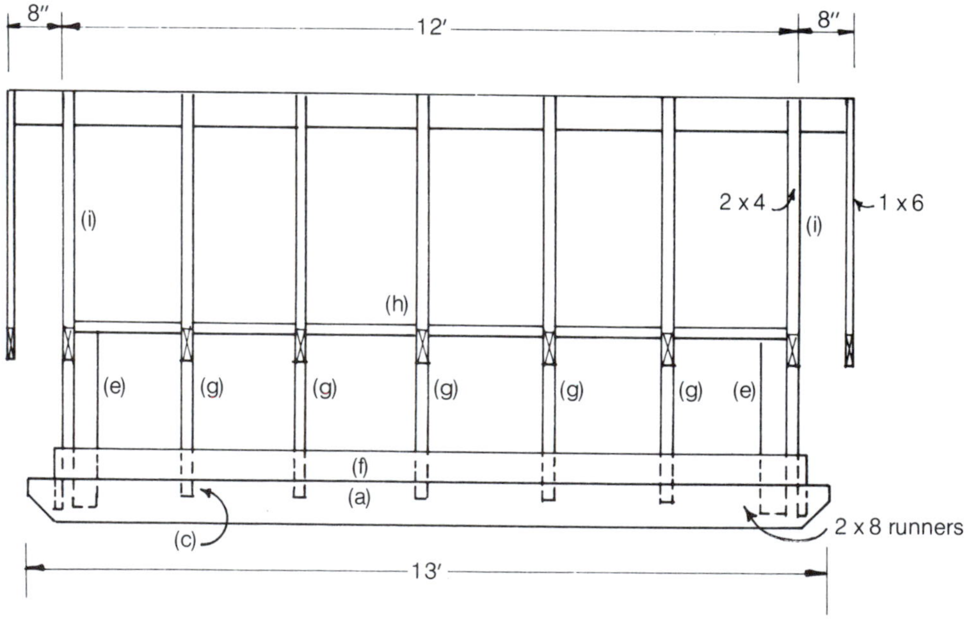

The range shelter has long 2 x 8 runners to enhance mobility. Letters identifying parts are keyed to text, p. 192.

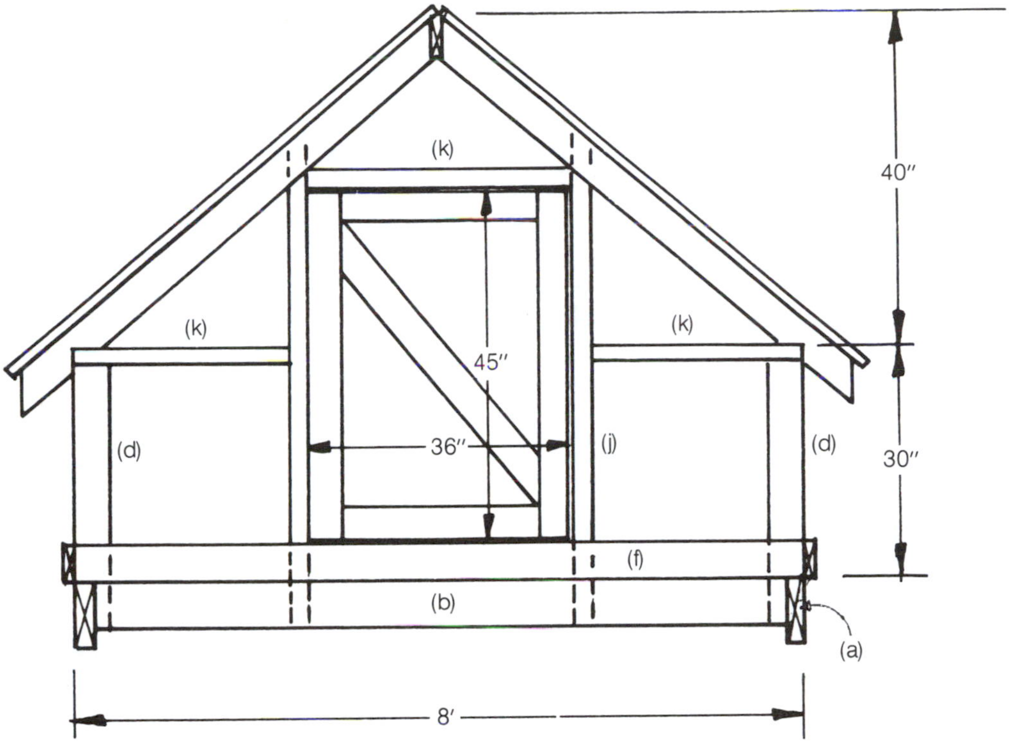

A front view of the range shelter, showing framing details.

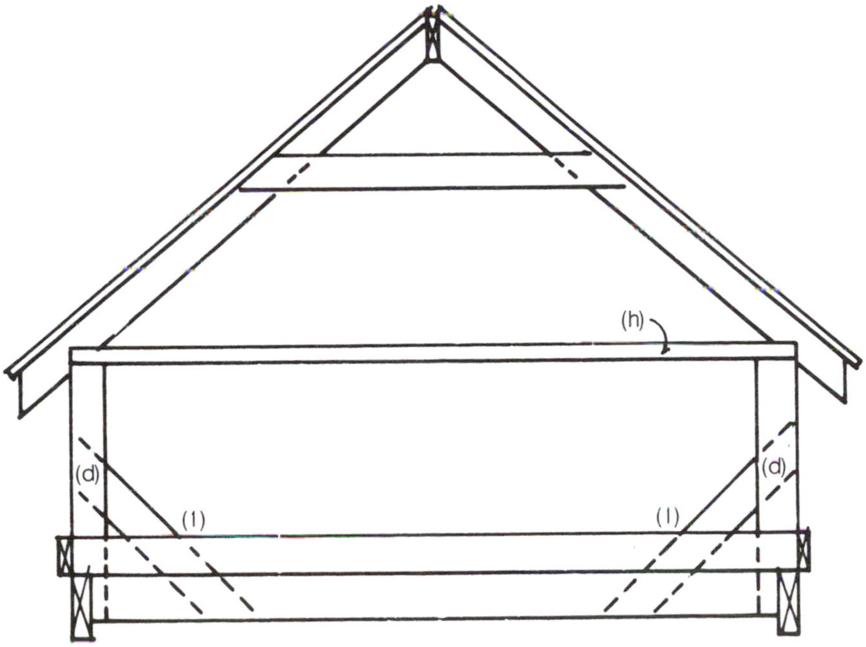

Rear view of the range shelter. Corner braces are used in the front of the shelter as well as the rear.

birds with health and vigor. They can feed on fresh grasses, and the shelter can be moved from place to place to insure sanitary conditions. Inside a range shelter, the birds are up off the ground away from diseases, and they are well protected from bad weather.

There are disadvantages to this type of poultry housing, though. Usually, range shelters must be used inside a fenced-in area, for protection from dogs, cats and predators. And even then, the protection may not be adequate. Hawks and owls can be a problem.

You can build a simple, 8- x 12-foot range shelter to raise a flock of turkeys or to raise fryers after they are old enough to be taken from a brooder house (eight weeks). In addition, the shelter may be used to house game birds. Twenty-five to thirty turkeys would be comfortable in this shelter. It may be scaled down to suit your needs.

To build the shelter, first cut the 2 x 8 runners (A) to shape, then nail the two front and back 2 x 6 cross members (B) to the runners. Cut the 2 x 4 floor joists (C) and install them between the runners, spacing them 2 feet o.c. Cut the front and back 2 x 4 uprights (D), notch them to fit around the runners, then nail them in place. Cut the side uprights (E) of 2 x 4s and nail them to the 2 x 4 joists already fastened in place.

Cut the 1 x 6 aprons (F) for the sides, front and back and nail them in place, then cut the remainder of the side uprights (G) and nail them to the apron. Toenail the uprights into the runners as well. Nail the sides' two top plates in place as well as the back top plate of 2 x 2s (H). Cut the rafters to the proper size and nail the end rafters (I) to the 1 x 6 ridge board, then install the remaining rafters. Cut the door support uprights (J) for the shelter front and notch them to fit around the rafters as well as the bottom runner. Nail securely. Cut the upper plates for the front (K) from 2 x 2s and nail them in place. Construct the door and hang it on hinges. Install the corner braces (L).

Cover the inside bottom with either ⅝-inch hardware cloth or 1 x 1s spaced 1 inch apart. If you have a good supply of rough lumber, these work better than the wire because they won't sag as badly as the wire mesh under the weight of the heavy birds.

Cover the roof with reinforced, green-tinted fiberglass panels or with solid exterior plywood sheathing. Then paint the entire structure and staple 1-inch poultry netting over the back and front sides including the door. In harsh climates, add sheathing to the side facing prevailing winds. For those working around the shelter, fascia boards on the outer edges of the rafters will provide some protection from the rafters' sharp edges.

Game-bird Housing

A 12- x 12-foot portable pen, covered by poultry wire, is suitable for raising game birds on a small scale. This pen is typical of those used on hunting preserves. Built on 2 x 8 runners, it may be towed from place to place with a tractor.

This pen holds about six pheasant hens and one cock for breeding purposes. If you're raising birds for the table, or to be turned loose, you can house about a dozen pheasant or two dozen quail.

To construct this shelter, first cut the runners (A) to shape and bore 1-inch holes in their ends. (The ends extend 6 inches beyond the edges of the shelter, so the total length of each runner is 13 feet.) Nail the front and back 2 x 8s (B) to the runners. Nail two 2 x 6 joists to runners (C).

Cut and notch two pairs of uprights (D and E); these support the 2 x 4 top cross members (F). Cut the remaining uprights, and nail all 12 uprights to the 2 x 8s. Cut the upper side plates (G) and nail them to the uprights. Cut the top cross members and nail them in place; do the same for the diagonal bracing at the corners (H). Cut and nail the top plates (I).

Game birds do best with as little human contact as possible, so cut feeding slots in the 1 x 12s (J). One vertical 1 x 4 slot is for access to a watering pan; another horizontal slot is for a two-compartment feed bin. Place a high-protein feed such as turkey mash in one side of the bin and charcoal, grit and oyster shells in the other.

(You can leave the floor uncovered, so the birds can feed from fresh greens below. Rotate the pen's location about once a week.) Cover the pen sides and top with poultry wire. Use 1-inch mesh for quail and 2-inch mesh for pheasants. Nail the 1 x 12s in place to form a protective apron.

Finally, construct the door of 1 x 4s, cover it with wire and hinge it in place. It may be secured shut with a hook and eye, or any other useable door hardware. To provide a shelter for the birds, simply make a small, 3-sided box-like house, using

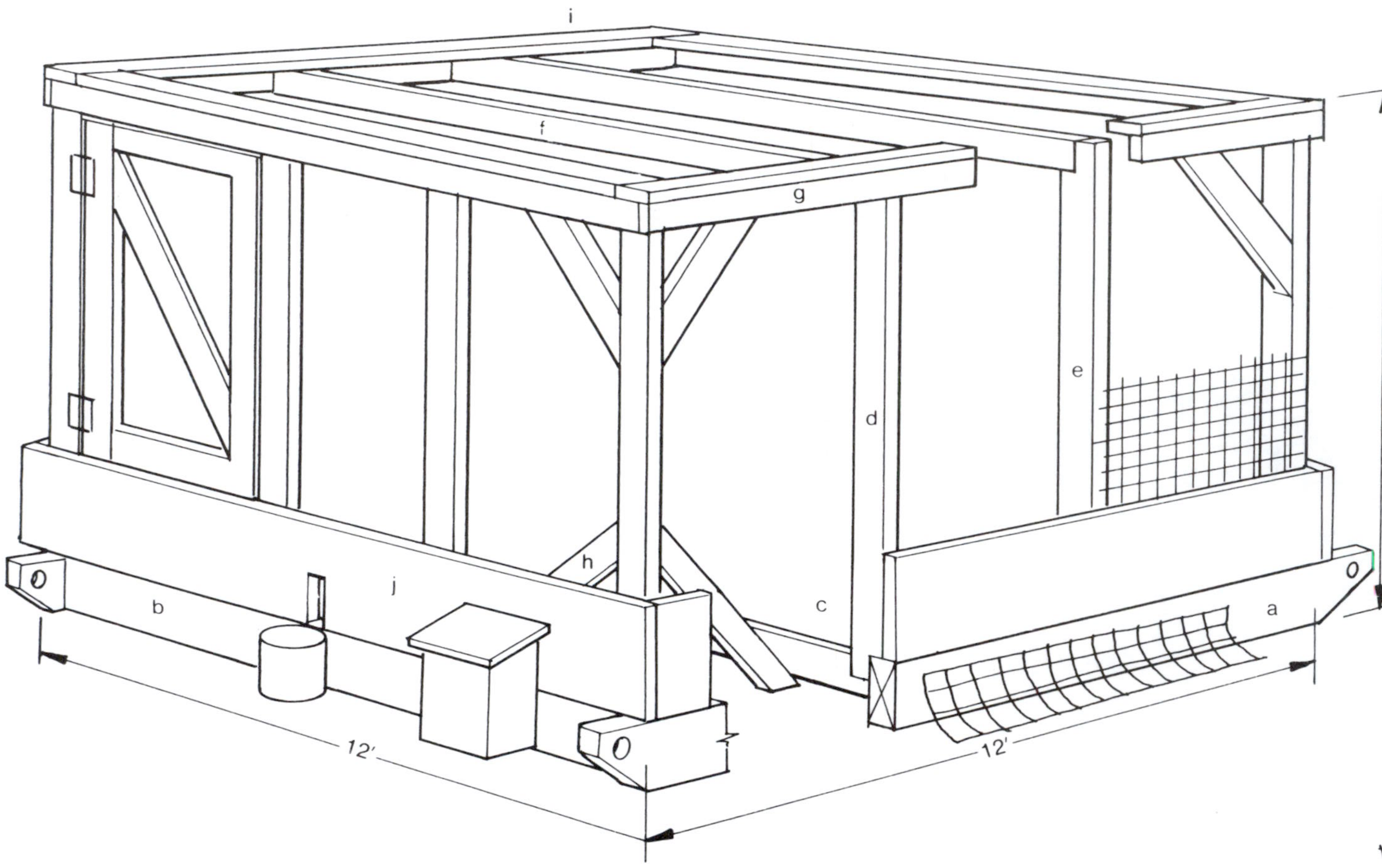

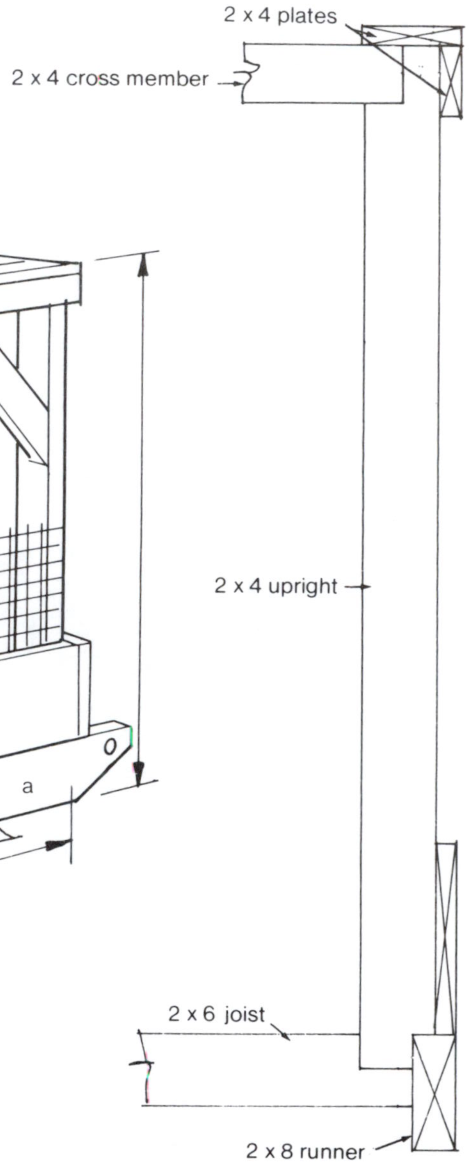

This game-bird pen is designed so feed and water may be provided from the outside. Also, the floor may be covered with hardware cloth, or it may be left open, so the birds may feed on grasses inside the pen. At right is a cross section of a corner upright.

3- x 3-foot pieces of exterior plywood. Place it near one of the corners. Game birds also like pine boughs or corn stalks to hide in.

If you have problems with predators, staple a 2-foot strip of screen wire or 1-inch chicken wire to the 2 x 8s around the outside of the pen. To prevent the outside edges of the wire from curling up, you can place boards or rocks on the outer edge.

CHAPTER 13

Rabbit Housing

Borg

NEXT TO THE right breeding stock, the right housing and equipment will do the most to insure success with rabbits.

Too often, a prospective rabbit raiser looks at some old wooden hutches, obtains lumber and chicken wire, hammer and nails and begins to build a similar structure. And too often the results are disastrous. A wooden hutch absolutely will not do. There is only one kind of hutch worth using, whether you build it yourself or buy it. That is the all-wire hutch.

The all-wire hutch is easy and economical to build or buy. And it is self-cleaning. Only an occasional wire brushing is needed, plus periodic disinfecting. Droppings and urine fall right through to the ground or to pans. Complete ventilation is afforded.

The All-wire Hutch

If you use it outdoors, the hutch will need some protection from the weather, depending upon your climate, but indoors it is fine the way it is. The hutch may be hung from joists in a barn, garage or beneath a south-facing shed. Mount the hutch so the animals receive plenty of light, but are well shaded.

To make the wire hutch, you will need only pliers and wire cutters, and your cost will be less than if you used lumber and screws, hinges and assorted hardware.

For the front, back and sides, purchase in a hardware, farm supply or department store a length of 1- x 2-inch, welded, 14-gauge galvanized wire fencing (sometimes called turkey wire). It should be 18 inches wide, so the hutch will be high enough to let the rabbits stretch up on their hind legs. For a cage with a floor area of 2½ by 3 feet, you will need an 11-foot piece. Giant breeds need an extra foot or two. (If you anticipate raising rabbits, purchase welded wire with a close mesh at the bottom, to prevent young rabbits from slipping out of the hutch.)

Don't cut this wire, but lay it flat on the floor and bend four corners by hammering it around a length of 2 x 4 lumber. Don't bend against the welds. Fasten it into a rectangle using hog rings, which can be purchased in auto supply stores, strangely enough, because they are used to fasten auto seat covers, not hogs. They come with a special pair of pliers and usually cost a few dollars including the pliers.

You may do without these rings if you twist the cut ends of the wire with pliers. Or, you may purchase inexpensive C-rings or J-clips from suppliers of rabbitry equipment. These rings and

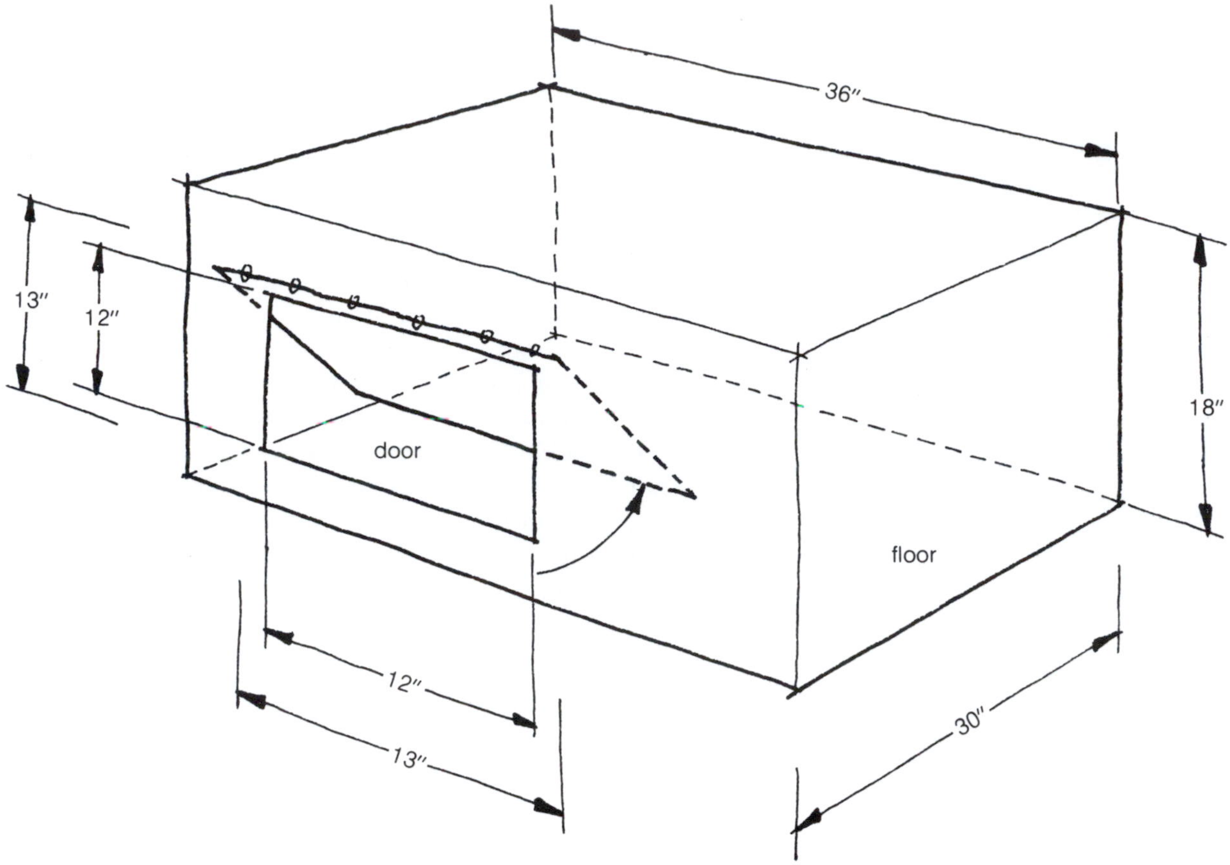

The wire hutch is simply an 18-inch wide piece of fencing, folded into the form of a box. Then the square floor and top pieces are added and secured with rings or clips.

clips may be bought by the pound or you can get enough for a hutch with only change. If you use these rings or clips, clamp them on the wire with pliers about every three inches or so. You've now got all four sides of the hutch assembled.

For the floor, buy a length of ½-inch mesh, 16- or 14-gauge (the latter is heavier and better, but initially more expensive) welded, galvanized wire (not hardware cloth—it doesn't have the necessary rigidity). For your 2½- x 3-foot hutch, you'll need a piece 30 x 36 inches. Fasten this to the bottom, using rings or clips or by twisting short pieces of wire around the sections. The top will be the same size and go on the same way, but use the 1- x 2-inch wire as you did on the sides.

Cut a door opening a foot square on one wide side (which will be the front), leaving ½-inch wire stubs. Bend the stubs back with pliers so there will be no sharp edges. The door, also cut from 1- x 2-inch welded mesh, should overlap at least 1 inch all the way around.

Many rabbit raisers prefer a door hinged at the top (with clips or rings) that swings up and into the cage. This way the door is inside the hutch when open and not extending out where it catches sleeves or otherwise gets in the way. Another good thing about this kind of door is that if you forget to latch it, it is still closed and prevents escape.

Latches and hangers for holding the door up in the cage can be purchased for just a few cents from rabbit equipment suppliers or you can make them yourself. A dog-leash snap fastener also makes a good latch.

If you have space available in a well-ventilated shed or garage, your rabbits will be safe and you and they will be comfortable in all kinds of weather. With the hutch indoors, you'll want to put a metal pan or cardboard box lined with plastic sheeting underneath it to catch the droppings. Use heavy-gauge wire to hang the hutch from rafters or joists. Connect the wire to the hutches with S-hooks. This makes removal for cleaning easy.

A Small Shed

Outdoors, the hutches need protection from weather. To provide this, you can build a small shed, using plywood and 2 x 4s. In windy spots, add 2 x 4 braces and anchor the legs with heavy pieces of wood or blocks that can be strapped on and buried beneath the soil. In well-protected yards, the shed can stand alone without anchoring. In either case, concrete footings are unnecessary. For good drainage, locate the shed over sand or gravel.

The legs should be high enough to put the hutch at a convenient level for you to feed and water the rabbits, and well above splashback from rain and drifting snow. On cold or rainy days, hang a heavy plastic or canvas from hooks mounted on the front. Canvas may be more durable, but plastic lets in light.

To build this small structure, first cut and assemble the 2 x 4 framing. Select pressure-treated, long-lasting 2 x 4s for the legs. Attach the plywood roof. If you live in a cold climate, add plywood to the back and two sides. Pound nails into the back piece only part way, so it may be removed easily on hot summer days when extra ventilation is desirable. If the roof is plywood, cover it with roofing paper. Another alternative is a corrugated metal roof.

Make sure your rabbits are out of the wind in winter and in the shade in summer. If your hutches are outside, try to locate them in a shady spot protected from dogs by a high, sturdy fence. All-wire hutches can also be suspended from a strong chain link or board fence, eliminating the need for legs and perhaps back-wall weather protection. Make every effort to provide shade.

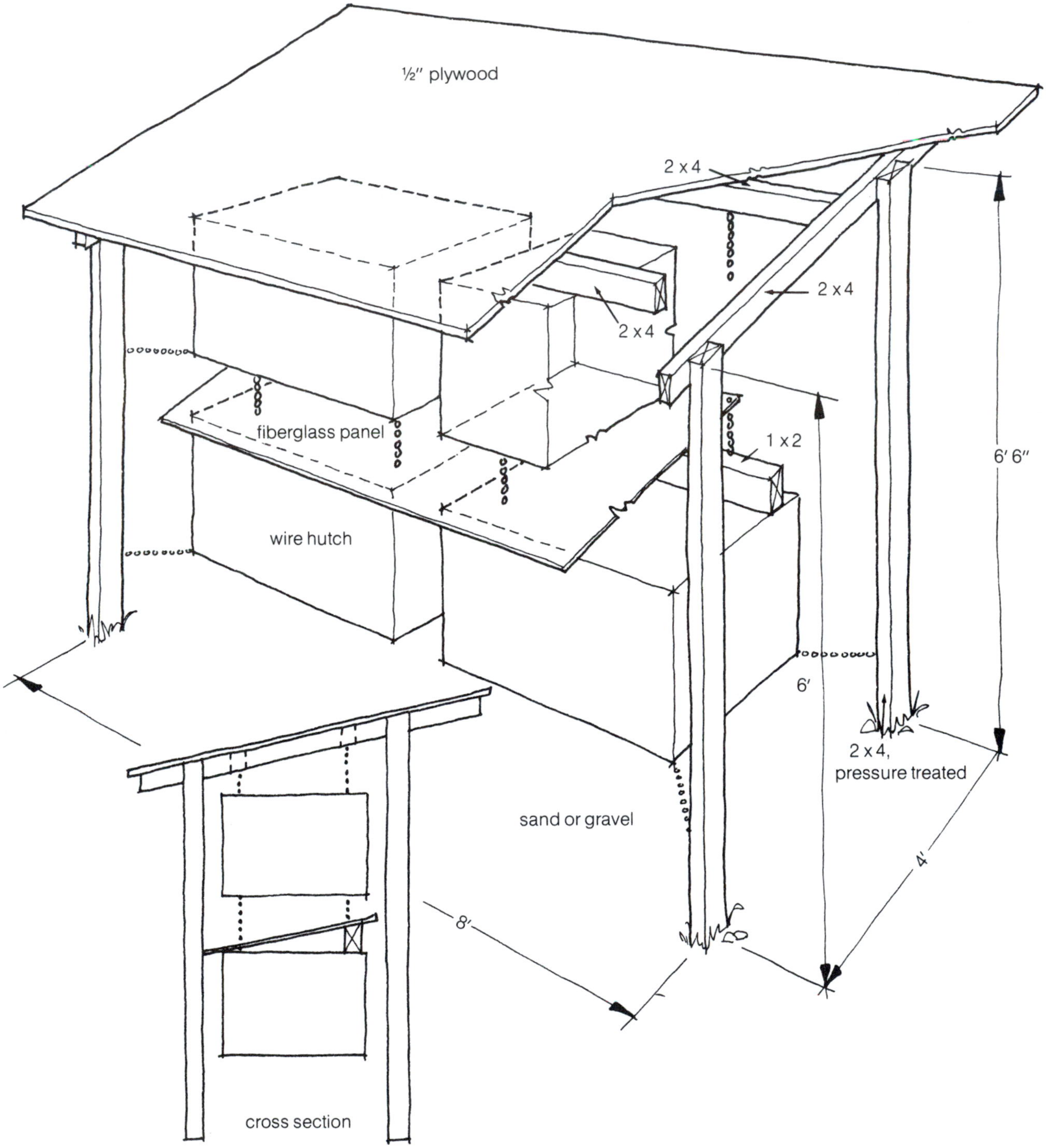

A cutaway view of the small shed suitable for wire-rabbit hutches. For more winter protection, add plywood to the rear and hook tarp or plastic to the front.

Use 14-gauge wire to suspend the hutches from the roof framing members. The two lower hutches should have a fiberglass panel slanted above them, to protect the rabbits from droppings. Cut the panel as necessary, to fit around the wires. Use a 1 x 2 piece to elevate the front of the panel, so it slopes toward the back.

The shed may be lengthened as necessary, to accommodate many more rabbit hutches.

CHAPTER 14

Hog Housing

Grant Heilman Photography

FARMERS AND RURAL homesteaders need efficient, multipurpose livestock buildings that can be constructed easily at very little cost. A good solution for the person who wants to raise a few pigs is a small, portable house. It's both economical and easy to construct.

Multi-purpose Pig Housing

One possibility is a portable, 12- x 8-foot shelter (p. 204). With a 12 x 12-foot pen (made of welded-wire panels and steel posts) added in front, the shelter will keep a couple of slaughter hogs. Or you can use it as a range shelter or housing for feeder pigs. And there's another possibility, too. When constructed with a center divider, the shelter serves as a two-sow farrowing house.

Regardless of the purpose of the house, follow these basic construction steps: First, cut the 2 x 8 runners, then nail the three 2 x 6 cross braces to them, as well as the two 2 x 4 side sills. Cut and nail together the pieces of the side-wall sections (2 x 4 uprights and 2 x 4 top plates), then nail these to the runners.

Cut the front- and back-wall uprights, notching those intended for the four corners so they fit around the side-wall sills. Then cut and notch the remaining uprights so they fit around the runners; nail uprights in place. Cut top plates for the front and back; nail these to the uprights.

Cut the six rafters and nail them to the top plates and 2 x 6 ridge board. To provide additional support for the roofing, add 2 x 4 blocking between rafters. Also, for extra rigidity, add 2 x 4 bracing in the corners.

Cover the shelter as required. I covered mine with 1 x 6, rough-sawn, native oak and metal roofing salvaged from an old barn. For a tighter building for farrowing, cover throughout with ¾-inch exterior-grade plywood.

Individual Farrowing House

If your aim is to raise a litter or two of pigs each year, the individual farrowing house is a good choice (p. 205). It is called *individual* because it houses one sow at a time. Within the house, the sow and pig areas are separated by 1 x 6 boards. The boards are mounted high enough (about 10 inches from the floor) so the pigs can crawl into the sow area for nursing.

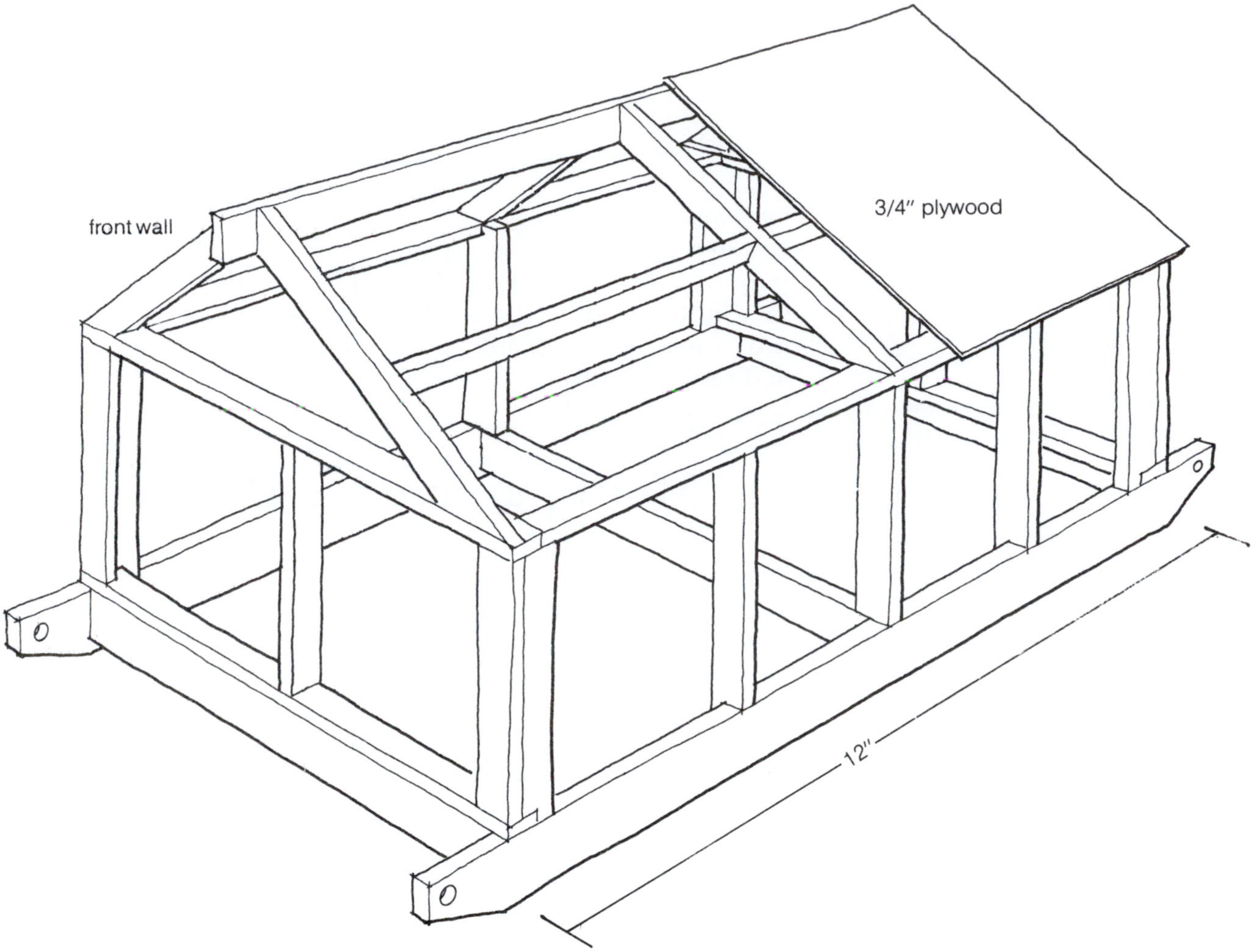

Multi-purpose, portable pig housing, seen in a cutaway view from the rear. The back wall studs are set 3 feet o.c. Fabricate plywood doors for the front wall.

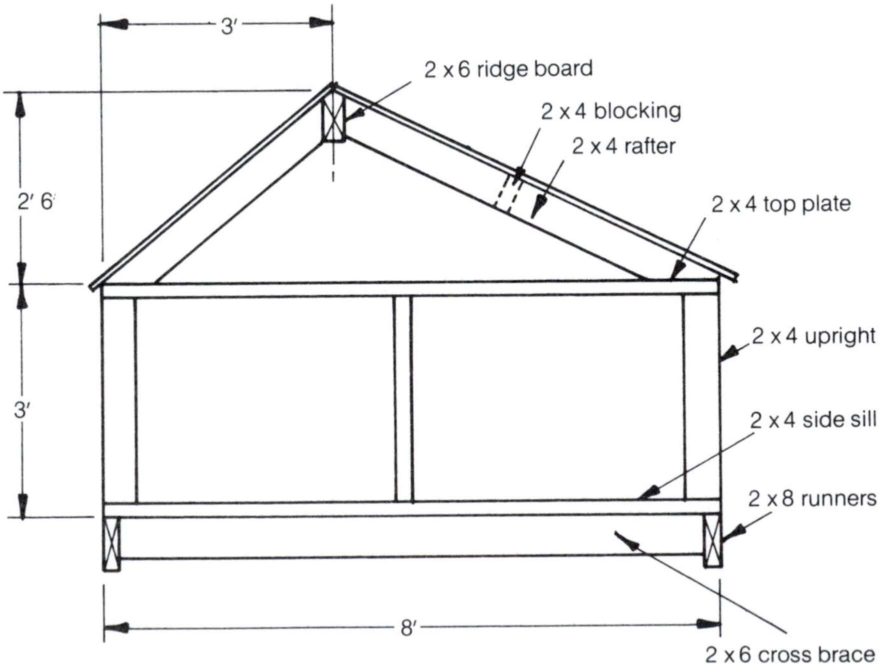

An end view of the portable pig house.

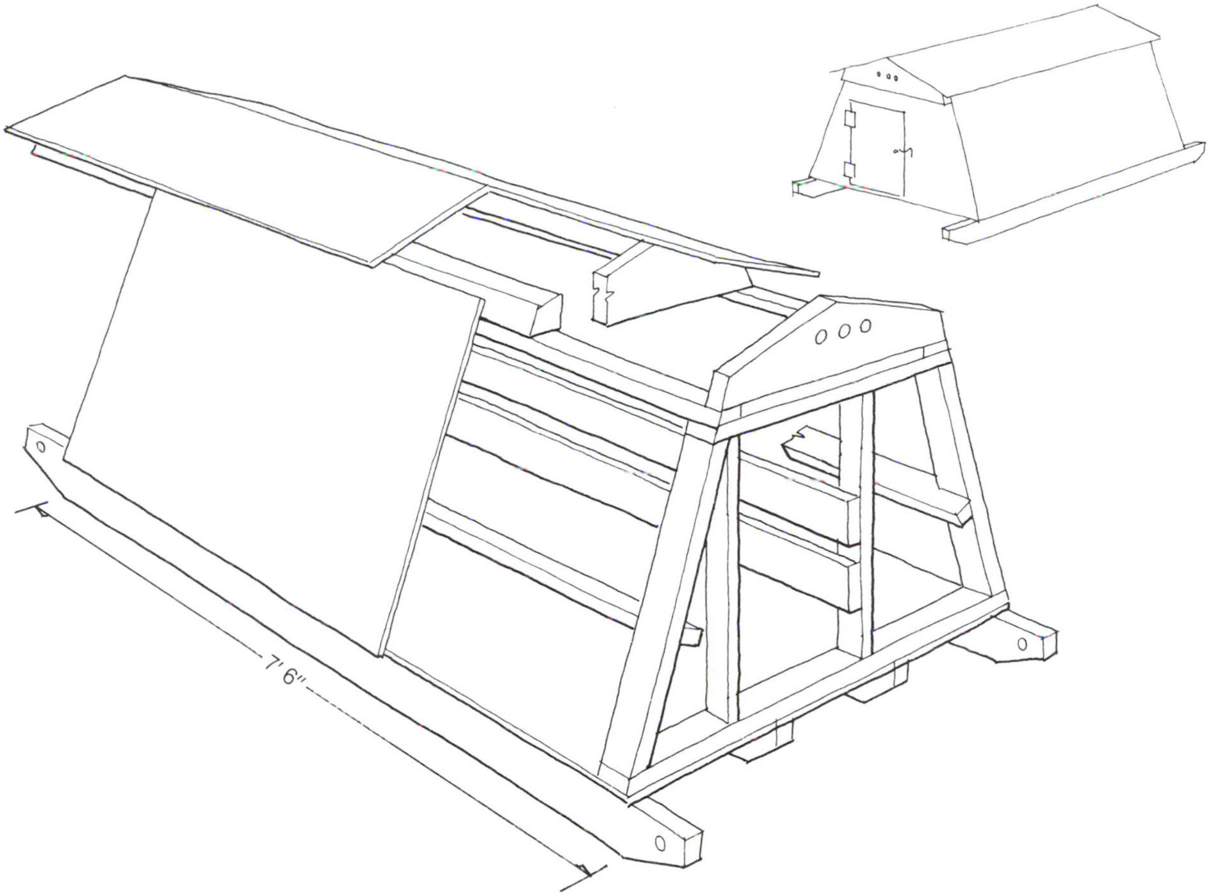

The individual farrowing house has a sliding roof, to permit access for cleaning or working with young pigs.

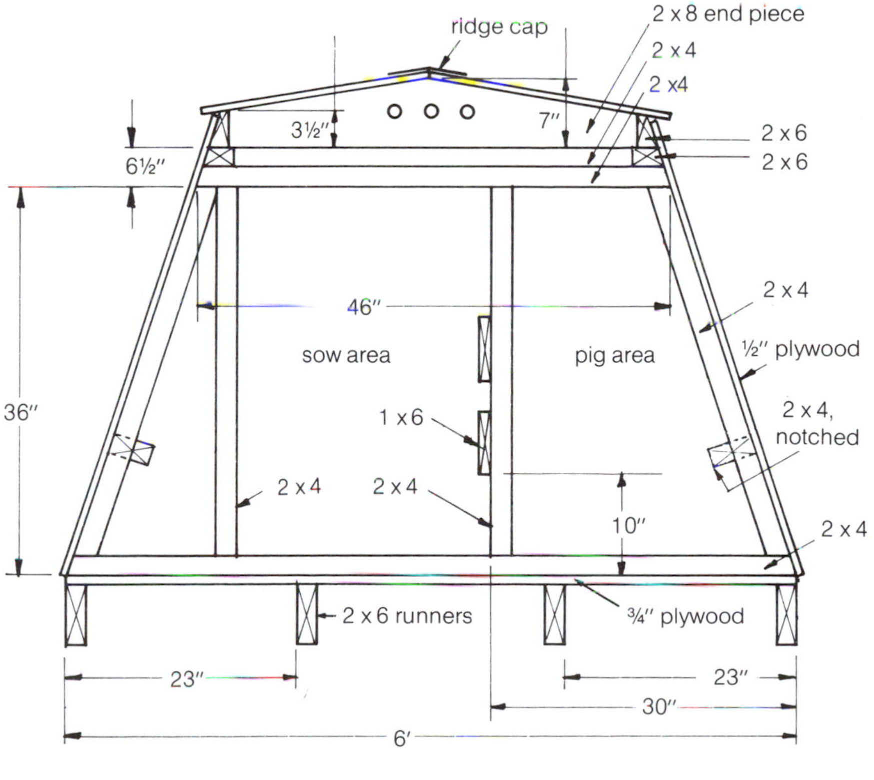

An end view of the individual farrowing house. Install the door so it opens into the sow area.

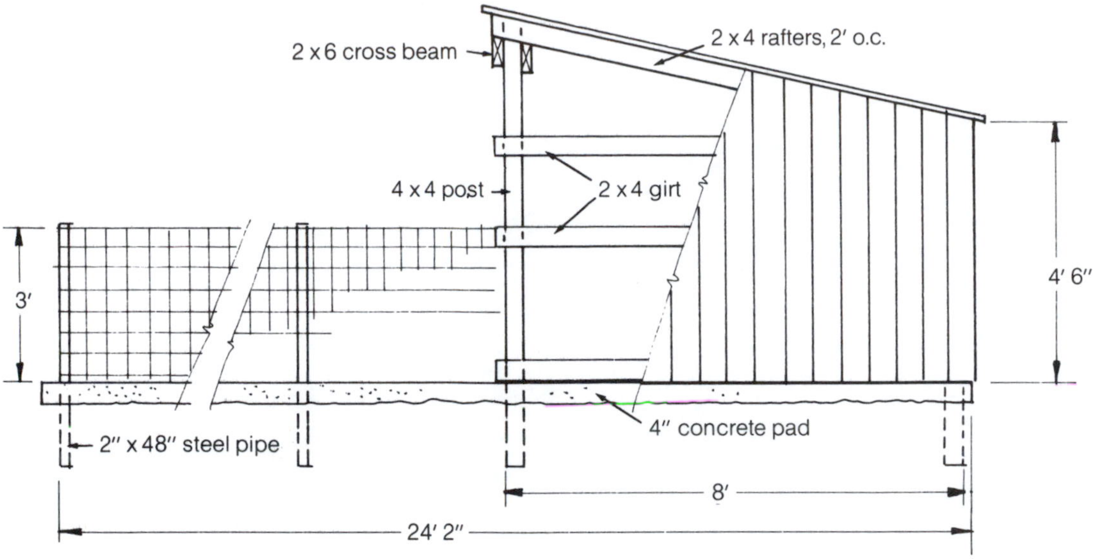

A finishing house for pigs may be built with a concrete pad and welded-wire pen. Design footings for the 4 x 4 posts that are appropriate for local soil conditions.

The house has simple framing, a plywood covering and runners for mobility. The roof rests on wooden slides, fashioned from 2 x 6s, so it can be slid back to allow access for cleaning or working with the young pigs.

To build this house, start by cutting the four 2 x 6 runners. Nail the flooring to the runners, making sure the entire base is square. I used native 1 x 12 oak flooring; ¾-inch exterior-grade plywood is a good substitute.

Cut the pieces for the front and back frames, then assemble them, using screws and waterproof glue. The front and back are identical. After completing the assembly, screw the end frames to the base of the shelter. Use a table saw to cut the lengthwise pieces to the right size and shape; glue and screw them in place. Cover the sides and ends with ½-inch exterior-grade plywood.

Cut the angled pieces for the top runners and the 2 x 8 end pieces to the correct size and shape and assemble the top unit. Then slide in place between the protruding upper edges of the plywood sides. Install the plywood top, using glue and screws. Bore 1-inch air vent holes in the upper 2 x 8 pieces.

The door is plywood, backed by 2 x 2 framing.

Finishing House

Hogs intended for butchering are often kept in any old pen nailed together hastily. As a result, they are usually out of the pen more often than in, and they don't do as well as they would in a better environment. The idea that hogs like to root around in filth and mud is a myth. Today's method of raising hogs and confining them to concrete is a welcomed farm technology.

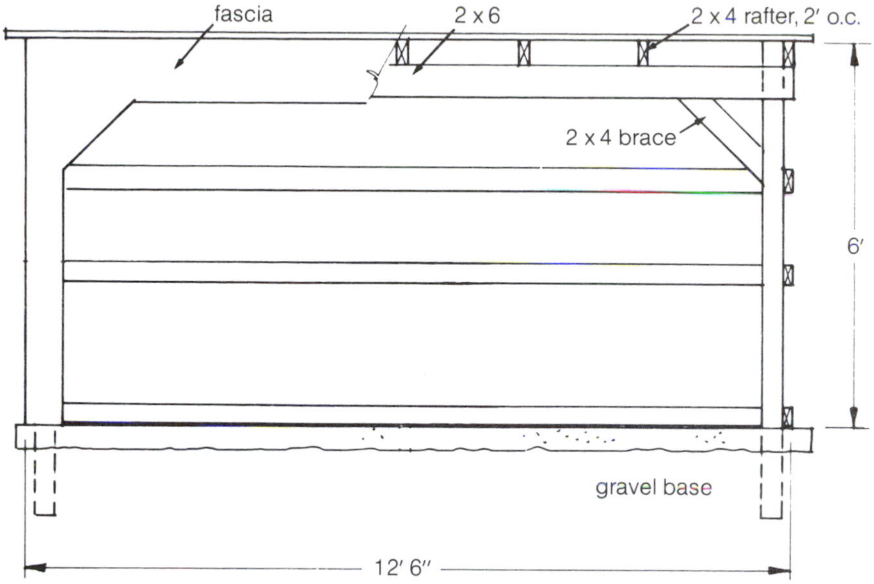

The open end of the shed-style house may be covered with fascia.

Here's an appropriate design suited for fattening two to four hogs each year. In areas with mild climates, it could also be a farrowing house for a young litter.

To begin, mark the outline of the concrete pad, then locate and dig all holes and set the posts. Brace them well; make sure they're plumb. Position the 2- x 48-inch steel pipes or posts so they will protrude 36 inches above the concrete. Lay the gravel base. Lay ½-inch rebar to reinforce the pad, then form the pad and pour the concrete. (See Chapter 3.) The pad should slope ¾ of an inch per foot toward the front. This slope makes cleaning easier. Sweep the concrete with a broom so there's a bit of "tooth" to the surface rather than a troublesome slickness.

(In cold climates, frost is likely to crack a large concrete pad. To reduce the chances of cracking, use expansion joints and pour the concrete in sections. Also, instead of embedding the building posts, you can mount them on top of anchor brackets, secured to the concrete with anchor bolts.)

After the concrete has cured, nail the cross beams and nailing girts to the shed posts. Position a rafter across the tops of the cross beams and mark the appropriate length to give the desired overhang. Then, using a plumb bob, mark the plumb lines on both ends. Cut the rafter to size, then use it as a pattern to cut the remaining rafters. Nail them to the cross beams, front and back. Nail the 2 x 4 braces in place. Cover the shed with the material you prefer, and nail fascia boards across the front and back of the rafters.

The concrete pad and pen is enclosed with welded-wire hog panels, wired to the steel posts and stapled to the corners of the shed. You will have to cut the panel for the 12½-foot wide front of the pen because the panels are only available in 16-foot lengths. Cut with a hacksaw or with a welder. Keep the panels about 2 inches above the level of the concrete to allow for easy cleaning of the pen.

CHAPTER 15

Smokehouse

Borg

BACK BEFORE ELECTRICITY, a prime method of preserving meat was to smoke it. Small smokehouses served this purpose and they also were used to store meats and other items during the off season. The one shown (p. 210, 211) is large enough to smoke eight to ten hams simultaneously.*

Naturally, smokehouses don't have any windows. They do have small doors for venting the smoke, as well as a main door fabricated from 1 x 6s or 1 x 12s nailed to a 1 x 6 frame. The smokehouse may sit on rocks or concrete piers, pads or walls. Six-inch diameter piers or 6-inch walls, extending below the frost line, would be suitable. Anchor bolts secure the framing to the foundation.

For the sides, you can use full dimension 1 x 12 lumber, covered with wooden siding or shingles. The result would be a tight house, requiring small door vents on both the front and back. All openings are covered with 30-mesh screen to prevent rodents and birds from entering the house. To facilitate cleaning smoke and soot from the house, line the inside with aluminum sheeting. This is sometimes available, at little charge, from newspapers where it's used to make page plates.

Firebox

Plan to situate your smokehouse with the firebox upwind and slightly below the smokehouse. In this position, smoke travels easily from the firebox to the smokehouse and access to the side of the firebox is possible.

Build the firebox inside a pit, about 2½ feet deep and 2½ feet wide. The firebox may be constructed of bricks or concrete reinforced with ¼-inch rebar. To facilitate firebuilding, make the firebox with a removable top and use a sliding plate for the front.

The 6-inch diameter smoke pipe is made of ceramic tile. The pipe should exit the firebox about 1 foot below the level of the smokehouse floor. Lay the pipe on 4 to 6 inches of gravel. Where the pipe enters the house through the concrete floor, use a drain fitting. And, when the house is not in use or you want to discontinue smoking, seal the pipe with a wooden plug.

Framing

Simple stud-frame construction is used throughout. First, construct the end walls and secure them to the anchor bolts, then

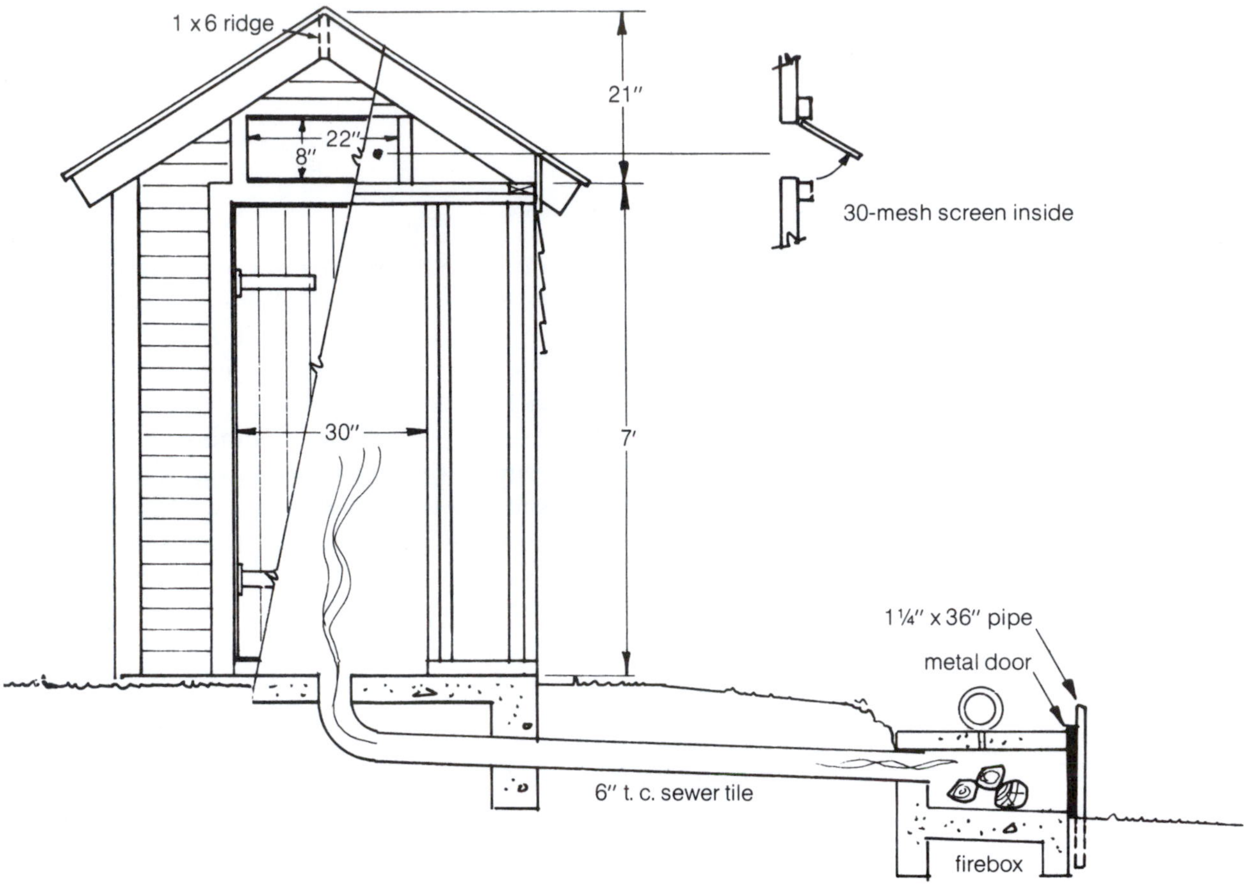

Build the firebox for the smokehouse slightly below and upwind of the smokehouse foundation.

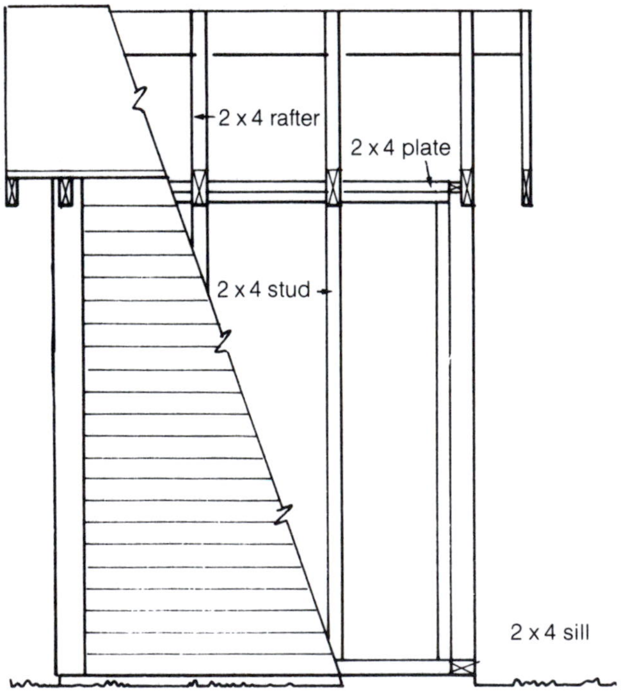

The smokehouse has simple platform frame construction, with 2 x 4 sills, studs, plates and rafters.

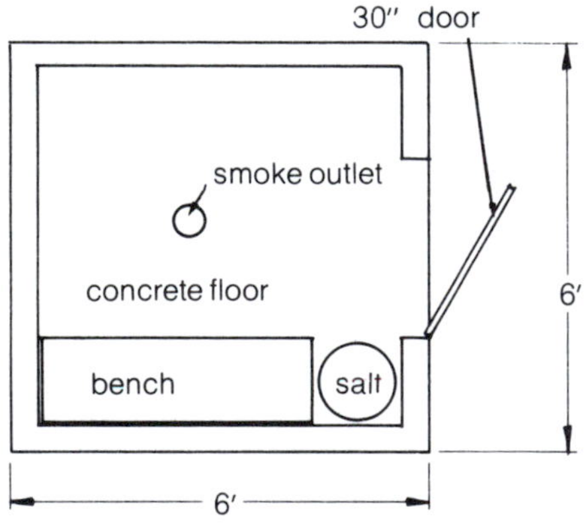

Here's a plan view of the smokehouse, showing overall dimensions and the door size.

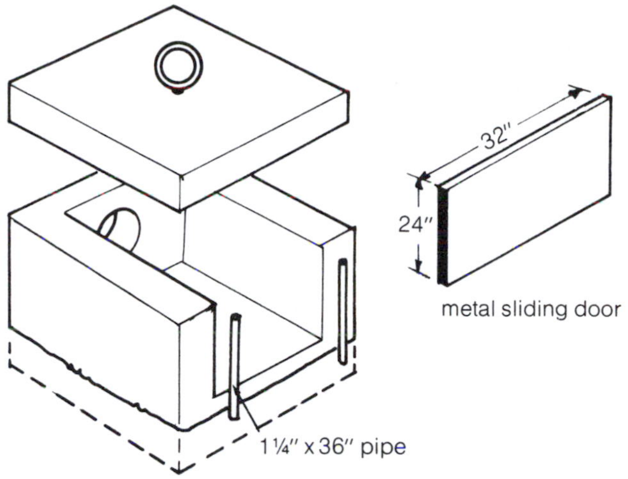

The firebox may be built with concrete, reinforced with No. 9 wire mesh and ¼-inch steel rebar, placed 6 inches o.c.

build the side walls and nail the corners together. Add the top plates to secure the walls solidly.

Cut the 2 x 4 rafters and nail them, spaced 24 inches o.c., to the top plates and ridge board. Note that the tails of the rafters are left square, without plumb cuts.

Place 2 x 6 joists, 24 inches o.c., so they rest on the top plates. These not only strengthen the smokehouse, they provide a place for hooks to hang meats. Trim the ends of the joists so they don't interfere with the roofing.

Construct the small 8- x 22-inch vent doors on both ends and the main door which is 30 inches wide. Hang the main door with two 8-inch Tee hinges; to secure the door shut, use a safety hasp or simple latch.

Apply sheathing and siding of your choice. Use blocking between the joists and rafters to create a tight seal where the top plates meet the roof.

Plan adapted from the United States Department of Agriculture Cooperative Extension Service.

CHAPTER 16

Fencing

Grant Heilman Photography

GOOD FENCING IS one of the most important features of a successful farmstead. If you have the right fencing in the right location, working with almost any animal can be easy and efficient.

Fencing varies from solid boards to welded steel panels or woven wire. Woven wire or a combination of woven and barbed wire is commonly used to confine livestock; for temporary enclosures, electric fencing is economical and effective.

Board fences are used for feed lots and corrals; welded steel panels are made for cattle feed lots and hog fencing; and chain link fences protect gardens and orchards.

Livestock Fencing

Farm animals are hard on fences in different ways. Horses will eventually break down a wire fence by reaching over it for bites of grass. They will do this even if they have good grass on their own side. They also ruin a board fence in this manner unless it is built especially to hold horses. For this reason, fencing for horses is often heavy-duty post and board fencing at least 5½ feet tall. A corral to hold a stallion should be even taller and built as strongly as possible.

Pigs can lift an amazing amount of weight with their rooters. Therefore their fences must be low to the ground and tightly constructed. Hog fencing is 3 to 4 feet high, and made of wire with a barbed strand stretched across the bottom; hog lots are also built of tough, welded-steel panels.

Cows push on fences until they create a weak spot. Eventually, they work their way through by brute strength. Consequently, cattle fencing is made of barbed wire or sturdy wire panels, at least 4½ feet high. For small calves, woven wire is often used.

Sheep may tear their fleeces on barbed wire. Therefore, woven wire is usually preferred, with barbed wire run along the top and/or bottom of the posts to discourage dogs and coyotes. The total height is about 4½ feet. An 18-inch apron of woven wire run along the ground will help prevent predatory animals from burrowing beneath the fence. Woven wire with a close mesh at the bottom is another alternative.

Posts

A fence is only as good as the posts which support it. Posts should be the correct size and set to the correct depth and height for the animal they confine.

Steel posts are often chosen for livestock operations. They are light, durable and easily driven. They are available in lengths between 5 and 8 feet.

A second type is the pressure-treated wooden post. It is used for everything from corrals to hog fencing. These posts are 2½ to 6 inches thick or larger and up to 10 feet long. Usually, their ends are pointed and they are driven in place with power equipment.

Home-cut, home-treated wooden posts are another less expensive alternative, although they won't last as long as the pressure-treated posts. They can be anything from huge corner posts cut from a locust, hedge or other hardwood to smaller pine posts.

Preserving Posts

If pressure-treated posts are not available, or you find they are too expensive, you can prepare a home preservative. When properly treated, the life of a post is doubled or even tripled. Coal-tar creosote is the best preservative. Since heat and special equipment are required, this is a difficult process for the homeowner. There are many other ways of treating wooden posts to extend their usefulness, but only two of them are practical for the small farm: cold soaking on seasoned posts and end soaking on green posts, which is not as satisfactory.

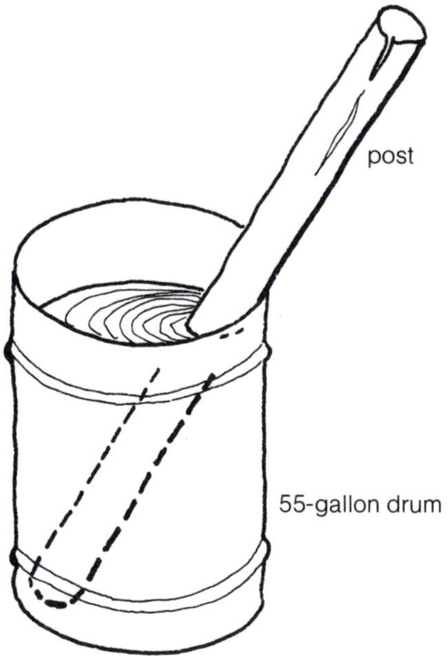

Soaking a post in preservatives can extend its life significantly.

End soaking. If you need a fence right now and don't have time to cut your posts ahead and season them before treatment, end soaking is the method to use. Cut round posts and leave the bark on the posts. Then use a 15 percent to 20 percent solution of zinc chloride, or chromatized (chromated) zinc chloride in water. Allow about 5 pounds (or about a half-gallon) of the solution for each cubic foot of post to be treated. The chromated zinc chloride is sold in a granular form that is easy to use, and is less subject to leaching from the posts than the plain zinc chloride. Often it is difficult to find a source for this chemical. If you can't find it, you usually can buy its two ingredients from a chemical company, and mix them. Use 80 percent zinc chloride and 20 percent sodium bichromate. For the solution, mix a 20 percent chemical, 80 percent water solution (each by weight). Thus 100 gallons of water weigh about 830 pounds and would require 166 pounds of zinc mixture.

Stand the posts bottom down in a tub or drum of this solution until they absorb about three-fourths of it, which takes from three to ten days. Then stand them on their tops, and let

> **CAUTION**
>
> Do not use paint containing lead on livestock equipment or on parts of buildings accessible to livestock. Poisoning may result when animals constantly lick or chew objects covered with paint containing lead.
>
> Also, use extreme caution when applying preservatives. They vary in terms of toxicity, but most are poisonous to animals and humans. Use of some preservatives may be restricted by regulations. Contact your state agriculture department for the latest information and regulations.

them absorb the rest. They should be seasoned for about a month before using, to allow the treated wood to dry.

Either green or seasoned posts may be soaked, covered completely and steeped, in a 5 percent solution of this same zinc chloride, for from one to two weeks, but this is not as effective as a pentachlorophenol solution treatment.

Cold soaking. The best absorption and penetration are obtained by first seasoning the posts. This lets the sap dry out to make room for the preservative. Peeled posts should be open piled, so that the air can circulate around each one, and the bottom of the pile should be at least a foot above the ground. The best place for piling would be an exposed area on well-drained ground.

While posts cut in the spring will peel more easily, posts cut in the fall will have a chance to dry more slowly, which prevents some cracking and checking. This is more important with oak posts than with wood from cone-bearing trees.

The seasoning of posts adds little to their life *unless* they are also treated with preservatives.

Cold soaking of seasoned posts consists of soaking in a solution of *pentacholorophenol* (called *penta* for short), or in *copper naphthenate,* diluted with either fuel oil or diesel oil. Wear protective clothing and rubber gloves, for these solutions irritate the skin. Penta can be purchased in a concentrated solution, or in a ready-to-use solution (more expensive) or in flakes. If you use penta in dry flakes, wear goggles and a dust mask when mixing,

Table 16-1 Estimated Life of Untreated Wood Posts[1]
(diameter, 5 to 6 inches in size)

Over 15 Years	7 to 15 Years	3 to 7 Years	
Black locust	Cedar	Ash	Honey locust
Osaga-orange	Red cedar	Aspen	Maple
	Red mulberry	Balsam fir	Pine
	Redwood	Beech	Red oak
	Sassafras	Box elder	Spruce
	White oak	Butternut	Sycamore
		Douglas fir	Tamarack
		Hemlock	Willow
		Hickory	Yellow poplar

[1] Split posts, which have more "heartwood," will last longer than the time listed, and larger sized posts also last longer.

to avoid irritating the eyes and throat. Penta is made in several strengths, calling for dilution by mixing with two to twelve or more parts of oil, to make a 5 percent solution. The label should specify strength and amount of dilution. Fifty gallons of a mixture of oil and penta will treat 50 posts of 6-inch diameter and 6-foot length. For convenience, they can be soaked in upright drums, soaking the bottom portion longer, for the tops are less subject to decay. (This is also a good treatment for seasoned boards to be used in a board fence.)

After removal from the solution, posts should be stacked so that the excess solution dripping from each post will be absorbed by the post beneath it. Wear gloves when handling them until they are dry. Penta is highly toxic and can be absorbed through the skin and by way of the lungs. Animals should be kept away from newly treated wood. Do not treat wood for feed troughs. Store penta carefully, away from the reach of children, for there is no antidote for penta poisoning.

Building a Fence

Karl Schwenke, author of *Successful Small-Scale Farming*,* suggests a few good principles for solid fence building that I think are worth repeating: Use plenty of fence posts and space them, on the average, about eight to ten feet apart. Brace and secure all corner and gates posts; where possible, dig or cement them 3 to 4 feet in the ground. Be sure to use the correct type and gauge of wire for your purposes. Neatly stretch the wire to a tautness that will resist probing pressures by livestock and pests, but not so taut as to prohibit contraction in cold weather. Set all corner and line posts before applying the wire or boards. Affix the wire or boards on the outsides of the posts only for appearance's sake or when the sole intent is to keep pests out. If the fence crosses a contour, it may be a good idea to put the wire outside, to achieve the best strength and balance.

Erecting the Corner Posts

Let's start with fencing a typical pasture using woven wire for the bottom portion and barbed wire for the upper. The follow-

* Schwenke, Karl, *Successful Small-Scale Farming* (Charlotte, Vt.: Garden Way Publishing Co., 1979).

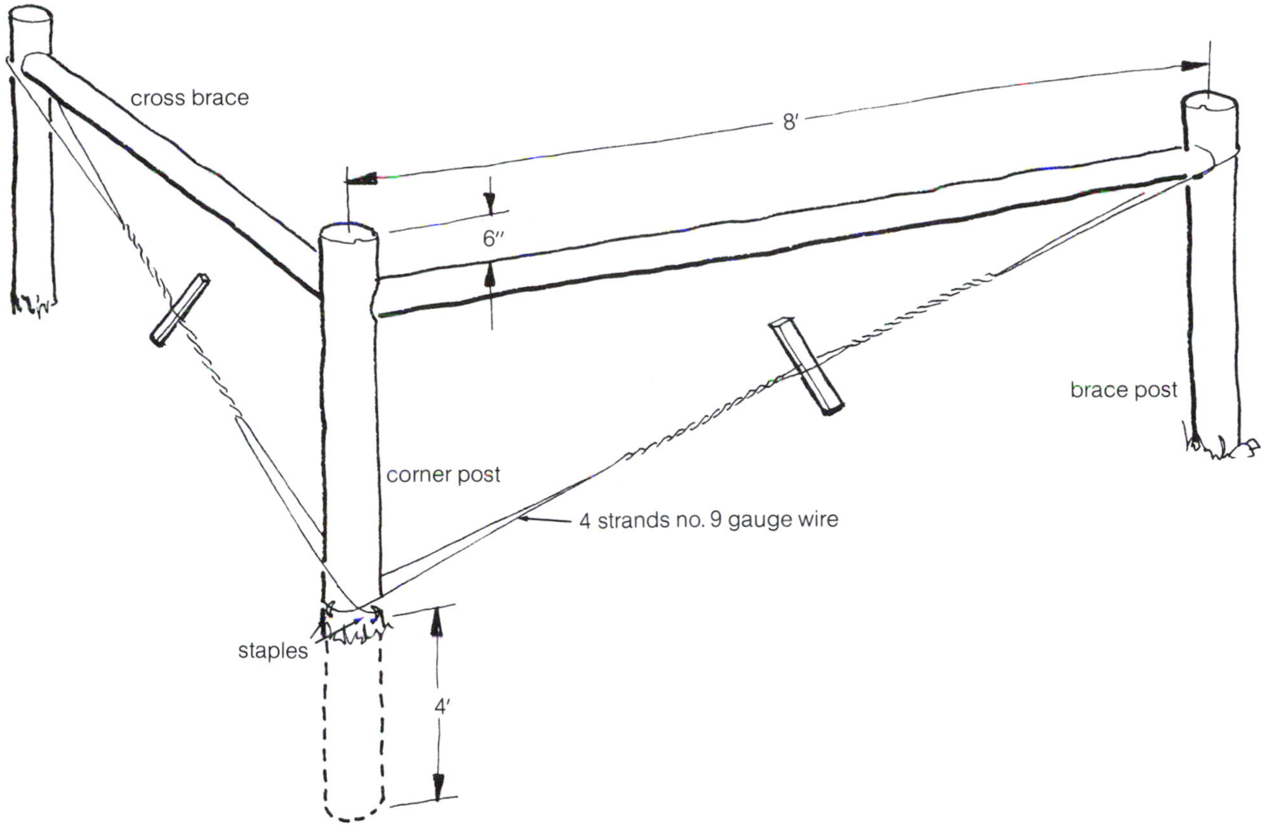

Once the posts have been installed, tighten the corner with No. 9 gauge wire and a small stick or piece of wood.

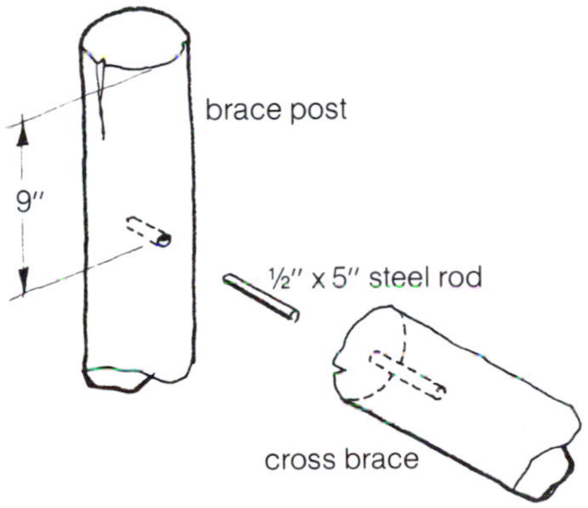

Cross braces may be notched into the corner posts and brace posts, or fitted on steel dowels.

ing post installation may also be used for barbed wire alone. Keep fence lines as straight as possible. Curved fences can be built to fit land contours, but they never stay as tight and snug as straight-line fencing.

The first step is to construct the corner or end-post sections of the fence. Clear away all brush, and level the area as much as possible.

A fence corner consists of the corner post, a sturdy post (8 to 10 feet long, 6 inches in diameter) set 4 feet in the ground, and two notched brace posts of similar size set 8 feet in either direction from the corner post. The posts can be set in pre-dug holes or driven into the ground. Power post-hole diggers or a powered driver on the back of a tractor can save much time and labor in setting the posts.

Posts may be set in concrete, or they may be secured by tamping dirt and stones around them. If posts are set in concrete, use pressure-treated wood. Setting the posts in concrete is recommended if the fence will be subjected to heavy pressure or to wind strain.

Plumb the corner post and "sight" it to suit the fence line. This means, in the case of a crooked post, turning the post so the best side is presented to the fence line.

Place several small stones around the post and tamp down solidly with a wooden tamper. Add a layer of earth, preferably

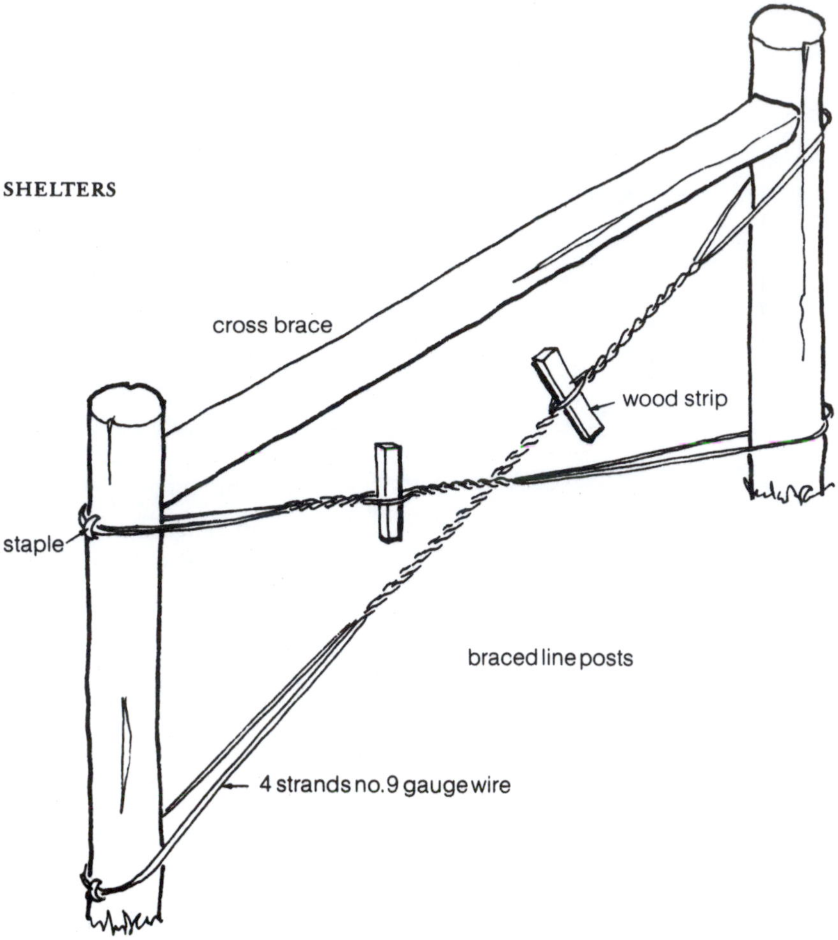

Approximately every 20 yards, line posts should be braced as shown.

clay taken from the bottom of the post hole, on top of the stones and tamp solid. Continue until the hole surrounding the corner post is filled and tamped, and the post feels solid.

Place the brace posts in their holes, but don't pack them in yet. Secure the cross braces. These can be notched into the corner and brace posts or fitted on steel dowels made of ½- x 5-inch steel rod.

Replace the soil around the brace posts and tamp firmly. Extend four strands of No. 9 gauge wire from corner to brace posts. Twist the wire with a stick or rod and leave in place so you can adjust the tension when necessary.

You may wish to anchor the corner posts further by securing them with wire to a "deadman," a log buried in the ground. This keeps the corner post from being pulled toward the fence.

Line Posts

Most fence builders like to set the corner posts, then temporarily stretch a single strand of barbed wire, cord or rope between the corners. This establishes a good straight line to mark the location of the other posts called *line posts.* Again, these posts may be installed in holes or they may be driven in place.

Line posts are set 2 to 3 feet in the ground (1 to 1½ feet for steel posts) and are spaced 16 feet apart. Closer spacing may be necessary if the ground is uneven or if you need a stronger fence. Braced line posts are recommended every 20 yards; these are posts set 8 feet apart and braced in the same manner as corner posts except that additional strands of No. 9 wire are added to form an X. Posts for feed lots and corrals are spaced 6 feet apart when heavy use is anticipated.

Avoid setting posts in a gully or stream where they could be washed out by heavy flows. If the water can't be avoided, stretch the fence between two well-secured posts. Or you may wish to use a fence-like swinging water gap. The rising water forces the gap to open up and swing out with the current, but allows it to drop back down as the water recedes.

Stretching and Attaching the Wire

Woven wire can be stretched only from one anchored post to another, but an anchored post can be a gate, corner, end or just a braced line post—any but unanchored line posts. What follows is a description of stretching woven wire from one corner post to another. If woven wire is used in combination with barbed wire, attach the woven first. Fasten woven wire to the posts using 1½-inch galvanized staples. Drive each staple in snugly, but don't allow it to crimp or bury the wire. Angle the staple slightly so both prongs don't split the same grain.

Wear heavy leather gloves, boots and tight-fitting clothing for protection. Carry staples or other metal fasteners in an apron, not on your person or in your pockets. Stand on the opposite side of the post from the wire and stretcher unit. If you are handling preservative-treated posts, do not rub your hands or gloves in your face; you may be allergic to the wood preservative.

After the corner and line posts have been solidly placed, the woven wire is fastened to the corner post. Remember for best appearance, the fencing should be on the outside of the posts, but if the fencing will be subjected to pressure from livestock, the wire should be on the inside. Here are the steps to follow when setting your wire fence:

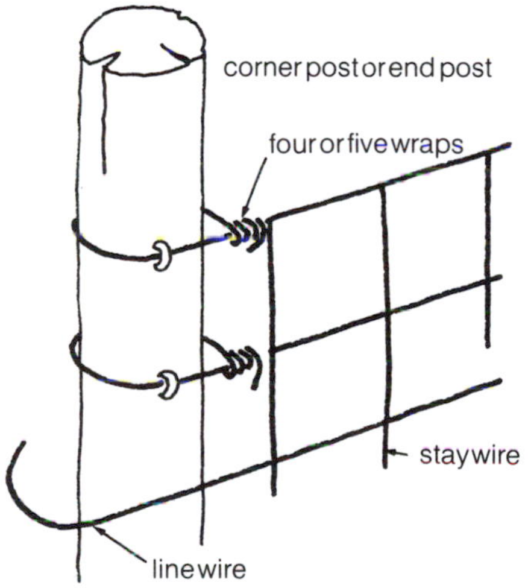

To attach woven wire, remove two or three stay wires, then staple and wrap line wire around post.

1. Set the roll of fencing on end, with the end having the closer wire spacing at the bottom. Unroll enough wire to make a wrap around the corner post. Remove two or three stay (vertical) wires, depending on the circumference of the post, and position the next stay wire a few feet from the post. Adjust the

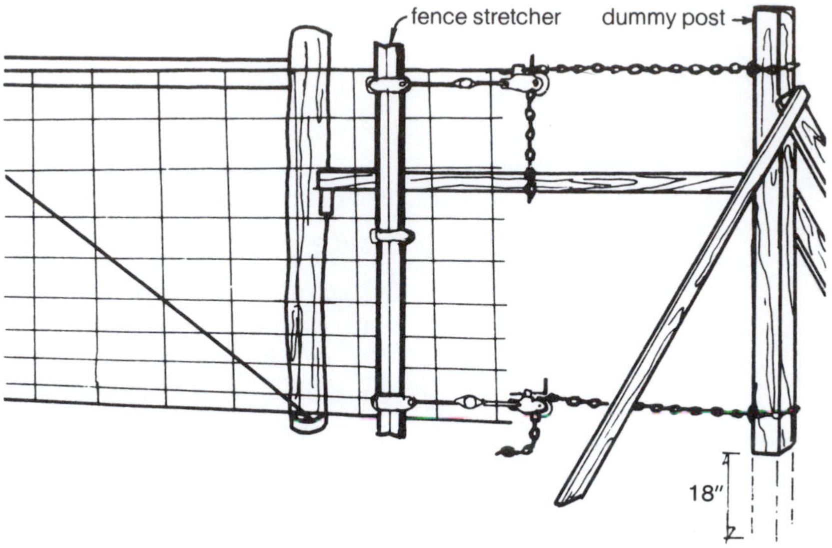

You can use a braced dummy post to support a fence stretcher. Attaching the stretcher to a vehicle is another possibility.

fence to the desired height. Starting with the center line wire, first staple then wrap each line wire around the end post and back around the wire itself, four or five wraps.

2. Resume unrolling the fencing, keeping the bottom wire close to each post. Before stretching the wire, tie it loosely to the line posts with baling twine, or prop it up with temporary stakes. Place the clamp of the fence stretcher on the wire so that when the fence is stretched, the clamp will pass well beyond the other end post, enabling you to staple the fencing to the post. The stretcher may be secured to a dummy post or a tree.

Before you start stretching the woven wire, note the shape of the tension curves (the little crimps in the line wires). The fence is stretched properly when these tension curves are about half straightened out. This provides for expansion and contraction with temperature changes. To avoid sagging, wire should be set out during warm weather.

3. Stretch the wire evenly so that the top and bottom wires are not pulled more tightly than the others. Also try to keep the stay wires as vertical as possible If your terrain is uneven, it may be necessary to fasten the fence to the posts on the ridges and low places first. Staple line wires one at a time, starting at the top or bottom, whichever is tighter.

4. When fastening the woven wire to the end post, secure each line wire with two staples angled in opposite directions to prevent slippage. Measure the circumference of the post. Cut the fencing, allowing that many inches plus at least 6 inches more. Remove a couple of stay wires, wrap the fence around the end post and secure it by wrapping the ends of the line wires back around the corresponding line wires of the fence. If the wire continues around a corner post, there is no need to cut it. Just fasten it and continue around.

5. Secure the fence to line posts, as necessary starting with the post nearest the first corner post.

6. The last step is to stretch and fasten the barbed wire strand to each post. Be careful when working with any type of

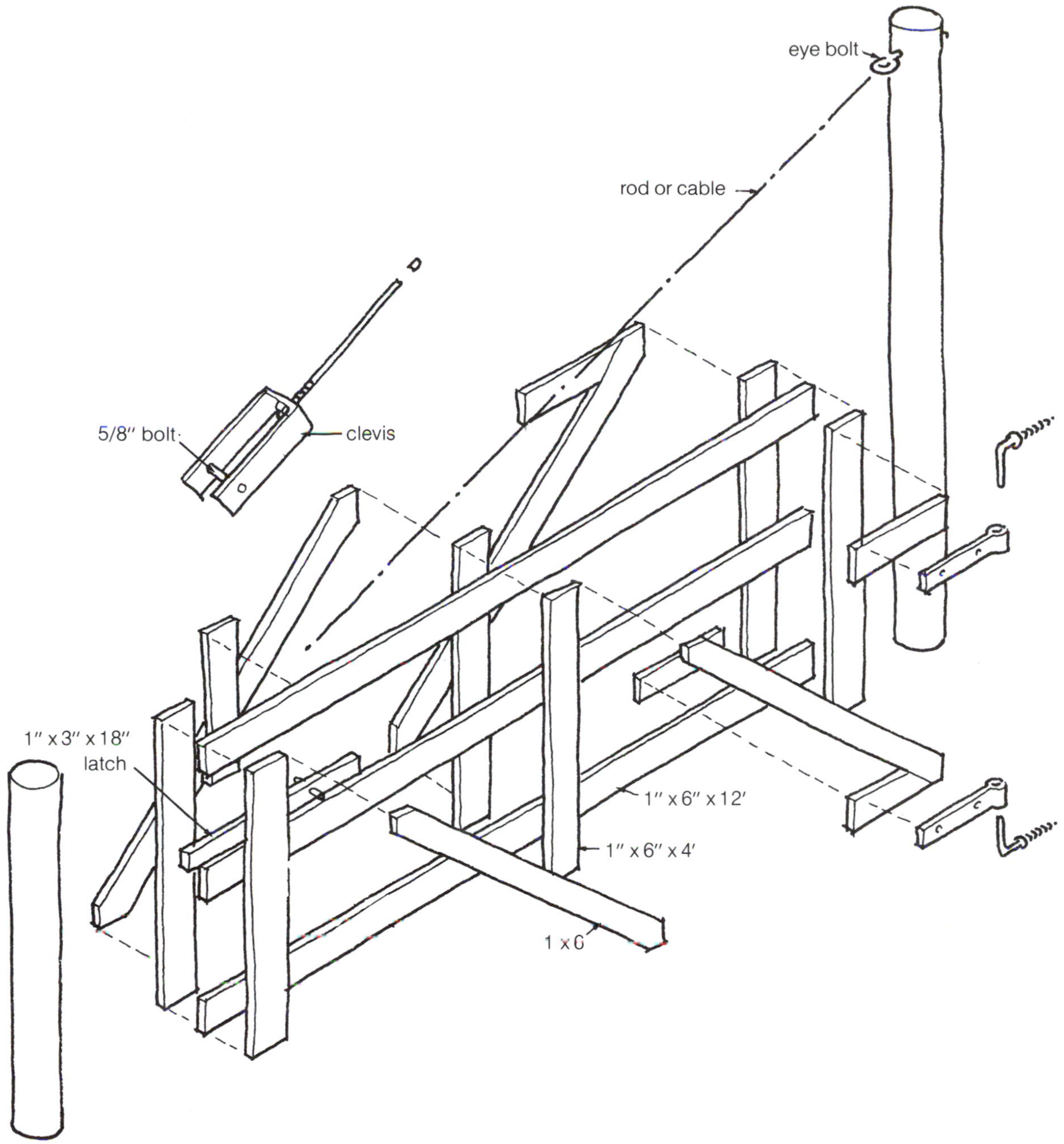

An exploded view of a gate made of 1 x 6 lumber and supported by a cable attached to a 12-foot post (8 feet above surface). The post should be set in concrete or securely braced.

wire, but particularly with barbed wire and fence stretchers. Fence stretchers may "let go" of the wire, causing it to suddenly roll up. You can stretch barbed wire with a block and tackle secured to a dummy post.

Gates

One of the most practical gates is the board gate, made from 1 x 6 lumber. You can hang this with hardware fitted to the gate and into holes bored in the post. Gates should be wide and

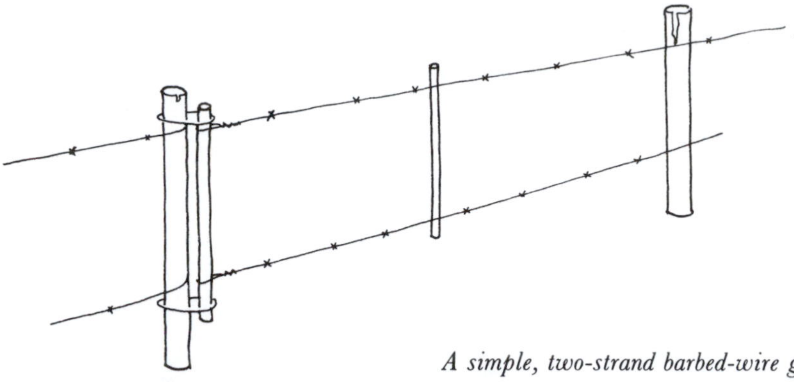

A simple, two-strand barbed-wire gate.

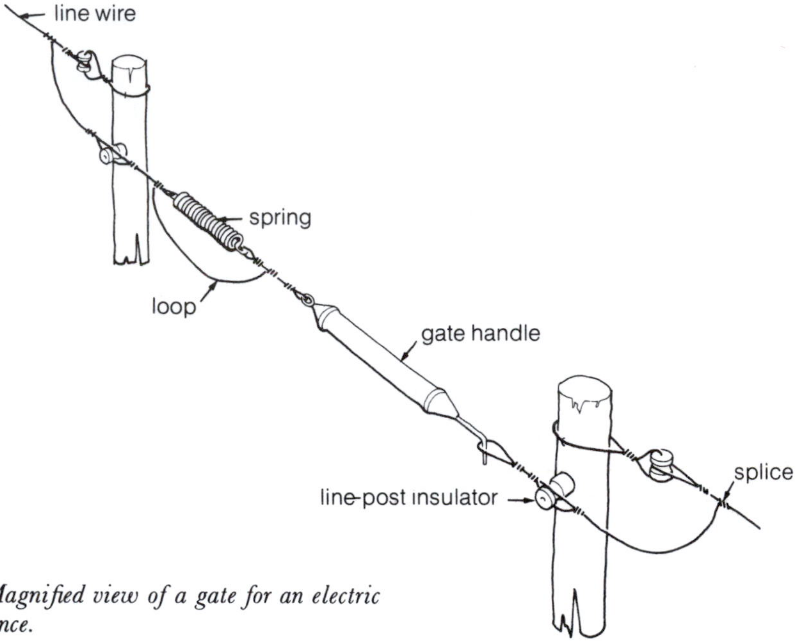

Magnified view of a gate for an electric fence.

well braced to withstand constant wear and frequent bumping by animals. For extra strength, the gate posts may be braced with wire and brace line posts, set 8 feet away. Another good alternative is the metal gate that may be assembled from pipe or purchased as a finished unit.

Gates may also be made from wire. Shown are two options, one using barbed wire, the other suitable for an electric fence.

Rigid Welded-steel Panel Fencing

Today many corrals, cattle feed lots and hog areas are fenced with welded-steel panels made of ¼-inch diameter steel rod. These are 16 feet long and available in two different heights; hog panels are 34 inches high; and combination hog-and-cattle panels are 52 inches high. When coupled with sturdy posts, they can withstand a lot of abuse.

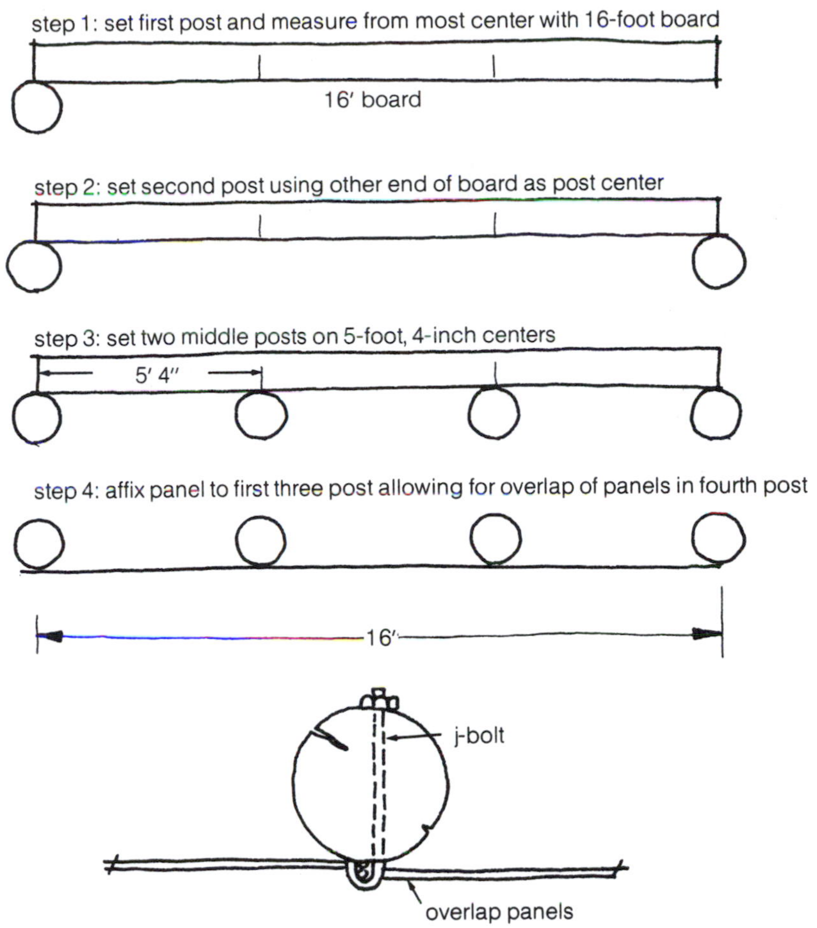

step 1: set first post and measure from most center with 16-foot board
16′ board
step 2: set second post using other end of board as post center
step 3: set two middle posts on 5-foot, 4-inch centers
5′ 4″
step 4: affix panel to first three post allowing for overlap of panels in fourth post
16′
j-bolt
overlap panels

These are recommended steps for neat installation of welded-steel panel fencing.

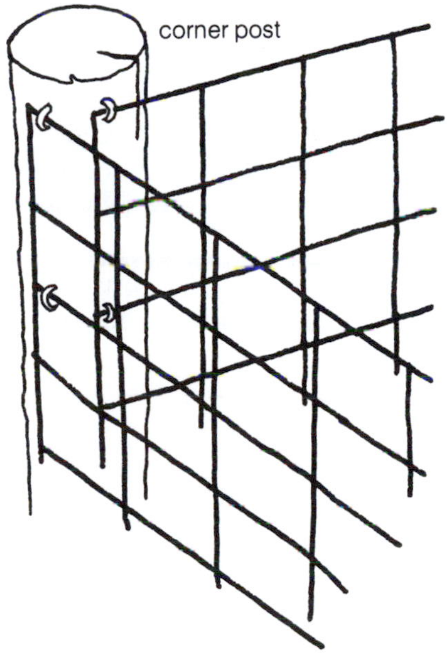

corner post

At corners, welded-steel panels are installed this way.

Typically, posts should be about 6 inches in diameter by 8 feet long for cattle panels and 3½ inches in diameter by 6 feet long for hog panels. Suggested post spacing is on 5-feet-4-inch centers.

A good way to achieve even spacing quickly is to use a 16-foot-long board, marked at 5-feet-4-inch intervals. Locate the first post and position the board. Then locate the end post and the two center posts. Fasten the panel to the first three posts and allow for overlap of the next panel on the fourth post.

The panels can be stapled to the posts or wired. A more permanent arrangement is to use special J-bolts that hold the panels securely in place. Three of these bolts are adequate for most installations. Drill ½-inch holes in the posts for the bolts. In most cases, the fencing panels are secured to the inside of the lot or confinement area, except in the corner where one end must be fastened outside.

Other Types of Fencing

Another fence for keeping out varmints and restraining livestock is the electric fence, powered by a battery or plugged into a 110 to 120 volt circuit. The wires deliver a small electrical charge

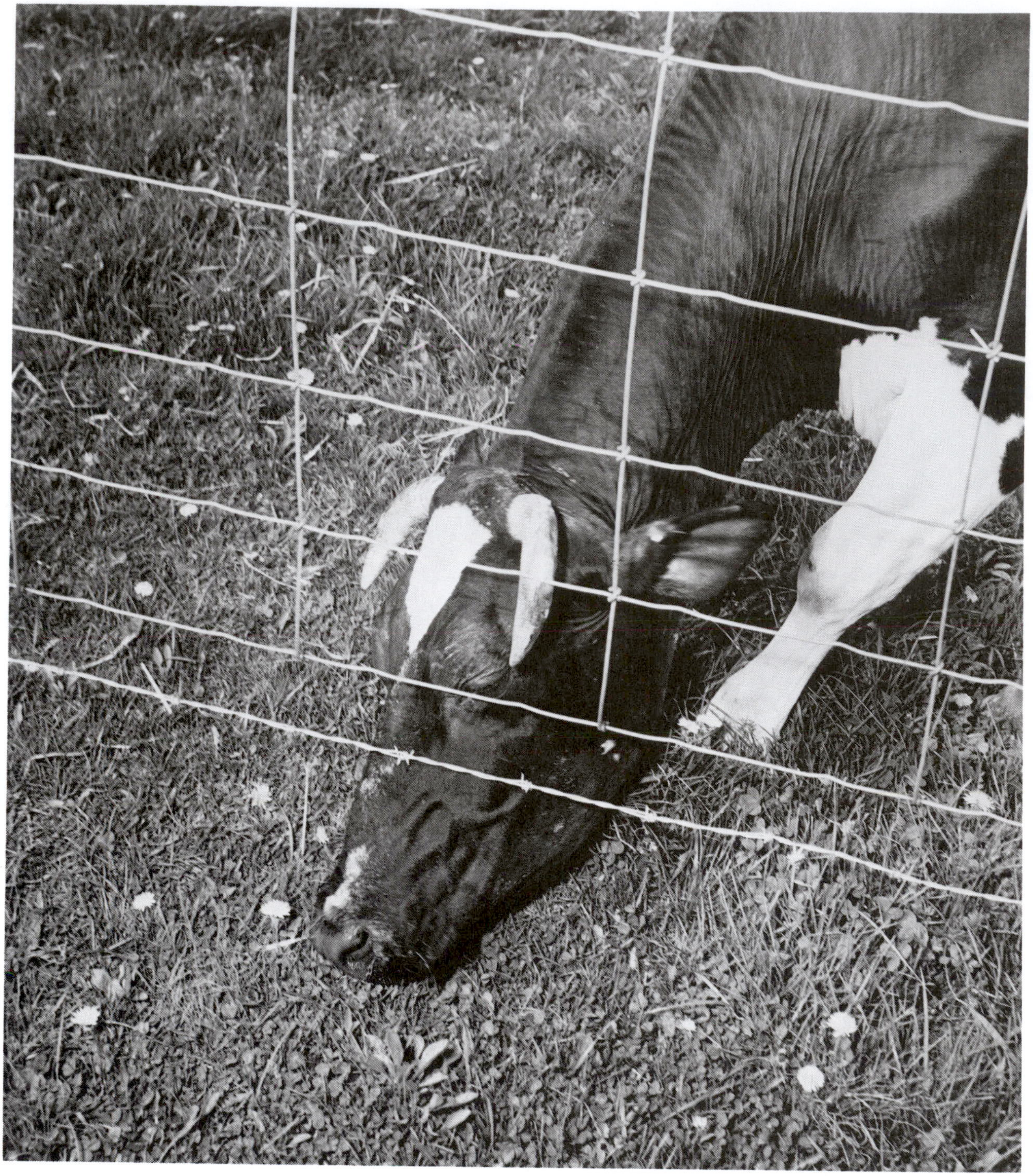

Welded steel panels may be combined with barbed wire to confine livestock and prevent dogs and pests from slipping through.

which most animals learn to respect. The charging unit may be placed inside or outside; it is grounded with a small wire attached to a steel post driven into the ground. One of the main advantages of electric fencing is that you can set it up temporarily wherever desired.

For an electric fence you'll need plastic insulators, lightweight posts or stakes (set at any interval up to about 30 feet) and a grounded charging unit. First, attach the insulators to the posts at the desired height (two-thirds the height of the animal to be restrained), install posts, then attach charging unit. To protect the fence from lightning, install a lightning arrestor. Be sure to keep the entire fence free of weeds and grass; if these touch the wires, the unit will be grounded.

Poultry fencing, or netting, is made of fine mesh, galvanized steel with 1- or 2-inch hexagonal mesh openings. It is available in heights from 1 to 6 feet.

Poultry wire can be run under the ground to form an apron. Bury it about 4 inches to keep varmints or dogs from digging under the fences; a 4-barbed strand of barbed wire about 3 to 4 inches under woven-wire fencing could be used instead.

One of the biggest concerns of orchardists, gardeners and flower growers is adequate protection against rabbits, opossums, deer and raccoons. To exclude these varmints, there are many different styles of fencing; each serves a slightly different purpose. One of the most effective, though expensive, is the chain-link fence.

Chain-link fencing is available with kits that include poles, top rails and hardware. Chain-link fencing is galvanized, and therefore does not rust or corrode. It is available in heights of 39 to 72 inches.

Appendices

Appendix A

Outbuilding Space Requirements for Livestock, Beddings and Feeds*

Space Requirements for Beef

Feedlot, sq. ft./head	
20 in barn and 30 in lot	Lot surfaced, cattle have free access to shelter
50	Lot surfaced, no shelter
150–800	Lot unsurfaced except around waterers, along bunks and open-front buildings, and a connecting strip between them
20–25	Sunshade
Building with Feedlots, sq./ft. head	
20–25	600 lb. to market
15–20	Calves to 600 lb.
½ ton/head	Bedding
Cold Confinement Buildings, sq. ft./head	
30	Solid floor, bedded
17–18	Solid floor, flushing flume
17–18	Totally or partly slotted
100	Calving pen
1 pen/12 cows	Calving space

Space Requirements for Sheep

Shelter space

Open-front building with lot:
- 10–12 sq. ft./ewe
- 12–16 sq. ft./ewe and lambs
- 6–8 sq. ft./feeder lamb

Lot:
- 25–40 sq. ft./ewe
- 25–40 sq. ft./ewe and lambs
- 15–20 sq. ft./feeder lamb

Solid floor (confinement):
- 12–16 sq. ft./ewe
- 15–20 sq. ft./ewe and lamb
- 8–10 sq. ft./feeder lamb

Space Requirements for Horses
Dimensions of Stalls Including Manger

	Box Stall Size		Tie Stall Size
Mature animal (mare or gelding)	10′x 10′	........small	....
	10′x 12′	........medium	5′x 9′
	12′x 12′	........large	5′x 12′
Brood mare	12′x 12′	........or larger........................	
Foal to 0-year-old	10′x 10′	........average	4½′x 9′
	12′x 12′	........large	5′x 9′
Stallion*	14′x 14′	........or larger........................	
Pony	9′x 9′	........average	3′x 6′

* Work stallions daily or provide a two to four acre paddock for exercise.

Space Requirements for Swine

Building floor space
Sows and boars: 8 sq. ft. indoors, 18 sq. ft. outdoors
Sow and litter: 26 sq. ft. slotted floors
or 32 sq. ft. indoors, 42 sq. ft. outdoors
Pigs to 60 lb.: 3 sq. ft.
60 to 125 lb.: 6 sq. ft.
125 and up: 8 sq. ft.
or 5 sq. ft. indoors, 13 outdoors

Pasture space
10 gestating sows/acre
7 sows with litters/acre
50 to 100 growing-finishing pigs/acre depending on fertility

Shade space
15 to 20 sq. ft./sow
20 to 30 sq. ft./sow and litter
4 sq. ft./pig to 100 lb.
6 sq. ft./pig over 100 lb.

Space Requirements for Poultry

Type of Bird		Floor Space[1] (square feet)
Chicks	0–10 weeks	.8 — 1.0
	10-maturity	1.5 — 2.0
Layers	brown egg	2.0 — 2.5
Layers	white egg	1.5 — 2.0
Layers	meat-type breeders	2.5 — 3.0
Broilers	0–8 weeks	.8 — 1.0
Roasters	0–8 weeks	.8 — 1.0
Roasters	8–12 weeks	1.0 — 2.0
Roasters	12–20 weeks	2.0 — 3.0
Turkeys	0–8 weeks	1.0 — 1.5
Turkeys	8–12 weeks	1.5 — 2.0
Turkeys	12–16 weeks	2.0 — 2.5
Turkeys	16–20 weeks	2.5 — 3.0
Turkeys	20–26 weeks	3.0 — 4.0
Turkeys	breeders (heavy)	6.0 — 8.0
Turkeys	breeders (light)	5.0 — 6.0
Ducks	0–7 weeks	.5 — 1.0
Ducks	7 weeks maturity	2.5
Ducks	breeders (confinement)	6.0
Ducks	breeders (yarded)	3.0
Geese	0–1 week	.5 — 1.0
Geese	1–2 weeks	1.0 — 1.5
Geese	2–4 weeks	1.5 — 2.0
Geese	breeders (yarded)	5.0

[1] Many factors determine the floor space requirements including the type of management system, the type of house, the number and kind of bird, the climate and even the management the birds receive.

Hay, Beddings and Feeds
Weights and Storage Space Requirements

Material	Weight Per Cubic Foot in Pounds	Cubic Feet Per Ton
Hay—loose in shallow mows	4.0	512
Hay—loose in deep mows	4.5	444
Hay—baled loose	6	333
Hay—baled tight	12	167
Hay—chopped long cut	8	250
Hay—chopped short cut	12	167
Straw—loose	2–3	1000–667
Straw—baled	4–6	500–333
Silage—corn	35	57
Silage—grass	40	50
Barley—48# 1 bu.	28	72
Corn, ear—70# 1 bu.	28	72
Corn, shelled—56# 1 bu.	45	44
Corn, cracked or cornmeal—50# 1 bu.	40	50
Corn-and-cob meal—45# 1 bu.	36	56
Oats—32# 1 bu.	26	77
Oats, ground—22# 1 bu.	18	111
Oats, middlings—48# 1 bu.	39	51
Rye—56# 1 bu.	45	44
Wheat—60# 1 bu.	48	42
Soybeans—62# 1 bu.	50	40
Any small grain[1]	Use 4/5 of wt. of 1 bu.	
Most concentrates	45	44

[1] To determine space required for any small grain use wheat (60# = 1 bu.) for example.
Then: 60 (4/5) = 48# wheat per cubic foot volume. To find number cubic feet wheat per ton, Then:

$$\frac{\text{2000\# (Wt. of one ton)}}{\text{48\# wheat per cubic foot volume}} = 42 \text{ cu. ft.}$$

* Sources: American Plywood Association; *Midwest Plan Service Structures and Environment Handbook* (Aimes, Iowa: Midwest Plan Service, 1976); Len Mercia, *Raising Poultry the Modern Way* (Pownal, Vt.: Garden Way Publishing, 1975).

Appendix B

Common Wood Characteristics

SPECIES	WORK-ABILITY	SHRINK-AGE	STRENGTH[a] (BENDING STRESS AT PROP. LIMIT)	WEIGHT[b] (LBS.PER CU. FT.)	DECAY RESIS-TANCE	INSULA-TION[c] (R-FACTOR PER INCH)	USES
Balsam Poplar	easy	low	very weak (5000–6000 psi)	26	low	1.33	walls
Northern White Cedar	easy	very low	very weak	22	high	1.41	walls, posts
Hemlock	mod.	low	weak	28	low	1.16	walls
Black Spruce	mod.	low	weak	28	low	1.16	walls
Basswood	easy	high	weak	26	low	1.24	walls
Red Cedar (east)	easy	very low	weak	33	high	1.03	walls, shingles
Red Cedar (west)	easy	very low	weak	23	high	1.09	walls, shingles
Redwood	easy	very low	weak	28	high	1.	walls, shingles, trim
Cypress	mod.	low	weak	32	high	1.04	walls, posts
Aspen	mod.	low	weak (6000–7000 psi)	26	low	1.22	walls
Cottonwoods	mod.	med.	weak	24–28	low	1.23	walls
Balsam Fir	mod.	med.	weak	25	low	1.27	walls
White Pine	easy	very low	fair (7000–9000 psi)	25	mod.	1.32	general, trim
Ponderosa Pine	easy	low	fair	28	mod.	1.16	walls, trim
Jack Pine	easy	low	fair	27	mod.	1.20	walls
Red Pine	easy	low	fair	34	low	1.04	walls, joists
Tamarack	fair	med.	fair	36	mod.	0.93	general
Yellow Poplar	easy	med.	fair	28	low	1.13	general
Elm, soft	hard	high	fair	37	low	0.97	fuel, floors

Continued

Common Wood Characteristics

SPECIES	WORK-ABILITY	SHRINK-AGE	STRENGTH[a] (BENDING STRESS AT PROP. LIMIT)	WEIGHT[b] (LBS.PER CU. FT.)	DECAY RESIS-TANCE	INSULA-TION[c] (R-FACTOR PER INCH)	USES
Maple, Soft	hard	med. high	fair	38	low	0.94	fuel, floors
White Birch	hard	high	fair	34	low	0.90	fuel, floors
Black Ash	hard	high	fair	44	low	0.98	fuel, floors, furn.
Douglas Fir	mod.	med.	strong (9000–11,000 psi)	34	mod.	0.99	general
Yellow Pines	hard	med. low	strong	36–41	mod.	0.91	floors, joists
White Ashes	hard	med.	strong	38–41	low	0.83	fuel, furn.
Beech	hard	very high	strong	45	low	0.79	fuel, furn.
Rock Elm	hard	high	strong	44	low	0.80	fuel
White Oaks	hard	high	strong	47	high	0.75	fuel, floors
Red Oaks	hard	very high	strong	44	low	0.79	fuel, floors
Sugar Maple	hard	high	strong	44	low	0.80	fuel, floors
Black Locust	hard	low	very strong (11,000–13,000 psi)	48	high	0.74	fuel, posts
Yellow Birch	hard	high	very strong	44	low	0.81	fuel, floors, furn.
White Ash (2nd growth)	hard	high	very strong	41	low	0.83	fuel, floors, furn.
Hickory, Shag	hard	very high	very strong	51	low	0.71	fuel, floors, furn.

[a] Stress at which timber will recover without any injury or permanent deformation.

[b] At 12 percent moisture content.

[c] Calculation for 12 percent moisture. Value *varies* greatly with moisture: variation is 43 percent for softwoods, and 53 percent for hardwoods (see USDA FPL handbook No. 72) for moisture ranging from 0–30 percent. Values given *per inch* of thickness in direction of heat flow; normal to grain.

Appendix C

Insulation Types*

Form and Type	R-Value Per Inch	Cost	Characteristics[3]
BLANKET AND BATT			
Fiberglass	3.2	low	non-combustible except for facing
(spun glass fibers)			difficult with irregular framing
Rock wool	3.4	low	non-combustible except for facing
(expanded slag)			difficult with irregular framing
LOOSE-FILL[1]			
Fiberglass attic	2.2	low	non-combustible
wall	3.3		good in irregular spaces
Rock wool attic	2.9	medium	non-combustible
wall	2.9		good in irregular spaces
Cellulose attic	3.7	low	combustible—specify "Class I, non-corrosive"
(paper fiber)	3.3		can be damaged by water
Perlite	2.5	high	non-combustible
(glass beads)	3.7		expensive
Vermiculite	2.4	high	non-combustible
(expanded mica)	3.0		expensive
RIGID FOAM BOARDS[2]			
Molded polystyrene	4.0	medium	combustible
("bead board")			permeable—do not use below grade
Extruded polystyrene	5.0	high	combustible
("Styrofoam")			impermeable—best below grade
Polyurethane/	6.0	high	combustible
Polyisocyanurate	7.2		used outside framing, protected by siding or sheathing
("Thermax," etc.)			

1) All loose-fill insulation must be installed at manufacturer's recommended densities as shown on bag to insure proper performance.
2) All rigid foams are combustible and must be covered with ½-inch drywall or equivalent 15-minute fire-rated material when used on interior.
3) Data taken from *An Assessment of Thermal Insulation Materials and Systems for Building Applications,* Brookhaven National Laboratory, June 1978, GPO Stock No. 061-000-00094-1.
Source: Cornerstone Energy Audit.

* Adapted from *Rodale's New Shelter* (Emmaus, Pennsylvania: Rodale Press, Inc., September, 1980).

Appendix D

Rafter Lengths*

Size	Spacing	Live Load	Groups A & B	Groups C & D
in.	in.	p.s.i.	ft. in.	ft. in.
2 x 4	16	20	9— 9	7—10
		40	7— 4	5—11
	24	20	8— 0	6— 5
		40	6— 0	4—10
2 x 6	16	20	14—11	12— 0
		40	11— 4	9— 1
	24	20	12— 4	9—11
		40	9— 4	7— 6
2 x 8	16	20	19— 8	15— 9
		40	15— 0	12— 0
	24	20	16— 4	13— 1
		40	12— 4	9—11
2 x 10	16	20	29— 6	19— 9
		40	18—10	15— 1
	24	20	24— 7	16— 5
		40	15— 7	12— 6

* National Building Code

Maximum allowable lengths for rafters that are sloped greater than 3 in 12. Length is the distance from the plate to the ridge.

Index

Other Storey Titles You Will Enjoy

BE YOUR OWN HOUSE CONTRACTOR: 3rd Edition. Carl Heldmann. How to save 25% without lifting a hammer, how to pick the right subcontractor, and more. 144 pages, 6x9, sample documents, charts, construction notes. Paperback. ISBN 0-88266-266-X.

BUILD YOUR OWN LOW-COST LOG HOME: Updated Edition. Roger Hard. A classic book that includes complete construction details, site selection, complete list of kit manufacturers. 208 pages, 8½x11, 100 illustrations, photos, tables, charts. Paperback. ISBN 0-88266-399-2.

HOMEMADE. Ken Braren and Roger Griffith. 101 easy-to-make projects for your garden, home, or farm. 176 pages, 8½x11, 150 drawings. Paperback. ISBN 0-88266-103-5.

LOW-COST POLE BUILDING CONSTRUCTION. Ralph Wolfe and D. Merrilees and E. Loveday. Learn all the basics in this one-of-a-kind book. Extensive tool lists and specific project materials list. 192 pages, 8½x11, 290 black and white photos, illustrations, plans. Paperback. ISBN 0-88266-170-1.

TIMBER FRAME CONSTRUCTION. Jack Sobon and Roger Schroeder. All about post-and-beam construction. How to design and build using post-and-beam methods. 208 pages, 8½x11, photos and detailed drawings. Paperback. ISBN 0-88266-365-8.

HOW TO BUILD SMALL BARNS & OUTBUILDINGS. Monte Burch. Step-by-step fundamentals of general construction. Over 20 projects offer complete plans and instructions. 288 pages, 8½x11, heavily illustrated, photos, detailed drawings, index. Paperback. ISBN 0-88266-773-4.

Photography Editor: PETER ENSENBERGER
Book Designer: MARY WINKELMAN VELGOS
Copy Editors: PK PERKIN McMAHON, EVELYN HOWELL
Book Editor: BOB ALBANO

Text and Photographs: *Arizona Highways* Contributors

Map: KEVIN KIBSEY

Library of Congress Control Number: 2003108599
ISBN-10: 1-932082-17-4
ISBN-13: 978-1-932082-17-3
Printing History: 1st & 2nd - 2004; 3rd & 4th - 2005; 5th - 2006; 6th - 2007.
Printed in China.

Published by the Book Division of *Arizona Highways* magazine, a monthly publication of the Arizona Department of Transportation, 2039 West Lewis Avenue, Phoenix, Arizona 85009.
Telephone: (602) 712-2200
Web site: www.arizonahighways.com

Publisher: WIN HOLDEN
Managing Editor: BOB ALBANO
Associate Editor: EVELYN HOWELL
Director of Photography: PETER ENSENBERGER
Production Director: MICHAEL BIANCHI
Production Assistants: ANNETTE PHARES, RONDA JOHNSON

ARIZONA HIGHWAYS
BOOKS

MOUNT HAYDEN, RIGHT, VIEWED FROM THE NORTH RIM'S POINT IMPERIAL, EMERGES FROM AN EARLY MORNING MIST.

THE GRAND CANYON

THE GRAND CANYON

THE DESERT VIEW WATCHTOWER

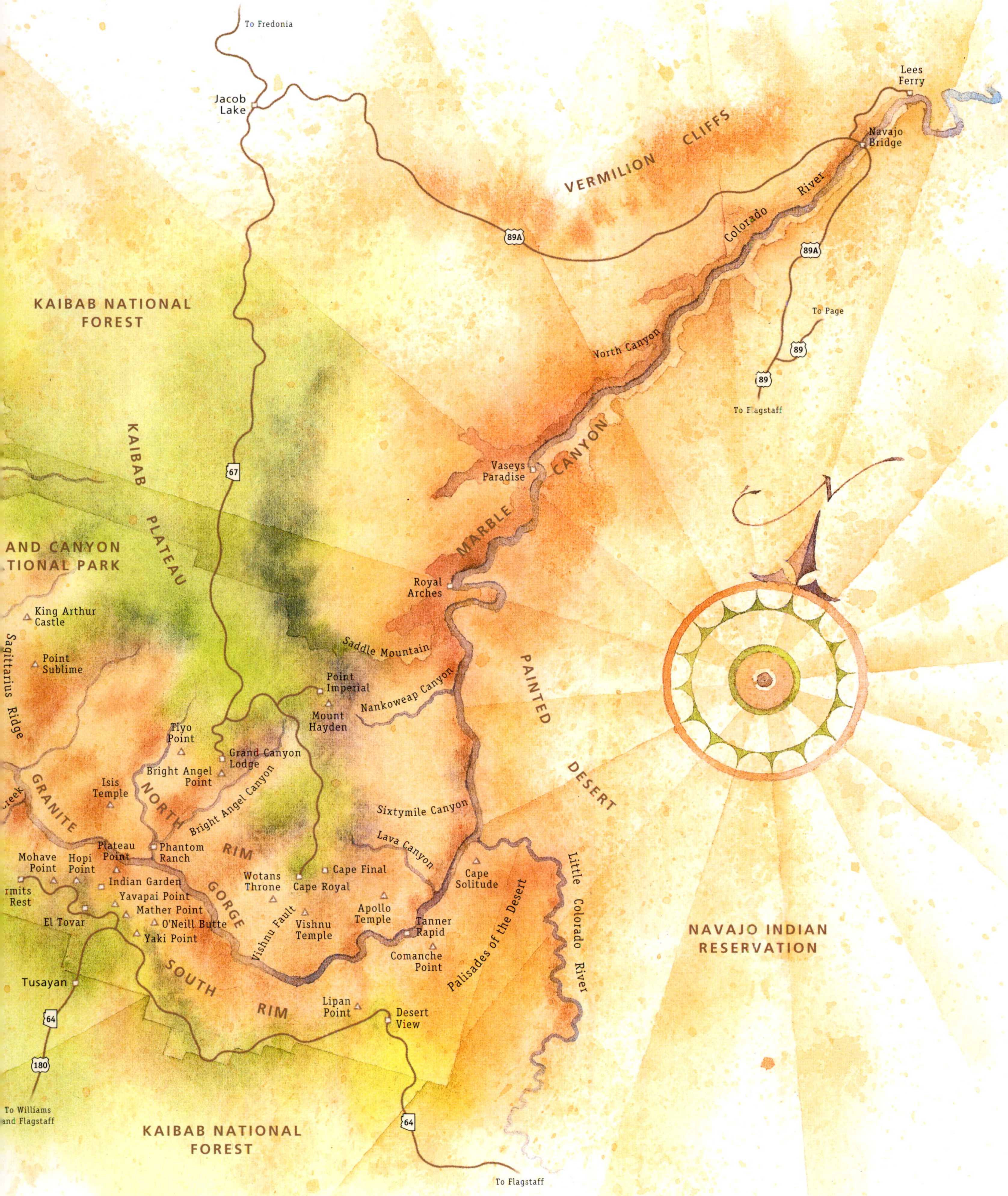

To Fredonia
Jacob Lake
Lees Ferry
Navajo Bridge
VERMILION CLIFFS
Colorado River
89A
89A
To Page
89
89
To Flagstaff
North Canyon
KAIBAB NATIONAL FOREST
KAIBAB PLATEAU
67
Vaseys Paradise
MARBLE CANYON
AND CANYON
TIONAL PARK
Royal Arches
King Arthur Castle
Point Sublime
Sagittarius Ridge
Saddle Mountain
Point Imperial
Nankoweap Canyon
Mount Hayden
PAINTED DESERT
Tiyo Point
Grand Canyon Lodge
Bright Angel Point
Bright Angel Canyon
Isis Temple
NORTH RIM
GRANITE
reek
Sixtymile Canyon
Lava Canyon
Plateau Point
Phantom Ranch
Mohave Point
Hopi Point
rmits Rest
Indian Garden
Yavapai Point
Mather Point
O'Neill Butte
Yaki Point
El Tovar
GORGE
Wotans Throne
Cape Royal
Cape Final
Vishnu Fault
Vishnu Temple
Apollo Temple
Tanner Rapid
Cape Solitude
Palisades of the Desert
Little Colorado River
Comanche Point
NAVAJO INDIAN RESERVATION
Tusayan
64
SOUTH RIM
Lipan Point
Desert View
180
To Williams and Flagstaff
KAIBAB NATIONAL FOREST
64
To Flagstaff

INTRODUCTION

After reading about the Grand Canyon's formation, you'll begin your photographic tour at Navajo Bridge, just a few miles south of Lees Ferry, the start of the Canyon, River Mile 0 in the language of river runners. From there, you'll move westward, or downriver, stopping at various viewpoints until you reach Hermits Rest. Then you'll swing over to the North Rim and again work your way westward to select viewpoints. As you look, keep this in mind: The Grand Canyon consists of some 600 major canyons and hundreds of smaller ones.

STEWART AITCHISON

The Earth has fallen apart. Instead of rolling highland hills feathered with pines, a complicated gash cuts through the planet. A metropolis of chasms plunges into shadow after shadow. At the chasms' bottom, a river tumbles, crashing back and forth through the planet's underbelly. If you walk to the edge of some points on one of the Grand Canyon's two rims, you might glimpse this river as it peeks around bends, but you will experience only fleeting looks. Elusive, overlapped by cliffs upon cliffs, the bottom of the Grand Canyon lies a mile below the South Rim. Even if you walk to the bottom, following trails that scale through palisades, and touch the cold river water with your hands, you still will experience only a glimpse as you stand on an isolated bank with towers of rock rising mile-high around you. The Colorado River goes through the Grand Canyon for nearly 300 miles.

The Grand Canyon is not one place, nor is it a single canyon. Six hundred major side canyons lace through it, aiming downward for the Colorado River. Branching from these side canyons, thousands of lesser canyons form a web-like maze, and thousands more branch from the lesser canyons until the terrain no longer looks like land, but instead like the endless reaching roots of a tree.

Many options allow you to witness the Grand Canyon — in person, movies, videos, books, and photographs taken during vacations. You may return to it year after year. Or, you might see it only once, standing mute at its rim. You may never go there in person, knowing it only from images and stories. However you witness the Grand Canyon, it will leave a mark in your imagination.

It did so on García Lópes de Cárdenas. In the fall of 1540, he became the first Spaniard to reach the Grand Canyon. A conquistador, Cárdenas was led by Hopis through the rising forests southeast of the Grand Canyon. They had told him that ahead was a great division within the earth, an unbridged gash that would stymie his travels. But even having heard these stories, like people coming here today, he could scarcely imagine what this place might be. He rode through thickets of piñon pine and juniper trees, hot in his armor, body swaying rhythmically to the motions of his horse, head lulled by days of travel. The land seemed to go on forever. Suddenly, he could see sky through the trees ahead — sky reaching all the way to the ground. As he rode out of the forest, he came to a great opening, an absence, as if the world had been eaten. There, he could feel his own breathing. Every movement he made in his saddle leather seemed deafening.

As others from his train of horses arrived beside him at the unexpected edge of the Grand Canyon, what were his first words to them? Did he gaze mutely for a long time? Did he utter the name of God and shake his head?

You will know in your soul what Cárdenas said when you come to the Grand Canyon. You will stand at one of its rims, forgetting all that you have heard of this place, and new words of awe will fall from your mouth.

MAPLE TREES, LEFT, STAGE AN OCTOBER COLOR SHOW IN THE TRANSEPT, GENERALLY CONSIDERED ONE OF THE KAIBAB REGION'S MOST PICTURESQUE GORGES. HIKERS, ABOVE, TRAVERSE THE TANNER TRAIL IN THE CANYON'S EASTERN AREA.

RANDY PRENTICE

SCIENCE OF THE CANYON

Since Maj. John Wesley Powell, an American Civil War veteran, first led a group of explorers through the Grand Canyon in 1869, scientists have studied it and formed theories on its origins. They still are revising and forming theories. They describe uplifts that began some 65 million years ago, at first furiously and then slowing about 30 million years ago. Theories tell of the river cutting through the terrain, of floods depositing material ranging from sand to boulders. Now it is thought that most of the Canyon was created in the last 5-6 million years. In short, the river did it.

GARY LADD

Every landscape on this planet poses an unlikelihood, a coming together of diverse and ancient elements to form a single moment, a place that will soon be gone. The Grand Canyon seems permanent with its colossal cliffs and temples — buttes with the grandeur of temple spires — that gathered around each other, more everlasting, certainly, than our quick lives and even quicker visits. But permanence does not explain the Grand Canyon's existence. Change does.

This deep cleft called the Grand Canyon came to be after the land beneath it lifted up. Because of the simple fact that water runs downhill, the river riding across the surface of the rising land refused to go up with it. The river cut its way down through the land. Imagine holding a hot butter knife. Now lift a stick of butter through it. The butter is the land. The knife is the river.

Of course, that is too simple. This land melts and cuts open at completely different rates depending on what kind of rock is being lifted. And some places are not lifting at all, but are falling. There is not only one knife, either. Over tens of millions of years, different rivers have probably come through this region, starting in different places, heading toward different destinations. The Colorado River happens to be the one here now.

If all that was going on was a single river cutting a solitary gash into the earth, then the Grand Canyon would be nothing but a slice. Instead, thousands of lesser canyons riddle it as they stretch down toward the slash where the Colorado River flows. These other canyons exist because, again, water flows downhill. The low gorge of the Colorado River has created a sink in the landscape. Streams and waterfalls and thunderstorm floods fall into this sink; as they fall, they carve the surrounding land, etching out buttes and canyons and mesas.

The forces that created the Grand Canyon also riddled much of the West with chasms and mountains and towers. The earth is buckling, lifting in some places but falling in others. Rivers cut through the lifting back of the continent. Sand dunes gather in the low basins between mountains. Floods carry boulders and throw them across the countryside. This entire landscape, the Grand Canyon and everything around it, results from the forces of a tumultuous planet.

Tiers of rock visible in the Grand Canyon first formed from instability. The Earth's surface rolled like someone trying to get to sleep, and, like a great storm, environments shifted over hundreds of millions of years — glaciers upon rain forests upon deserts upon deep oceans. All in this one location.

Each rock layer represents some change in the world. The gray, limestone caprock on the Grand Canyon is the most recent formation: the low, hardened floor of a sea that was here 260 million years ago. Below that, layer by layer, the Grand Canyon reveals what happened in the past. The Hermit shale, its rock as deeply red as a Japanese maple tree in the fall, is what is left of reptile-filled swamps and fern jungles. Just above that stand the white walls of the Coconino cliffs, the cemented remnants of coastal dunes when this part of the world was once the sea-edge of a continent.

Every past environment left a mark. It is as if the floor of a house had never been cleaned: Coats of shoe-mud are covered by spilled kitchen food, covered by furniture dust, by mud again, by pieces of paper, by tracked-in asphalt. If you cut down into the thick veneer of this filthy floor, you would be able to see every event, each change that happened in the house. You might be able to differentiate winter-snow mud from summer-rain mud. You could analyze the kitchen refuse to

RUNOFF IN NORTH CANYON, LEFT, EVENTUALLY FLOWS INTO MARBLE CANYON, NAMED FOR ITS POLISHED CLIFFS REACHING AS MUCH AS 2,500 FEET HIGH AND UP TO 4 MILES WIDE.
A LATE AFTERNOON RAINBOW, ABOVE, DECORATES THE MATHER POINT AREA NEAR GRAND CANYON VILLAGE.

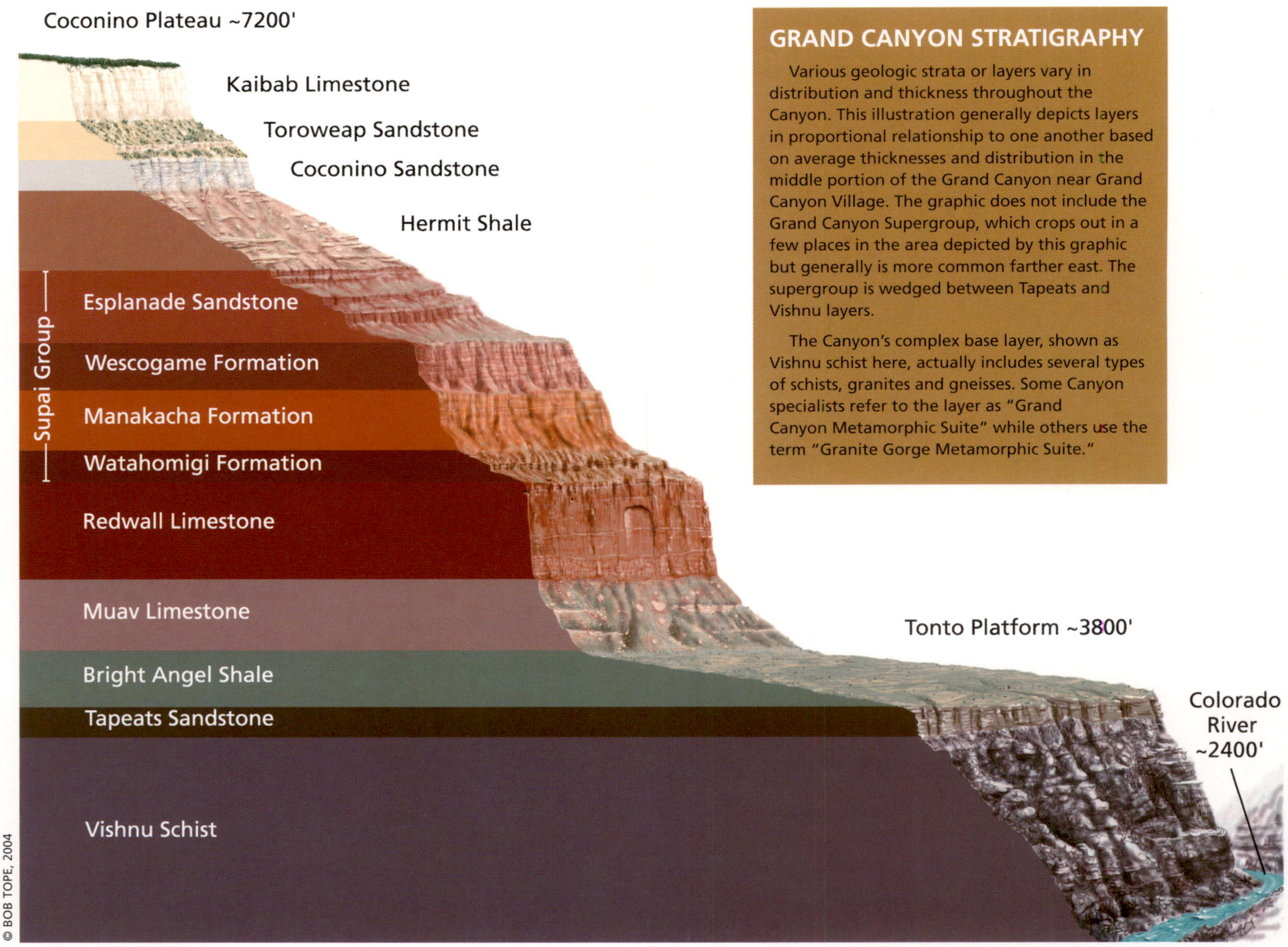

GRAND CANYON STRATIGRAPHY

Various geologic strata or layers vary in distribution and thickness throughout the Canyon. This illustration generally depicts layers in proportional relationship to one another based on average thicknesses and distribution in the middle portion of the Grand Canyon near Grand Canyon Village. The graphic does not include the Grand Canyon Supergroup, which crops out in a few places in the area depicted by this graphic but generally is more common farther east. The supergroup is wedged between Tapeats and Vishnu layers.

The Canyon's complex base layer, shown as Vishnu schist here, actually includes several types of schists, granites and gneisses. Some Canyon specialists refer to the layer as "Grand Canyon Metamorphic Suite" while others use the term "Granite Gorge Metamorphic Suite."

determine what kinds of diets came and went in the house. The cut you made in the floor would be the Grand Canyon. The layers would be a record of the world. This uncleaned house of geology dates back nearly 2 billion years.

As a side note, housecleaning has occurred in the Grand Canyon, leaving gaps, unknown layers of rock swept away by great floods and, to a lesser degree, ancient winds. In fact, the cleaning continues. Instead of material collecting, material erodes. The river carves downward, collapsing the land, rolling boulders down into rocks, wearing rock into sand. Once, before dams were built, this erosion flowed all the way to the Sea of Cortés.

Current erosion has been going on for only a few million years, a brief moment in the history of this landscape, a quick sweep of a broom. Possibly, the land will stop rising someday, maybe in 10 million or 15 million years. The swelling earth will quiet and again recede. The Sea of Cortés will back up several hundred miles to the Grand Canyon as this national park shrinks below sea level. Such occurrences will turn the Grand Canyon into a river delta, filled with mud. That mud eventually will again form a new rock, a record embedded into the future Earth where geologists may postulate about an eminent canyon that once lay carved here.

This is the future, and the future is hard to

tell. The one certain thing is that the Grand Canyon will be gone. If it is not packed to its rim with river mud, then the river will carve it away into nothing. The cliffs will cave in, as they do now to some extent every day. Flash floods teeming down side canyons will tear the Canyon apart.

The Grand Canyon is not here for long. Millions of years, and that is all.

If you are confused by all of this time, don't feel bad. We humans only live for 70, maybe 100, years. To grasp millions, much less hundreds of millions, of years exceeds our ability. Perhaps that is why people come here. The Grand Canyon reminds us that much lies beyond our vision. We can look into it and witness time in a way that our day-to-day lives do not allow. We can see in this place that a million years is as tangible as a single day. When we walk into the Grand Canyon, every foot down takes us 30,000 years into the past, even millions of years into the past depending on the particular rock where our feet land. Every step takes us into new environments — ancient environments preserved in a fossilized earth and modern environments ranging from forests at the North Rim and the South Rim to hot desert at the Canyon's floor.

The classic way of appreciating the variety of today's Grand Canyon environment is to realize that walking from a rim to the river, as far as ecosystems are concerned, you are walking from Canada to Mexico. This is just a useful simile, walking from forest to desert. In reality, viewing the Grand Canyon environment is much more rich and diverse than going from one country to the next. To say the bottom is desert oversimplifies the nature of that terrain. Along its course, the Grand Canyon cuts into three of the four North American deserts. It begins in the West's Great Basin Desert, falls into the Sonoran Desert, and finally it emerges at the west end into the Mojave Desert. Each desert has different plants, different rainfall, different temperatures.

The intricate geography of the Grand Canyon, all of its nooks and crannies, creates mazes of overlapping environments — Canadas within Mexicos within Canadas. At the desert floor there grow columbine flowers and even orchids around cascading springs. Dry, south-facing slopes look like deserts, yet when you turn the corner, you find yourself in a dense, north-facing stand of conifers.

Every turn in the Grand Canyon holds some dramatic change. Every step crosses thousands and millions of years and worlds of environments. But maybe this knowledge is unnecessary. There are people who will tell you that you cannot go to an art museum without having taken classes in art, that if you do not know the detailed history, you cannot appreciate what you are seeing. People might say the same about the Grand Canyon, that if you do not know the details of science, the Canyon holds no meaning.

Not true.

Come to the Grand Canyon with no knowledge of geology or environments. Stand at its edge and see a vastness that reaches beyond anything you could gather from a book. Remember, the Canyon projects a different look with every glance. Photographs do not hold it. Words only blow across it like wind. Forget what you have heard or read about this place, which really is thousands of places. Come to it with your eyes open.

Whatever comes to you in that first moment marks only the beginning.

Navajo Bridge

LARRY ULRICH

River Mile 4.5. Upriver lies Lees Ferry, River Mile 0, the start of the Grand Canyon. Look downriver, and you're peering toward Marble Canyon, so named because its walls are so polished.

Stretched 470 feet above the Colorado River, twin bridges join one side of the Grand Canyon to the other just downriver from the Canyon's head. This region, known as Marble Canyon, is the only place where a car can cross the Grand Canyon.

Look closely at the engineering of the new and of the original Navajo Bridge, at the cadence of bolts and steel-arch suspension beams. In them, human ingenuity is visible, a work of artistry and mathematics. Look also at the natural surroundings, shadows marking fracture lines, or geologic faults, crossing one direction and the next, making room for a river and erecting walls to keep it in.

The Canyon represents a feat of natural engineering. Pieces of earth have lifted skyward. Bedrock has buckled into great, continental fissures — not randomly like broken glass, but along exact crystalline planes within the rock. The Colorado River carves down through this lifting earth, building the Grand Canyon by cutting into the pre-existing angles of faults and fractures.

The first Navajo Bridge, which now only pedestrians use, was completed in 1929. Until the Arizona Highway Department built it, only a ferry service started in 1871 by John Lee and his wife, Emma, provided a river crossing for a 600-mile stretch. Designed to look like the original, the second, wider bridge was constructed in 1995 at a cost of $15 million.

However, dollars do not measure the currency of Earth; time and distance do. The Grand Canyon exposes nearly 2 billion years worth of rock. Several million years of uplifting have created canyons and buttes. A 2,340-mile river crosses through here. At this point, human and Earth technology meet, bridge and canyon, dollars and time, cars and stone.

NAVAJO BRIDGE, ABOVE, CROSSES THE COLORADO RIVER A FEW MILES SOUTH OF LEES FERRY. FLOWING PARALLEL TO THE VERMILION CLIFFS ALONG THE HORIZON, THE COLORADO, RIGHT, CHURNS INTO WHITEWATER AT BADGER CREEK RAPID ABOUT 8 MILES DOWNRIVER FROM LEES FERRY.

GARY LADD

LARRY ULRICH

TOM TILL

GARY LADD

ABOVE, LEFT TO RIGHT: NORTH CANYON MEETS THE RIVER IN MARBLE CANYON; MARBLE CANYON'S COLORS BOUNCE OFF THE RIVER; RAFTS APPROACH NAVAJO BRIDGE (NOT PICTURED) WITH THE VERMILION CLIFFS IN THE BACKGROUND.
AN EPHEMERAL WATERFALL, BELOW, FLOWS NEAR NAVAJO BRIDGE.
SOAP CREEK RAPID, FAR RIGHT, DROPS SOME 16 FEET ABOUT 11 MILES DOWNRIVER FROM LEES FERRY.

RANDY PRENTICE

Cape Solitude

River Mile 62. Just down from the confluence of the Colorado and Little Colorado rivers, the cape rests some 3,500 feet above the river. Serenity, the wind, and the lower end of Marble Canyon dominate. Nearby is the mound from which Hopi ancestors emerged from the Underworld.

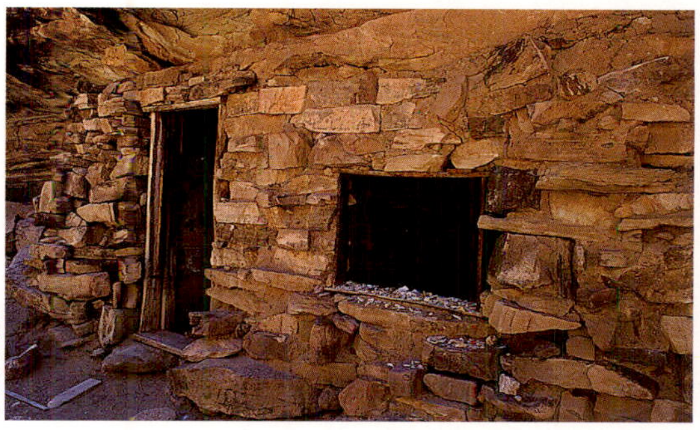
GARY LADD

Marble Canyon winds down into the Grand Canyon, taking into it the topaz blue waters of the Little Colorado River, waters that turn tomato-red with mud during storms. Here, some 60 miles below Navajo Bridge, the Grand Canyon reveals its high terraces — steps of cliff-and-slope for which the area is known.

Pouring in from the side, the Little Colorado amounts to another Grand Canyon, a long, narrow defile that works its way from the high country of central Arizona and western New Mexico and across a desert. Although the Colorado River now runs south and then west from here, some geologists once suspected that the Colorado River at one time headed southeast from this point. The Colorado's older passage would have carried it along the path of the Little Colorado, until the land changed and deflected the big river, turning it back to where it now lies deep within the Grand Canyon. Today, however, not many people subscribe to that theory.

The calm-water blue of the Little Colorado comes from minerals in a series of springs not far upstream from its confluence with the Colorado. If you were to walk up this tributary gorge, you would find deep patches of quicksand alongside the blue water. Eventually, you would arrive at the springs, from which sulfur-smelling water pours.

According to traditional Hopi belief, this is the area from which the Hopis — who now live not far from here — emerged from the Underworld onto the face of this world. Archaeologists have researched this story, finding for themselves that humans have lived in the Grand Canyon for at least the past 10,000 years. The last thousand years or so of occupation were primarily by people who farmed the high rims and canyon deltas. These people elaborately painted their pottery in styles that eventually evolved into what can be seen on modern Hopi ceramics, proof for archaeologists that at least some of the Hopi came from the Grand Canyon.

After the Hopi ancestors, and after the brief passage of conquistadors, came miners and explorers. In the 1800s, people of European origin erected buildings here and there, fleeting evidence of human lives in this massive landscape. After the prospectors and the government-sent surveyors, came river runners and rim-bound tourists and backpackers wandering along river shores and through canyon shadows.

Still, signs of humans are scarce. The Grand Canyon has only 33 miles of maintained trails. Rafts and dories slip down the river, quickly passing out of sight. Visitors centers perch at only a few high points. The rest consists of tumbled and eroding boulders, water, wind, plants, and animals.

The site called Cape Solitude stands in the eastern part of the Grand Canyon, 3,500 feet above the Little Colorado River. From the nearest road, the long walk to here passes through scrub brush and open, barren soils of the desert. From the cape, you can look across the river to Temple and Chuar buttes and to the sprawling arms and open delta of Lava Canyon.

From Cape Solitude, you will see that you have walked to only a single point among hundreds. This one has a name and a trail coming to it, but the majority of sites in the Canyon remain nameless, solitary monuments of rock.

NOTE THE COLORS, LEFT, AT THE CONFLUENCE OF THE COLORADO AND LITTLE COLORADO RIVERS.
ABOVE, A CABIN WITH ANCIENT ORIGINS REMAINS A FIXTURE IN THE CANYON. IN THE LATE 1800s PROSPECTER BEN BEAMER MODIFIED A PUEBLOAN STRUCTURE TO MAKE THE CABIN.

GARY LADD

THE LITTLE COLORADO RIVER, ABOVE, FORMED A MIGHTY GORGE. BELOW, RAFTERS CONTEMPLATE THE CONJOINED RIVERS.
FAR RIGHT, INTRICATELY ERODED SANDSTONE DECORATES THE COLORADO RIVER'S BANK, WITH DESERT VIEW WATCHTOWER BUT A SPECK ON THE HORIZON AT LEFT.

LARRY ULRICH

COLIN McKAY

Desert View, Lipan Point

River Mile 68-75. Here's the eastern entrance to the South Rim. Looking northeast from Desert View, you'll see the Painted Desert meet the Grand Canyon at the rugged Palisades of the Desert.

GEORGE STOCKING

Desert View and Lipan Point mark a dramatic change in the Grand Canyon. Until here, the Colorado River has flowed mostly south and southwest through Marble Canyon. More or less, the river has been following this course since leaving its headwaters in the ragged alpine peaks of Colorado. Suddenly, the Canyon bends and the river jumps westward, utterly out of character. It even turns north for a time, bursting through the hard, raised stone of the Kaibab Plateau. This is the turning point.

High above the river stands Desert View Watchtower, an enchanting amalgam of architecture and artwork. Architect Mary Elizabeth Jane Colter, whose South Rim career began with her designing the famous Hopi House built in 1905, designed the tower, which was completed in 1932.

Its windows frame various scenes of the metamorphosing Grand Canyon. To the north, the tower looks across the open stem of Marble Canyon. To the northeast, the Palisades of the Desert stretch in a dramatic line of cliffs separating the Painted Desert from the Grand Canyon. To the west, the tower looks into the growing Grand Canyon and its stone formations called temples. From this view, the rapids-spangled water vanishes into shadows.

From nearby Lipan Point (originally named Lincoln Point but renamed after a Texas Apache tribe) the river far below makes an S-curve into red shales. Dark, angular walls of schist are exhumed from the Canyon floor, revealing bright slashes of granite. From there the river aims into a tightly sealed interior of gorges, and the Grand Canyon becomes what it is famous for: a yawning pit of chasms.

A few miles west of Desert View, the Tusayan Ruin and Museum help tell the story of Indian life at the Grand Canyon some 800 years ago.

LEFT, TANNER TRAIL DROPS FROM LIPAN POINT TOWARD THE COLORADO RIVER. ABOVE, DESERT VIEW WATCHTOWER OVERLOOKS THE RIVER AND THE PALISADES OF THE DESERT.

ELIAS BUTLER

LOOKING NORTH FROM DESERT VIEW, LEFT, VIEWERS SEE THE PALISADES SEPARATING THE PAINTED DESERT AND THE GRAND CANYON. ABOVE, THE RIVER BENDS AT UNKAR RAPID. BELOW, THE COLORADO FLOWS CALMLY JUST UPRIVER FROM TANNER RAPID.

COLIN McKAY

Moran Point

GARY LADD

River Mile 76-79. The river appears glassy despite two major rapids. Lying unseen from here, in the folds of side canyons, are verdant areas. Coronado Butte lies just west of Moran Point.

JACK DYKINGA

The white roof of Kaibab limestone — a finely grained, thick-bedded sedimentary rock — breaks apart. Towers of rock stand from each other, leaving isolated points all over this edge of the eastern Grand Canyon. Below, the side canyons gather like threads bound together into a knot. What is down there?

From here, the land below looks absolutely dry, a desert of chalkboard cliffs and dusty canyon floors. However, down there, invisible from here, deep springs flow and grottoes of ferns create verdant settings.

The side canyon just below Moran Point falls into precipices of limestone where water seeps out, creating a stream. The stream plunges between narrow walls, cutting into saffron tiles of shale. Cottonwood trees rise like palms toward the sunlight, shading beneath them cold pools of water and strands of waterfalls. Eventually, the stream sinks into sand and the clutter of fallen boulders, no longer visible at the surface. It pours generally unseen into the river at Hance Rapid, about 72 river miles below Navajo Bridge.

What cannot be heard from Moran Point is the sound emanating from where the stream joins with the river, where boulders thrust against each other and the water breaks over them with a baritone rumble and a hissing roar. Each Grand Canyon rapid carries a different sound. Some, like Hance, are an orchestration of tones as the river bounds through waves and troughs. Others, like Lava Rapid far downstream, emit a single bass rumble where the river cascades over an edge.

Each Grand Canyon rapid aligns with a side canyon. Flash floods flush boulders and debris down side canyons into the river. The debris clogs the river, forming rapids. Hance is born from Red Canyon. Sockdolager, the next downstream, comes from Hance Creek, named for the genial John Hance, who insisted he was the South Rim's first Anglo settler.

Each rapid has its own demands. Boulders align like puzzles, giving each rapid a different solution, a different way through. Hance is a long maze, and running it in a raft or boat is like coming through a pinball machine. Sockdolager tightens down to a single punch near its end.

For the most part, though, the river through the Grand Canyon is not loud with the clash and strike of these roller coasters. Stretches between rapids are long, punctuated with ornate sounds of upwellings and harmless whirlpools.

The sound lulls river travelers, but ahead there always comes the low growl. Boatmen who have left oars sagging suddenly grip them and lean forward. Another side canyon appears and another rapid explodes into view.

MORAN POINT, ABOVE, OVERLOOKS THE DISTANT COLORADO RIVER AT DAWN.
BELOW, CORONADO BUTTE TAKES ITS NAME FROM A SPANISH EXPLORER WHO PROBABLY NEVER SAW THE CANYON.
FAR LEFT, A GEOLOGIST EXAMINES A MAGMA DIKE ABOVE HANCE RAPID.

LARRY ULRICH

COLIN McKAY

FAR RIGHT, RAFTERS APPROACH HANCE RAPID IN CALM WATER.
BELOW, BOATERS IN A DORY NEGOTIATE HANCE RAPID'S 30-FOOT DROP.
ABOVE, A LITTLE MORE THAN A MILE LATER, SOCKDOLAGER RAPID CONNOTES A BOXER'S KNOCKOUT PUNCH.

LARRY ULRICH

Grandview Point

RICHARD L. DANLEY

River Mile 78-82. Even a quarter-mile hike down Grandview Trail brings you to vistas not visible from the point. Below, lies Horseshoe Mesa, and the river flows in the upper Granite Gorge.

This is the last spot to look back. The Desert View Watchtower stands barely visible on the eastern horizon as cloud shadows graze the far cliffs of the Palisades of the Desert. Buttes, called temples because those who named them thought they resembled temples, dominate the vistas now as the open bends of Marble Canyon are swallowed by overlapping gorge walls. The tiers of Grand Canyon geology now stand out like pages in a book. Hard layers of sandstones and limestones form stark-edged cliffs. Between them, horizontal slopes consist of softer, crumbling materials like shales and mudstones.

At the bottom, an area difficult to see from here, the coal-black bedrock of Vishnu schist is exposed, a rock nearly 2 billion years old. It ranks by far as the Canyon's hardest formation, a solid basement through which the river cuts. The schist polishes smooth down in the floors of canyons, stricken by occasional white or pink bands of granite that stand out like bolts of lightning. This is Earth's bottom, the lowest, oldest point of rock exposed in the Grand Canyon — in fact, the oldest formation exposed anywhere in the Southwest.

From the South Rim at Grandview Point, all of this remains invisible, hidden deep in the Canyon's folds. And yet, Grandview Point, at about 7,400 feet above sea level, is among the highest points of the Coconino Plateau, on which much of the South Rim rests.

Near Grandview Point a hotel once housed Canyon visitors. From the point, miner Peter Berry built a 3-mile trail, dropping nearly 3,500 feet to Horseshoe Mesa, where he extracted copper. Hikers descending just a short distance on the trail readily can witness the dramatic change in the Canyon's nature as they look east and then west.

BOB AND SUZANNE CLEMENZ

ABOVE, RUINS OF COPPER MINER PETER BERRY'S CABIN REMAIN ON HORSESHOE MESA BELOW GRANDVIEW POINT.
RIGHT, GUESTS OF THE NOW-GONE GRAND VIEW HOTEL ENJOYED THIS VISTA.

BERNADETTE HEATH

ROCKS FROM DIFFERENT AGES FORM UPPER GRANITE GORGE, LEFT, AND VISHNU CANYON, BELOW.
REDBUD TREES, ABOVE, EMBELLISH GRAPEVINE CANYON NORTHWEST OF GRANDVIEW POINT.

GARY LADD

Shoshone Point

TOM DANIELSEN

River Mile 78-87. With binoculars, look north for buildings, including Grand Canyon Lodge, on the North Rim, and Zoroaster and Brahma temples below them. Vishnu Temple lies northeast.

CHARLES LAWSEN

In the beginning, the Grand Canyon had no name. Then, people came. Some Spaniards called the river *Rio Colorado* ("reddish river") because it ran red. Maps became catalogs of names, and the Grand Canyon took on its own complicated lexicon.

Geography here seems organized by subject matter. Names of Southwestern tribes are strung along the South Rim: Papago Point, Zuni Point, Shoshone, Yaki, Yavapai, Maricopa, Hopi, Mohave, Pima, Yuma, Cocopa, Mimbreno, Mescalero, Jicarilla, Piute, Walpai, Havasupai. At the far end of these points comes a cluster of Mesoamerican subjects: Toltec Point, Aztec Amphitheater, Point Quetzal. Just before them lie the gem canyons: Agate, Sapphire, Turquoise, Emerald, Ruby, and Serpentine.

Across, on the north side, lie the Middle Eastern and sub-Asian relations, named in 1880 by a geologist interested in Eastern religions: Ottoman Amphitheater, Zoroaster Temple, Buddha Cloister, Shiva Temple. Nearby is the Roman section: the temples of Jupiter, Venus, and Apollo.

Many of these labels were bestowed by early explorers such as John Wesley Powell, who led the first recorded run of the river in 1869. Whatever Powell's mood was on any particular day is reflected in the names eventually printed on the map (the great Powell Plateau is surrounded by the names of the Canyon's various surveyors: Wheeler Point, Walthenberg Canyon, Dutton Point, Ives Point, Beale Point, Thompson Point, Newberry Point, and finally, Powell Spring).

Even the colorful wooden dories that boatmen use to run the river carry names of the land painted on their sides: Nipomo Dunes, Hidden Passage, Music Temple, Dark Canyon. (The tradition for Grand Canyon dories calls for them to be named after places now covered by Lake Powell.)

Detailed maps of the Grand Canyon are so crowded with words that it seems obsessive, as if by naming places we have planted flags atop them. We claim them, then move quickly to a new spot, inventing a new name, and scurry off for more. Over time, these names have become embedded, fusing into the landscape. Perhaps Confucius Temple standing beside Tuna Creek aptly expressed the bizarre juxtaposition of a colossal, sunlit butte overshadowing an impenetrably deep and shadowed canyon.

Someday, though, the maps will become old. On the time scales suggested by the Grand Canyon, where a thousand years is only the tick of a second hand, these names will vanish. Again, the Grand Canyon will have no name.

A POINTED BUTTE NAMED VISHNU TEMPLE, ABOVE, LIES ON THE NORTH SIDE OF THE RIVER NORTHEAST OF SHOSHONE POINT. TWO OTHER TEMPLES, OR ROCK FORMATIONS, ARE SEEN BELOW: ZOROASTER (LOWER) AND BRAHMA. ZOROASTER, NAMED FOR A PERSIAN HOLY MAN, ALSO IS SEEN AT FAR LEFT.

TOM DANIELSEN

WHILE ZOROASTER TEMPLE IS VISIBLE FROM SHOSHONE POINT, THE NEARBY CLEAR CREEK WATERFALL, ABOVE, IS NOT.
RIGHT, BOATERS TAKE A BREAK IN GRANITE GORGE.

CHATTAHOOCHEE

Yaki Point

River Mile 77-89. The panorama embraces, east to west, Vishnu Temple and the flatter Wotans Throne. Cedar Ridge (a 4.5-mile roundtrip hike) yields dramatic views up and down the Canyon.

MARK LARSON

JEFF SNYDER

From Yaki Point, one of the most substantial trails in the Grand Canyon, the South Kaibab Trail winds down ridges and cliffs to be absorbed into the canyons below. This popular trail ranks second only to the Bright Angel Trail in the number of people who hike it.

Before the National Park Service paid Coconino County $100,000 in 1928 for the Bright Angel Trail, it was actually a toll road. Before that, when the county would not sell the trail to the park, the National Park Service built the South Kaibab as a new trail to the river from Yaki Point.

Standing at the South Rim's edge, you can see the South Kaibab Trail far beneath you, hints of a human-made line slashing back and forth like some ancient passage into another world. Barely visible from here, people walking this trail seem no larger than flies that shrink to ant size that then turn to nothing as the trail becomes too small to see.

In 1907, a cable car was strung across the Colorado River below Yaki Point. Before that, boats were used to carry people across, but there was no regular ferry service. In 1921, a footbridge was built where the South Kaibab Trail touches the river. Much of the construction materials for the bridge were hauled down by mules to make what then was the Canyon's only foot crossing of the Colorado River. Laborers, mostly Havasupai who live in the Canyon, carried the cables.

Later, a second bridge — often called the Silver Bridge — was built a little way downstream, primarily to hold a pipe carrying water from Roaring Springs on the North Rim to Grand Canyon Village, but a walkway was built on the structure.

When you look out from Yaki Point and see this hand-cut trail below, remember: We are not the first to come here and certainly not the last.

LEFT, SEEN FROM YAKI POINT, THESE FORMATIONS ARE NAMED FOR WOTAN, NORSE GOD OF WAR, AND THE HINDU DEITY VISHNU. ABOVE, PHANTOM RANCH ON THE CANYON'S FLOOR IS A COMPLEX OF CABINS AND OTHER BUILDINGS.

WILLIAM STONE

ABOVE, RAFTERS (FAR RIGHT) REST ON A BEACH UPRIVER FROM THE BRIGHT ANGEL SUSPENSION BRIDGE, OR "SILVER" BRIDGE, VISIBLE FROM THE SOUTH KAIBAB TRAIL BELOW YAKI POINT.
BELOW, COTTONWOODS STRUNG ALONG BRIGHT ANGEL CREEK SHELTER PHANTOM RANCH.
RIGHT, LOOKING DOWNRIVER TOWARD THE KAIBAB AND BRIGHT ANGEL SUSPENSION BRIDGES.

ELIAS BUTLER

Mather, Yavapai Points

River Mile 87-90. A short walk from Mather Point, or reachable via the park's shuttle, the Canyon View Information Plaza features an outdoor exhibit centering on a giant relief map. The complex's exhibits, staff, and programs bring the Canyon and its activities into focus for visitors.

GARY LADD

Mather Point's roomy parking lot makes it easy for visitors to park and stroll to the edge for a broad view of incised, creased, and corrugated earth spread to the horizon in waves of color. If you become impatient with the throng, go a little beyond to Yavapai Point for much of the same view.

The window-encased observation building at Yavapai Point hangs at the Rim like a glass-and-wood capsule protecting visitors from wind and vertigo. For being so close to a precipice, there is a strange sense of safety and comfort in here. But, again, there is no wind inside. The varied scents of the Grand Canyon — pine scents blowing from the forests and the astringent tang of juniper lifting from below — mostly are absent. The quickly shifting temperatures of winter breezes, of summer clouds passing by, are sealed outside, allowed in only when the heavy wooden doors swing open.

Out there, beyond the glass, is a tactile Grand Canyon, a place both rewarding and dangerous. Each year more than 400 search-and-rescue missions take place in the Canyon. More than $100,000 is spent each year for emergency response in the Grand Canyon, not including the astronomical cost of helicopter flights. That place down there has a long history of dramatic rescues and mysterious disappearances.

Perhaps the most quizzical of people-gone-without-a-trace stories is that of Glen and Bessie Hyde, who ran the river on their honeymoon in November 1928. This part of the Grand Canyon was where they were last seen, where famous Grand Canyon photographer Emery Kolb tried to talk them out of going any farther. Their skills were limited, their gear inadequate, their prospects bleak. Since then, many theories speculate on what happened to them after they left Kolb and the country beneath Yavapai Point. Witnesses said that Glen suppressed Bessie with his inflated ego and that Bessie feared running the river. Some suggest that Bessie ended up killing her recklessly arrogant husband to escape him and the river. Their boat was found upright and undamaged 150 miles downstream from here. No other clues were left besides a late-November etching of their names and the date on a piece of wood 11 miles upstream of their abandoned boat.

Bessie Hyde was the first woman to run the Colorado River through the Grand Canyon, even if she never made it out the other end. The next women to dare were two University of Michigan botanists, Elzada Clover and Lois Jotter, who came 9 years after the Hydes. With such a poor survival rate set already, many approached these two and begged them to rethink their 600-mile journey from Green River, Utah, to the end of the Grand Canyon.

But these botanists could not be swayed from their mission to collect plant specimens. By turns exasperated and amused at the media hoopla surrounding their trip, Clover and Jotter launched their expedition of three boats and four men on June 20, 1938. Their river ride hit its rough spots, but on Aug. 1, the women emerged from the Canyon's mouth with their crew, making history.

For the people peering from the Yavapai Observation Station, distance obscures these stories. It is impossible to touch the reality of the Grand Canyon through the glass, difficult to fully comprehend its stories. For some, like the Hydes, it may have been better if they had viewed the Canyon from this safe distance. For others, like Elzada Clover and Lois Jotter, it would have been a mistake to remain here.

You choose.

LEFT, THE SETTING SUN HIGHLIGHTS THE TOP OF VISHNU TEMPLE (TOP CENTER) IN THIS VIEW EASTWARD FROM YAVAPAI POINT.
ABOVE, LIGHTNING STRIKES NEAR O'NEILL BUTTE BELOW YAVAPAI POINT.

LARRY LINDAHL

STARTING WEST OF YAVAPAI POINT, THE BRIGHT ANGEL TRAIL TAKES HIKERS, ABOVE, TO INDIAN GARDEN, BELOW. RIGHT, FROM PLATEAU POINT NORTH OF INDIAN GARDEN, YOU CAN SEE PIPE CREEK RAPID.

RICHARD L. DANLEY

COLIN McKAY

Powell, Hopi, Mohave Points

River Mile 87-94. Lying just west of Grand Canyon Village, these points are favored by sunset watchers. Looking east from here gives a different perspective on formations seen from points farther east.

GARY LADD

This cluster of points juts farther out into the Grand Canyon than most others, like the prow of a ship cutting into deep water. Powell Point is named after John Wesley Powell, who led the first expedition through the Canyon. There is a memorial there commemorating his feat. And, from the point you can see the Orphan Mine, one of the last mines still operating in the Canyon.

Hopi and Mohave points are named after native peoples who live near or along the Colorado River. Between Hopi and Mohave points lies the Hopi Wall. From Mohave Point, you'll have an excellent view of the Abyss, where the Great Mohave Wall drops 3,000 nearly vertical feet to the Tonto Platform.

These points stand within a westward opening in the Canyon where sunset light casts shadows among the niches and gaps of far buttes. Of any of the roaded overlook points at the South Rim, these are ideal for watching the final light of day.

Come early, when the light is still clear but low in the late afternoon. Walk along the points to see which one suits you best, to find the very spot where you will want to stand as the western horizon eclipses the sun.

There will be other people there, perhaps hundreds, and on some days a thousand. Still, the viewing will be worth being in a crowd. Along the South Rim, one must make peace with the incredible numbers of people. They all have come for personal reasons, dragged along on a family vacation or journeyed tens of thousands of miles to see this place for themselves. Like you, they are here for the single purpose of absorbing the Canyon's magic.

The truly committed will be here at sunset. They've come from all around the planet for this very moment. Voices mingle — Vietnamese marked by French; curt German sentences weaving in and out of the Romance languages of Spanish and Italian; the high tinkling of Chinese; Navajo words mumbled like flowing water.

It means something that all people come here to watch the sun shed its final light through these canyons. If there is any true United Nations, if there is a land that promises peace, it is at these points at sunset.

Likely, wind will be pouring over the edges this time of the day, a fluid rush of air falling into the landscape below. In the fall and spring, a warm hat and coat are good to have along no matter how warm the days — the rim winds grow quickly cold. In the winter, snow will be wind-pelted into shapes around you, and the temperature will plummet into freezing and colder. In the summer, the stretching sunlight will be a relief from the day's heat.

If you are lucky, thunderstorms will be roving across the horizon, cutting the light into cinematic shafts. But even clear sunsets have their good points, cleanly banding the horizon with colors.

Choices must be made.

Standing with your back to the sun on the eastern side of the points, you will be able to see volcanic orange light wash the Canyon. Facing the sun on the western edges, you will watch the molten sun itself gliding across the horizon. Or, in the very center, at the far tip, you can watch both by merely turning your head.

Sitting beneath a juniper. Standing on rock with no railing in front of you. Leaning against the metal bars that keep you from falling. Make your choice and wait. The sunset is sacred. It will happen quickly. And then, if you wait long enough, the stars will come.

LEFT, A RAINBOW SLICES BY ISIS TEMPLE. ABOVE, VISITORS GATHER AT HOPI POINT FOR SUNSET.

JACK DYKINGA

GARY LADD

LEFT, THE VIEW FROM POWELL POINT REVEALS MUCH OF THE CANYON'S COLOR AND TEXTURES. BELOW, A FORMATION RESTS ON THE TONTO PLATEAU BELOW HOPI AND POWELL POINTS. DEEPER INTO THE CANYON FROM THIS POINT, ABOVE RIGHT, HORN CREEK RAPID MOVES THROUGH GRANITE GORGE AT RIVER MILE 91.
THREE MILES DOWNRIVER LIES MONUMENT CREEK, ABOVE LEFT, IN LINE WITH THE COMPLEX OF POWELL, HOPI, AND MOHAVE POINTS BUT NOT VISIBLE FROM THEM.

BERNADETTE HEATH

Pima Point, Hermits Rest

COLIN McKAY

These points lie at the end of the West Rim Drive. Below flows Hermit Creek, joining the Colorado at River Mile 95. West of Pima Point, across an expanse of Canyon, lies the South Rim's Eremita Mesa.

Most Canyon visitors stop here. Paved roads lead no farther along the South Rim. It is easy to imagine that the Grand Canyon ends here, that the show is over. But the Colorado River has come only 95 miles through the Grand Canyon to reach this point. It has 183 more miles to flow before leaving the final gates of the Canyon where the Grand Wash Cliffs open onto Lake Mead not far from Las Vegas. Hermits Rest, this last observation point along the South Rim proper, puts you nowhere near the end.

Of course, a hermit once lived out here, below Hermits Rest. His name was Louis Boucher, and in 1891, a decade before train service began hauling tourists to the South Rim, he certainly did not imagine himself living merely at the boundary of the Grand Canyon. Instead, he ventured deep within it, building a trail from the Rim down to a remote copper-mining claim in what would come to be called Boucher Canyon.

Boucher's copper mine never paid out, and 30 years later, after he had left the Grand Canyon, tourists were pouring into what was once a hermit's canyon. Hermits Rest became a staging point for a Fred Harvey Company hotel built below Hermits Rest. A tramway was constructed, carrying people 3,000 feet down to an encampment of tents covering wooden platforms in Hermit Canyon.

Then came the millions of people who gather at the Canyon's edges today, some thinking that at Hermits Rest they have reached the end because the road ends there, others realizing that the Grand Canyon is far from over.

The structure at Hermits Rest was designed by Mary Elizabeth Jane Colter, the visionary architect who also designed Hopi House in Grand Canyon Village and the Watchtower at Desert View.

STORM CLOUDS ENVELOP THE NORTH RIM, RIGHT, ACROSS FROM PIMA POINT.
ABOVE, A STATION AT SANTA MARIA SPRING PROVIDES A REST SPOT FOR HERMIT TRAIL HIKERS.

TOM DANIELSEN

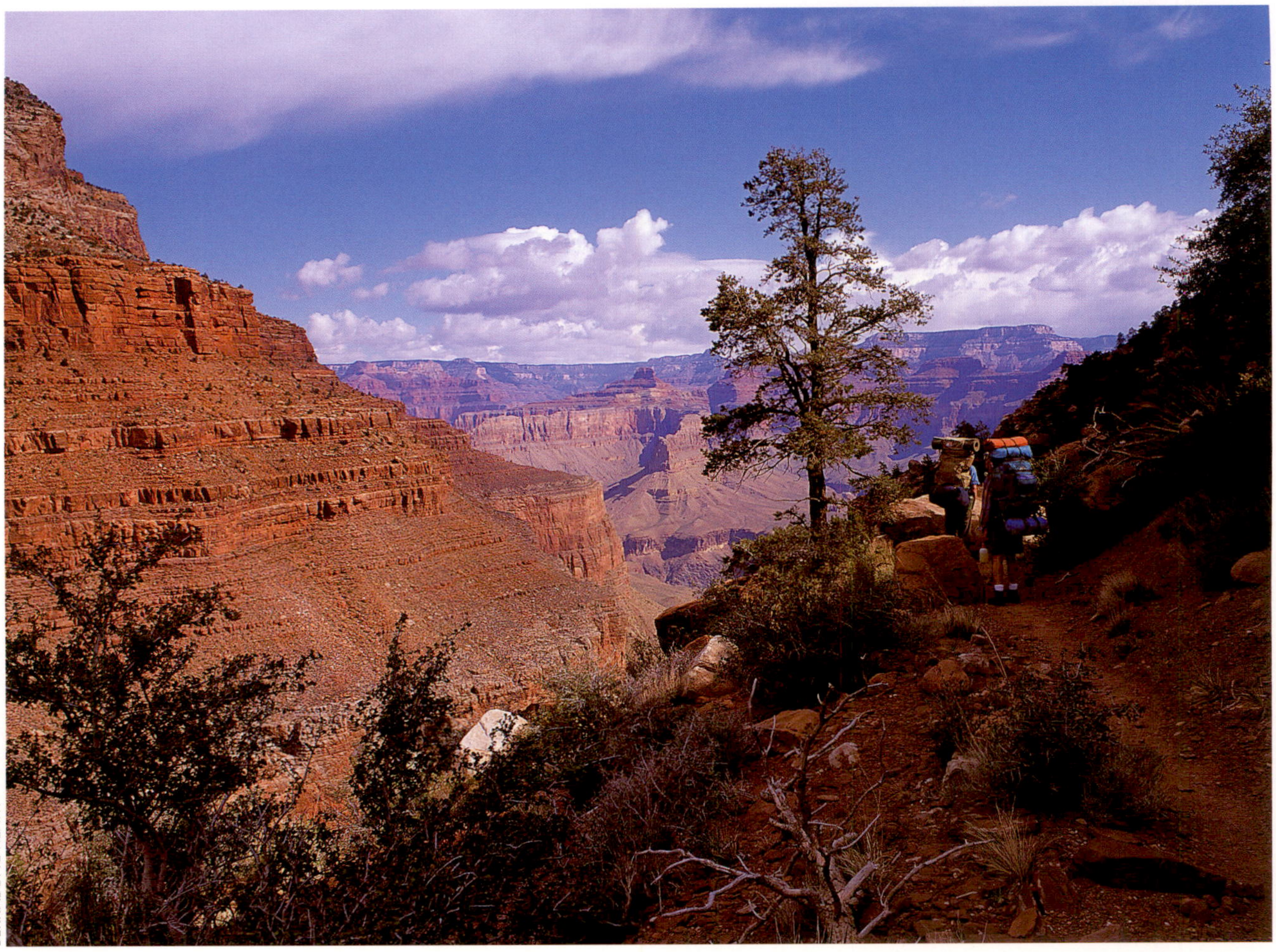

BERNADETTE HEATH

ABOVE, BACKPACKERS HIKE ON HERMIT TRAIL.
BELOW AND FAR RIGHT, HERMIT CREEK SPILLS INTO WATERFALLS ALONG THE TRAIL.

BERNADETTE HEATH

Point Imperial

GARY LADD

Looking east you see Nankoweap Mesa and the top edge of Marble Canyon at River Mile 55. South lies the Walhalla Plateau, Wotans Throne, and Grandview Point on the South Rim.

DAVID H. SMITH

The other side. The North Rim. The world here differs from the South Rim. Quieter, colder, higher. The forests are damper and loom heavier with trees. The side canyons that stretch down toward the river are longer, but a little less steep. The view, for those accustomed to the subtleties of the Grand Canyon, is utterly dissimilar. From here the buttes stretch much farther across the land than they do from the South Rim. Marble Canyon is completely obscured, veiled by butte after butte, far down into limestone walls where there lie caverns and arches.

The architecture of the North Rim is unlike that of the South Rim. The Grand Canyon's portion of Earth tilts southward with the Colorado River etching itself into the rock on the low, southern end of this dip. This means that the South Rim is perched just above the river, while the North Rim is drawn far back to the upsloping northern side of the land, far from the river, and Point Imperial's 8,803-foot elevation (6,600 feet above the Colorado River) makes it the Canyon's highest point. The side canyon that cuts from the North Rim below Point Imperial takes 14 miles to reach the river. On the opposite side, the nearest unnamed canyon carves from rim to river in less than half a mile.

The colors are deeper and richer on the north side, perhaps because of the greater abundance of plant life growing at this higher elevation or just because the light is seen from a different angle. The air smells fertile with Douglas fir and spruce and aspen trees, as opposed to the South Rim, where the butterscotch scent of ponderosa pines dominates.

Even though a paved road leads to Point Imperial, so far, the human population has only straggled in. Just reaching the North Rim takes longer — nearly a day's drive from Grand Canyon Village on the South Rim. Cities and airports lie farther away. All winter the North Rim is snowed in, closed to cars.

To say that the Grand Canyon is a single place obviously becomes, here, a fallacy. North Rim and South Rim speak different languages entirely.

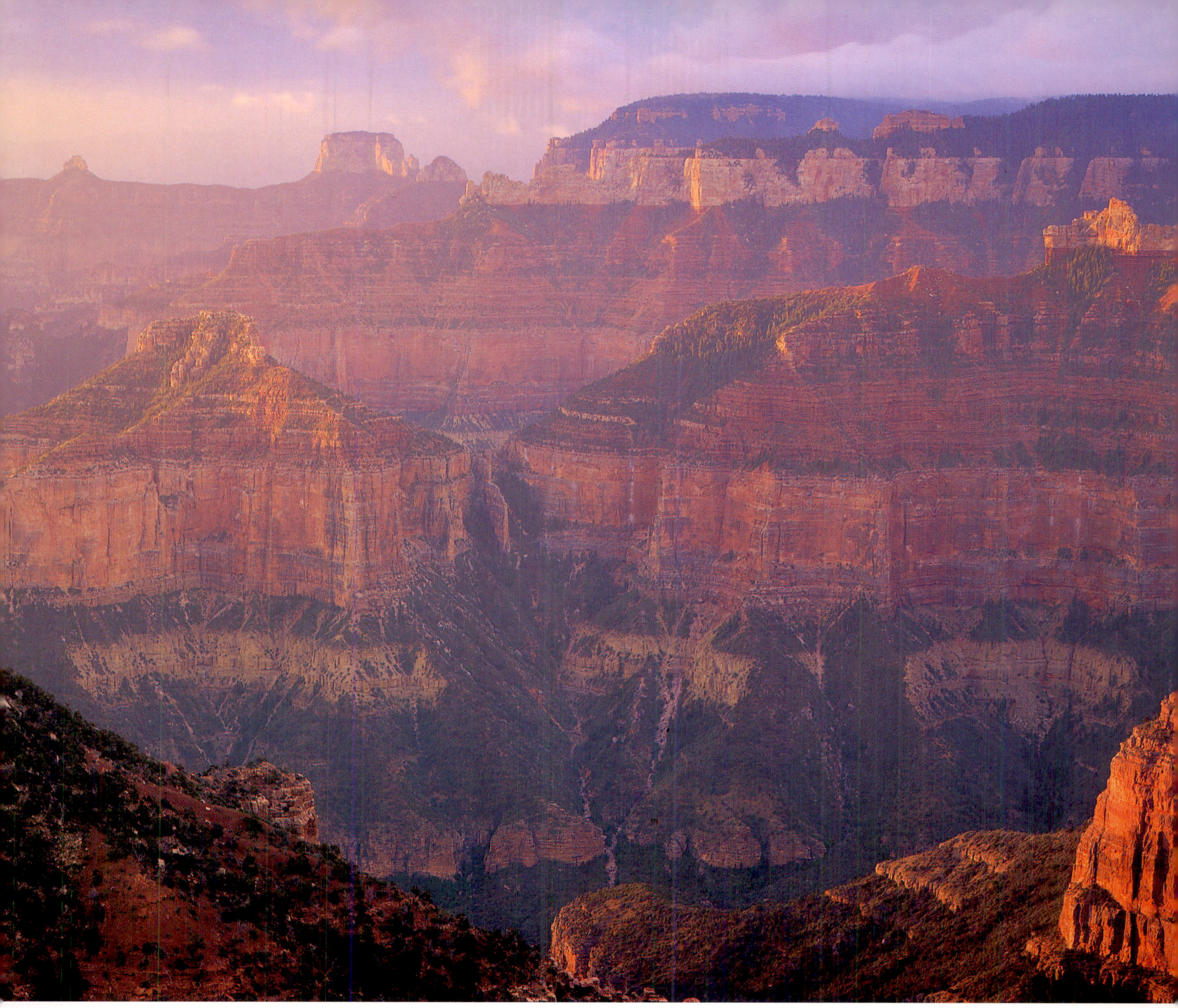

POINT IMPERIAL BRINGS VARIED SCENES TO VISITORS: ABOVE, MOUNT HAYDEN, WHICH RISES NEARLY 8,400 FEET ABOVE SEA LEVEL; AND BELOW, SADDLE MOUNTAIN (FOREGROUND), MARBLE CANYON, AND THE VERMILION CLIFFS IN THE DISTANCE. FAR LEFT, RIVER RUNNERS PLAY A PICKUP VOLLEYBALL GAME AT REDWALL CAVERN, ABOUT 30 MILES DOWNRIVER FROM NAVAJO BRIDGE.

WILLIAM STONE

KERRICK JAMES

GARY LADD

GARY LADD

BOATERS, ABOVE LEFT, DRIFT IN MARBLE CANYON BEFORE RUNNING PRESIDENT HARDING RAPID. EROSION CREATED A BARREN FORMATION CALLED THE BRIDGE OF SIGHS, ABOVE CENTER, AND VETERAN CANYON EXPLORER GEORGE STECK, ABOVE RIGHT, WORKS HIS WAY UP IT. AT FAR RIGHT, MONKEYFLOWERS THRIVE IN THE MICROCLIMATE AT VASEYS PARADISE, NOT FAR FROM THE BRIDGE OF SIGHS. BELOW, ANCESTRAL PUEBLOAN GRANARIES REMAIN AT THE JUNCTURE OF NANKOWEAP AND MARBLE CANYONS NEAR RIVER MILE 53.

JACK DYKINGA

Cape Final

The paved road that serves Point Imperial winds south to a 2-mile trail leading to Cape Final. East, down from River Mile 65, lies Lava Canyon Rapid, origins of which stem from volcanoes. From here, visitors catch a much closer view of Vishnu Temple and the Walhalla Plateau than they can from the South Rim.

RANDY PRENTICE

In the floor of the canyons below Cape Final lie the remains of long-dead volcanoes. Hundreds of millions of years before the Grand Canyon ever came to be, veins of molten rock broke their way through ancient bedrock, now seen as black, skeletal dashes in the canyon walls. Lava flows once burst onto the surface of the planet, dropping blankets of basalt 300 feet thick. This was so long ago that now only relatively thin pages of these dark rocks are visible in the Grand Canyon. The river cuts into canyon walls, scraping away these primeval histories of volcanoes, carrying them downstream.

The Grand Canyon is made of time. Here, you have to let go of notions of your own life span. Civilizations must be forgotten. Even life as we know it is almost out of the grasp of this place (the oldest rocks in the Grand Canyon bear fossils that date back to the very first single-cell life-forms). The black lava below Cape Final first came to the surface long before mammals, before dinosaurs, before insects, before fish, before algae. Standing at the edge, you are looking into a time machine, millions of years, billions even, stretched out beneath you.

Once volcanoes erupted here, even older than those in the canyons below Cape Final. About 2 billion years ago — on a time scale so monumental that the Grand Canyon itself is only a blink — underwater volcanoes covered a long-forgotten sea floor with lava. This lava has since been hardened, driven to the interior of the earth, and brought back out where it is now revealed in the bottom of the Grand Canyon as the rock called Vishnu schist.

Formation of the Canyon began 5-6 million years ago. The most recent set of volcanoes actually cropped up while the Grand Canyon was in place, some as recently as a few thousand years ago — just yesterday, really. Their lava flows can still be seen at the western end of the Canyon, where they poured into the side canyons, filling some to their brims.

This fresh lava spilled into the Colorado River, building vast dams of hard, black basalt. River water backed up behind these dams, creating reservoirs larger than anything humans have ever made. The lava reservoirs varied in size. The largest stretched from near the Nevada border all the way up the Grand Canyon, partially filling the Canyon with water, backing up through Marble Canyon and into Utah.

Eventually, the river broke through these basalt barricades, cutting through miles of solid rock to return to its channel.

Perhaps time is one reason that so many people come to the Grand Canyon. Whatever concepts we may have about our day-to-day lives are suddenly altered. Here time is measured by the wearing away of rock, by the rise and fall of entire landscapes. We look through the Grand Canyon and see in it what cannot be seen through normal vision. We see lava spreading across the floor of an ocean that was here long before North America. We see a time before living animals evolved into bones and teeth and eyes.

It is not necessary to come to the Grand Canyon knowing the details of the geologic past. You come here, peer inside, and immediately realize that you are in a place very old.

LEFT, VISHNU TEMPLE'S DISTINCTIVE SHAPE JUTS FROM THE CANYON WITH THE WALHALLA PLATEAU STRETCHING ACROSS THE HORIZON.
ABOVE, WILDFLOWERS ON THE NORTH RIM AT CAPE FINAL STAND OUT FROM THE NEARLY BARREN BUTTES AND RIDGES IN THE CANYON.

GARY LADD

LEFT, THE DISTINCTIVE BOULDER IN THE FOREGROUND IS DOX SANDSTONE LYING UPRIVER FROM TANNER RAPID.
ABOVE, LAVA CANYON (CHUAR) RAPID, FLOWS BELOW COMANCHE POINT, SEEN AT THE RIM'S CENTER .

BERNADETTE HEATH

Cape Royal

The paved road flanking the North Rim ends at Cape Royal. Just north are Angels Window and Walhalla Plateau. On the Colorado, River Mile 67 is east and River Mile 79 is south. Between these two points the river begins turning westward.

TOM TILL

The canyon that falls beneath Cape Royal is called Unkar, a Paiute word describing the blush color of the rock that exists down there. Where Unkar Canyon unfolds onto the Colorado River at about Mile 72.5, a broad delta has formed, and spread over the delta stands the Grand Canyon's largest known prehistoric settlement.

The remnants [see photo atop Page 62] at Unkar only hint at what once stood here. Single-story, roof-covered buildings that enclosed some 20 rooms have been whittled away by erosion to leave mere blueprints of stacked stones and bits of broken pottery scattered here and there. Wind-blown sand buries some artifacts and exposes others. The people who constructed this site lived here about a thousand years ago.

Archaeologists have surveyed little more than 3 percent of the Grand Canyon. Yet in that small portion, they have discovered more than 4,000 archaeological sites. Ancient pueblo ruins like those at Unkar and Hilltop Ruin [above] dot the river's length. Up in the side canyons, smaller structures, compact storage granaries built into the rock, still survive. And in the canyons above them are the ancient routes, handholds carved into the rock, wooden ladders and bridges plugged into cliff walls.

It is interesting to note that when the Hopi, some of whom may have descended from the early Grand Canyon inhabitants, first guided the Spaniards here in the 16th century, they led García Lópes de Cárdenas to a point on the South Rim where there were no visible ways down. However, the Hopi knew of many routes to walk from the Rim to the river. This leaves one to wonder if the Hopi wished for the interior of the Grand Canyon to remain unnamed and out of reach, an ancestral ground to be left untouched.

LEFT, LOOKING SOUTHWARD FROM WALHALLA POINT, JUST NORTH OF CAPE ROYAL, A VISITOR CAN SEE 7,533-FOOT VISHNU TEMPLE LOOMING OVER SUN-BATHED FREYA CASTLE. ON THE HORIZON LIES THE SNOW-CAPPED SAN FRANCISCO PEAKS.
ABOVE, THE HILLTOP RUIN LIES ABOVE THE COLORADO RIVER ON ITS SOUTH SIDE.

GARY LADD

RICHARD L. DANLEY

GARY LADD

ABOVE, LEFT TO RIGHT, UNKAR DELTA AND RAPID; CHEYAVA FALLS NORTH OF CAPE ROYAL IN CLEAR CREEK CANYON; ANCESTRAL PUEBLOAN RUINS ON UNKAR DELTA.
BELOW, VIEW TOWARD THE PALISADES OF THE DESERT FROM HILLTOP RUIN ACROSS THE RIVER FROM CAPE ROYAL.
RIGHT, FROM CAPE ROYAL, ANGELS WINDOW DOMINATES THE VIEW, WHICH INCLUDES VISHNU TEMPLE AND THE SLOPING SIDE OF FREYA CASTLE AT LEFT-CENTER. A HAZE DIMS THE OUTLINE OF THE SAN FRANCISCO PEAKS ON THE HORIZON.

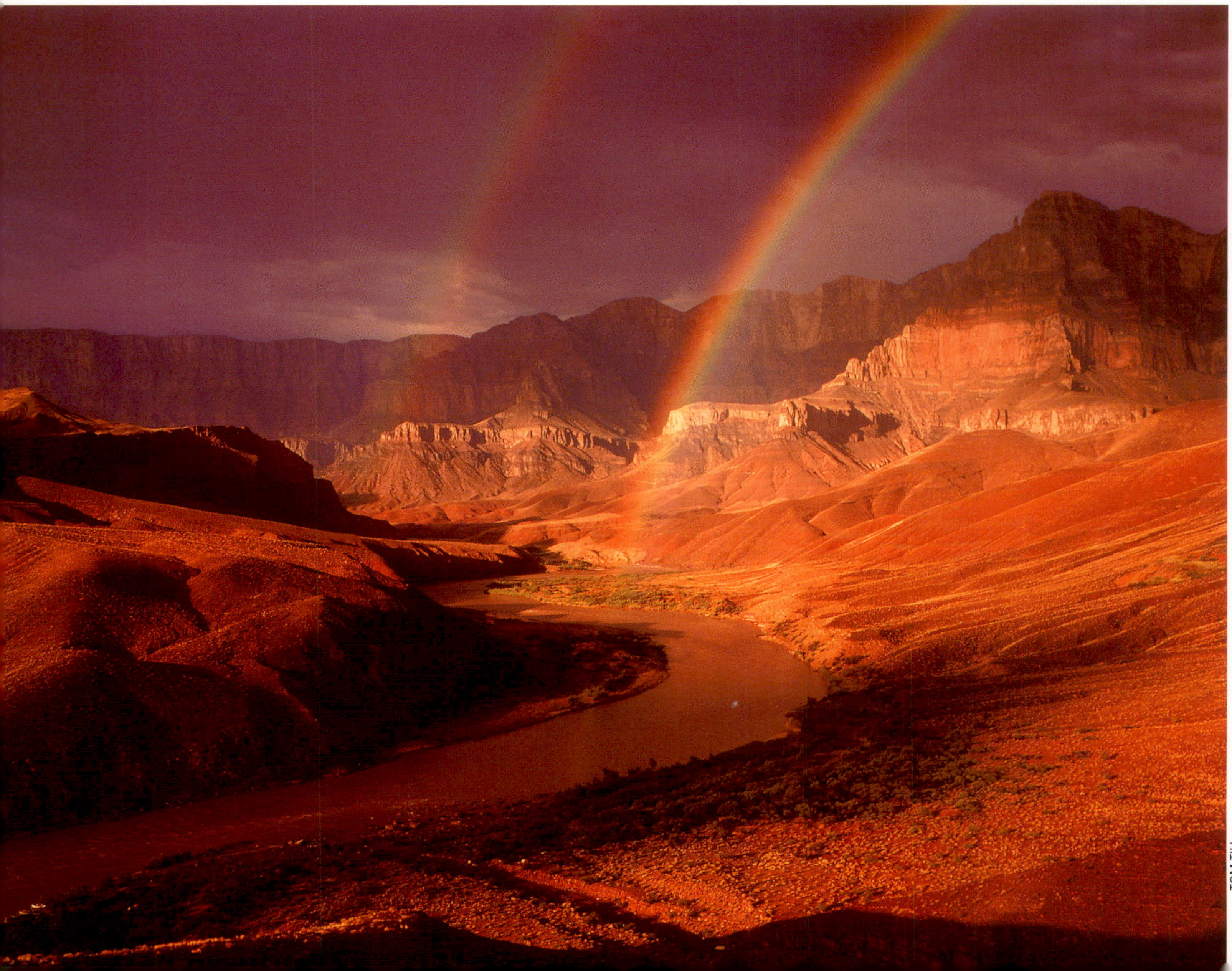
TOM TILL

Bright Angel Point

LAURENCE PARENT

Lying in the heart of the North Rim enclave, Bright Angel Point, at the head of Bright Angel Canyon, lies due north from the South Rim's Shoshone Point. The canyon's mouth empties into the Colorado just below Phantom Ranch.

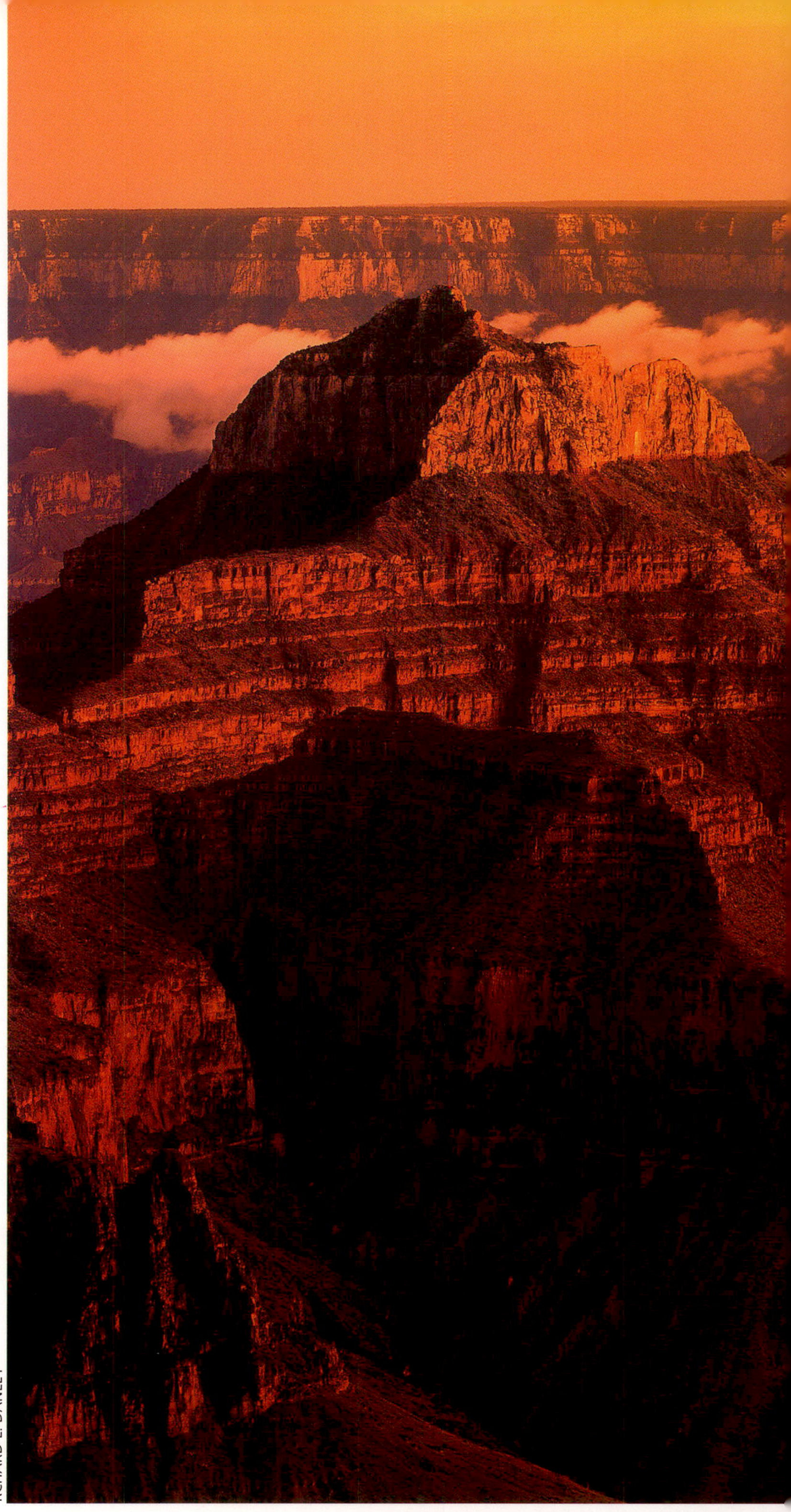

RCHARD L. DANLEY

The mules begin here. Trains of the long-eared animals start their arduous trek deep into the earth along the stairwell turns of a rock-and-dirt trail. From this side of the Grand Canyon, the mules carry strictly tourists, visitors swaying back and forth in their saddles. The ride ends at Roaring Springs, the water source for Grand Canyon Village on the South Rim, and Canyon visitors must hoof it themselves to the Canyon floor.

Bright Angel Point aims straight across the Canyon at the South Rim's South Kaibab and Bright Angel trails. Mules also ply these trails, hauling tourists and supplies all the way to the bottom, to Phantom Ranch.

In the predawn hours, before visitors start their rides down the trails, the pack trains of mules wind through the dark carrying supply loads for Phantom Ranch: lettuce, light bulbs, toilet paper, medical supplies, eggs, blankets . . . On the return, the sturdy animals pack out loads of garbage for disposal back on the South Rim.

Good-tempered mules are reserved for those visitors wanting to ride, while the rough and temperamental ones handle the less socially demanding task of lugging goods.

Below Bright Angel Point, when the first footbridge crossing the Colorado River was built in 1921, mules carried most of the materials into place. The trails they use for all of this transportation are in turn maintained by equipment carted down by mules. A second-generation Grand Canyon mule wrangler once said, "I personally think it's beautiful, this working relationship between people and animals. We work together to maintain that trail. Pack mules carry the food, fuel, tools, and dirt used to keep the trail up. We keep each other in business."

Other than going in by foot or floating the river, mules are the only way to reach the Canyon's interior. Besides the occasional historic tram, it has long been this way here. Mule tales pepper the Grand Canyon's history, the first commercial mule rides going down in 1901, and mules assisting as the Birdseye Expedition that floated through in 1923. Mules carried supplies down to mining claims, brought fingerling trout to stock Bright Angel Creek, and ushered in Teddy Roosevelt on a number of Grand Canyon treks he took in the early 1900s.

BRAHMA AND ZOROASTER TEMPLES AT SUNSET, ABOVE, AND WITHIN A BROADER PANORAMA, BELOW.
THE GRAND CANYON LODGE, FAR LEFT, ON THE NORTH RIM.

LAURENCE PARENT

CHARLES LAWSEN

THE TRANSEPT, ABOVE.
BELOW LEFT, CLEAR CREEK INDIAN RUIN; BELOW RIGHT, HIKERS AND A MULE TRAIN SHARE THE NORTH KAIBAB TRAIL.
PIÑON AND JUNIPER TREES, FAR LEFT, FIND PURCHASE IN KAIBAB LIMESTONE.

STEWART AITCHISON

DAVID ELMS JR.

Tiyo, Widforss Points

GEORGE H.H. HUEY

Tiyo Point, west of Bright Angel Point, rests directly across the Canyon from the South Rim's Grand Canyon Village. Tiyo's vantage provides excellent Canyon views in all four directions.

Tiyo and Widforss points are nearly surrounded. Canyons are constantly carving back into the land, clawing at the high platform on which these points stand. Eventually, the heads of these distant canyons will meet. Tributaries to Bright Angel Creek will wrap around and touch the arms of Dragon Creek, leaving this piece of land as a stranded island, a new solitary temple within the Grand Canyon. Each of the points at the edge of the Grand Canyon is destined for this future. The Walhalla Plateau, which supports Cape Royal and Cape Final, is soon to be cut away from the mainland as the head of Bright Angel Creek reaches to join the head of Nankoweep Creek.

The Grand Canyon is an active place. You are seeing only an instant of its evolution when you stand and look at it. Where we install roads, there will someday be nothing but empty space between points and plateaus. We build visitors centers and restaurants on what seems like solid ground, but erosion is quickly pulling down cliffs as if they were curtains. The ground is on its way out. This is the whole principle behind the Grand Canyon: It is a land falling apart. The fingers of Haunted Canyon reach up toward Widforss Point as if grabbing at it. Phantom Canyon does the same to Tiyo Point, grasping upward, scratching back the cliffs. Eventually, the canyons will take these points and we will have to find new places to stand.

JACK DYKINGA

ABOVE, CLOUDS CAST A SURREALISTIC TONE OVER OZA BUTTE, EAST OF WIDFORSS POINT AND ADJACENT TO THE TRANSEPT.
TIYO POINT, RIGHT, WITH ISIS TEMPLE IN THE DISTANCE.

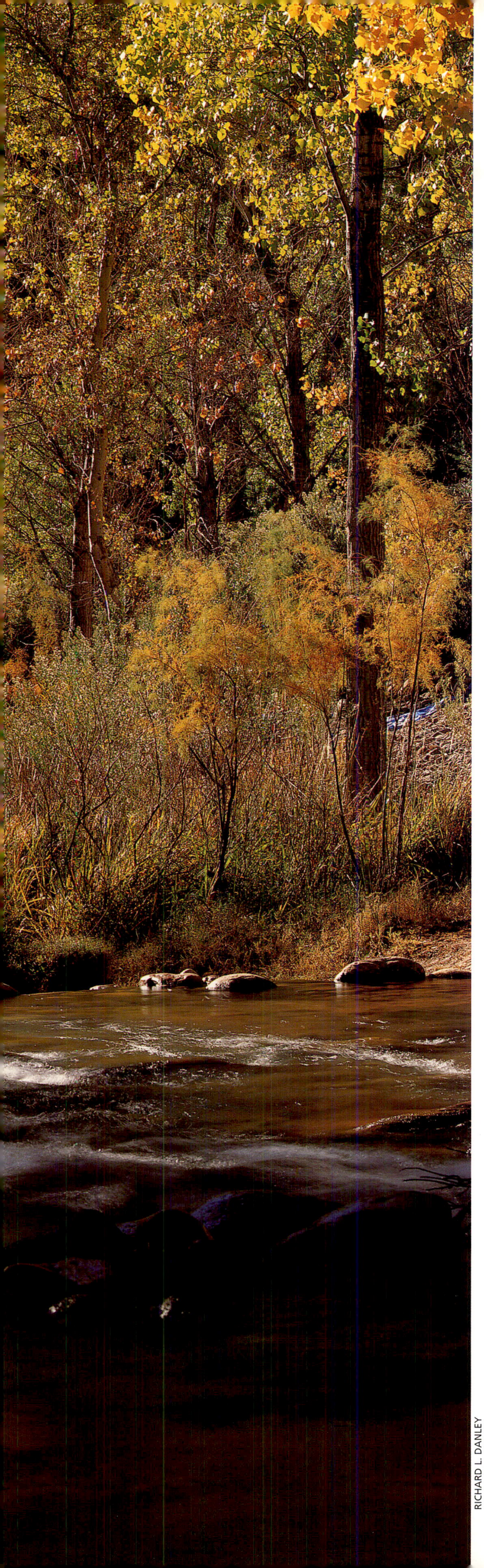

RICHARD L. DANLEY

LEFT, BRIGHT ANGEL CREEK FLOWS BY ITS NAMESAKE CAMPGROUND.
BELOW, SUMMER HIKERS COOL OFF IN THE CREEK.

GARY LADD

Point Sublime

Reachable only by foot, the point lies on a ridge providing views in all four directions. Looking south from here puts your view in line with Hermits Rest on the South Rim. Sublime is a landmark for hikers on the Bass Trail. Just west, the Colorado River swings northward again.

LARRY LINDAHL

The North Rim is a sponge. Winter snows melt and soak into the ground, sending water into subterranean fractures to be spit back out at springs down in the canyons.

The South Rim tends to be much drier. In fact, public water for the South Rim is pumped from springs that blossom straight out of a canyon wall below the North Rim.

In some places, such as this pumped-water source, caves are exposed in cliff sides and out of them pour graceful, white streamers of waterfalls. These caves lead back into hidden areas deep inside the planet. Some of these caves have been explored. In one case, an intrepid group carried rafts through a North Rim cave after climbing its waterfall. Once inside, they found long, quiet reflection pools. Water dripped from the ceiling, tapping at the pool's surface. They found in there an unexpected Grand Canyon, a place unknown.

Most long canyons on the north side carry sizeable streams, clear water babbling over boulders and down falls. Ferns and cottonwood trees gather around the running water. Now and then, though, the water turns dark. Brief and powerful floods streak down these canyons, driven by passing storms, sometimes pushing 50-ton boulders at speeds up to 50 miles per hour.

Such was the case in a storm that struck Crystal Creek in December of 1966. Crystal Creek reaches up the peninsula of land on which stands Point Sublime. Its arms are fed by alpine forests far above the Canyon's points and rims. When this winter storm hit, a flood more than 40 feet deep bellowed down Crystal Creek.

At the river below, there was once a small rapid at the mouth of Crystal Creek. It never received much attention from river runners, who considered it just a light riffle, easy to navigate in a boat.

Suddenly, though, this 1966 flood unloaded hundreds of tons of debris into the river, wholesale pieces of canyon chewed into boulders and sand. All of this material acted as a dam, causing the river to narrow by 180 feet. The once docile little rapid that graced the meeting of Crystal Creek and the Colorado River turned into one of the largest white-water obstacles in the entire Grand Canyon.

Now it is famous among river runners. They speak with awe about "the hole at Crystal."

The hole is just that, a deep pit of water on the backside of a massive wave. The wave itself has been known, at especially high river levels, to reach a height of 20, even 30 feet. River runners gather at the nearest shoreline before working their way around this hole. They stand and stare, talking among themselves about how it should be run, pointing at their anticipated routes. There is palpable fear.

All of this came from a single event, a debris flow that lasted maybe a few hours. Now Crystal Creek, as usual, runs clear and lovingly among its boulders and young cottonwood trees. It takes water from rains and snowmelt, gently delivering it to the river far below.

VISTA FROM POINT SUBLIME, LEFT, AND RELICS FROM THE BASS TOURIST CAMP IN SHINUMO CANYON, ABOVE.

BERNADETTE HEATH

CHARLES LAWSEN

ABOVE LEFT, VIEW ABOVE BOUCHER RAPID ACROSS THE RIVER FROM POINT SUBLIME; ABOVE RIGHT AND BELOW, CRYSTAL CREEK.
RIGHT, AN ANCESTRAL PUEBLOAN RUIN SANDWICHES CRYSTAL RAPID AGAINST THE DISTANT SOUTH RIM.

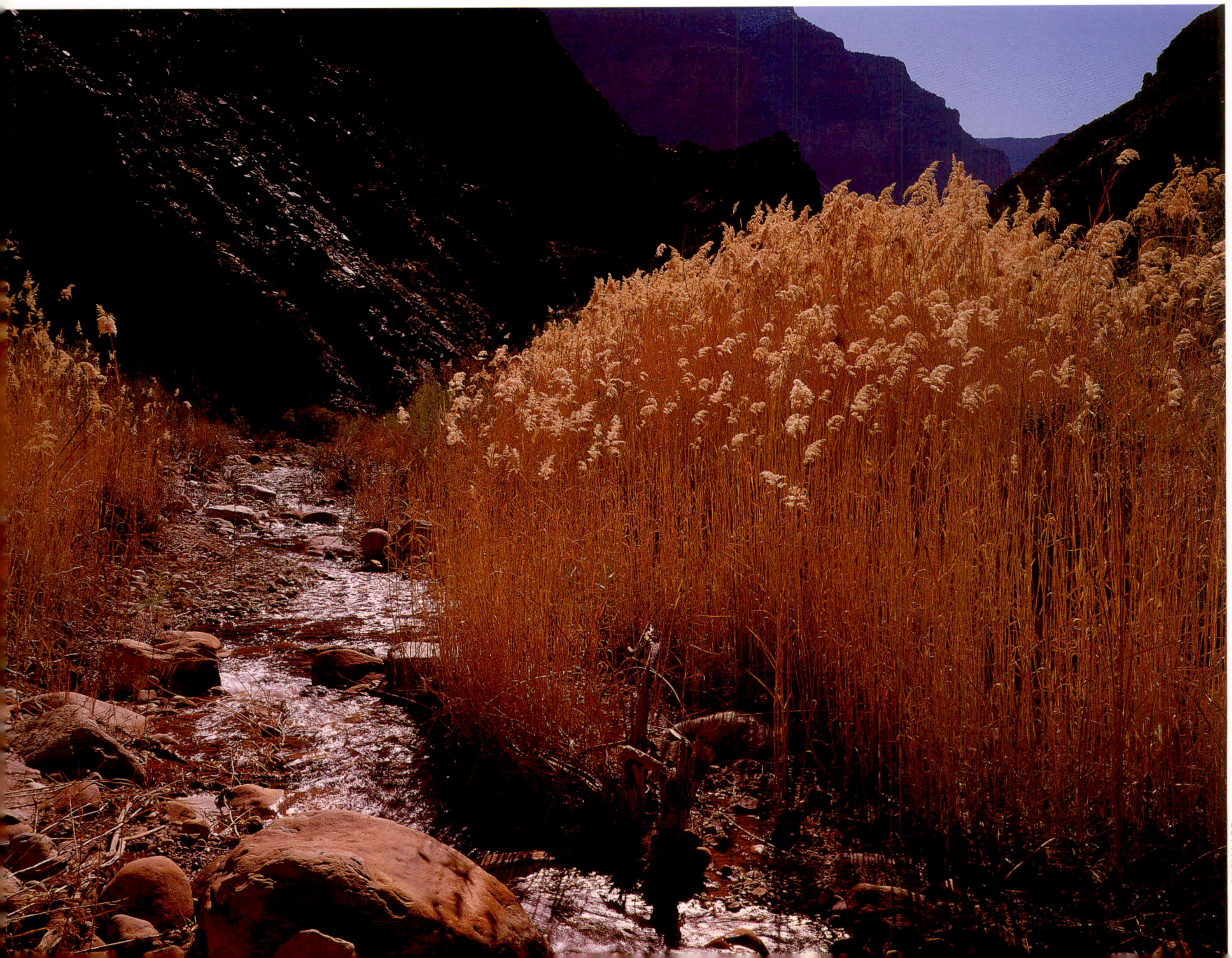

GARY LADD

Toroweap Point

PETER ENSENBERGER

Some people who peer into the Canyon at Toroweap crawl to the precipitous edge on hands and knees. Less daring folks approach the edge only after tethering themselves to their car's bumper.

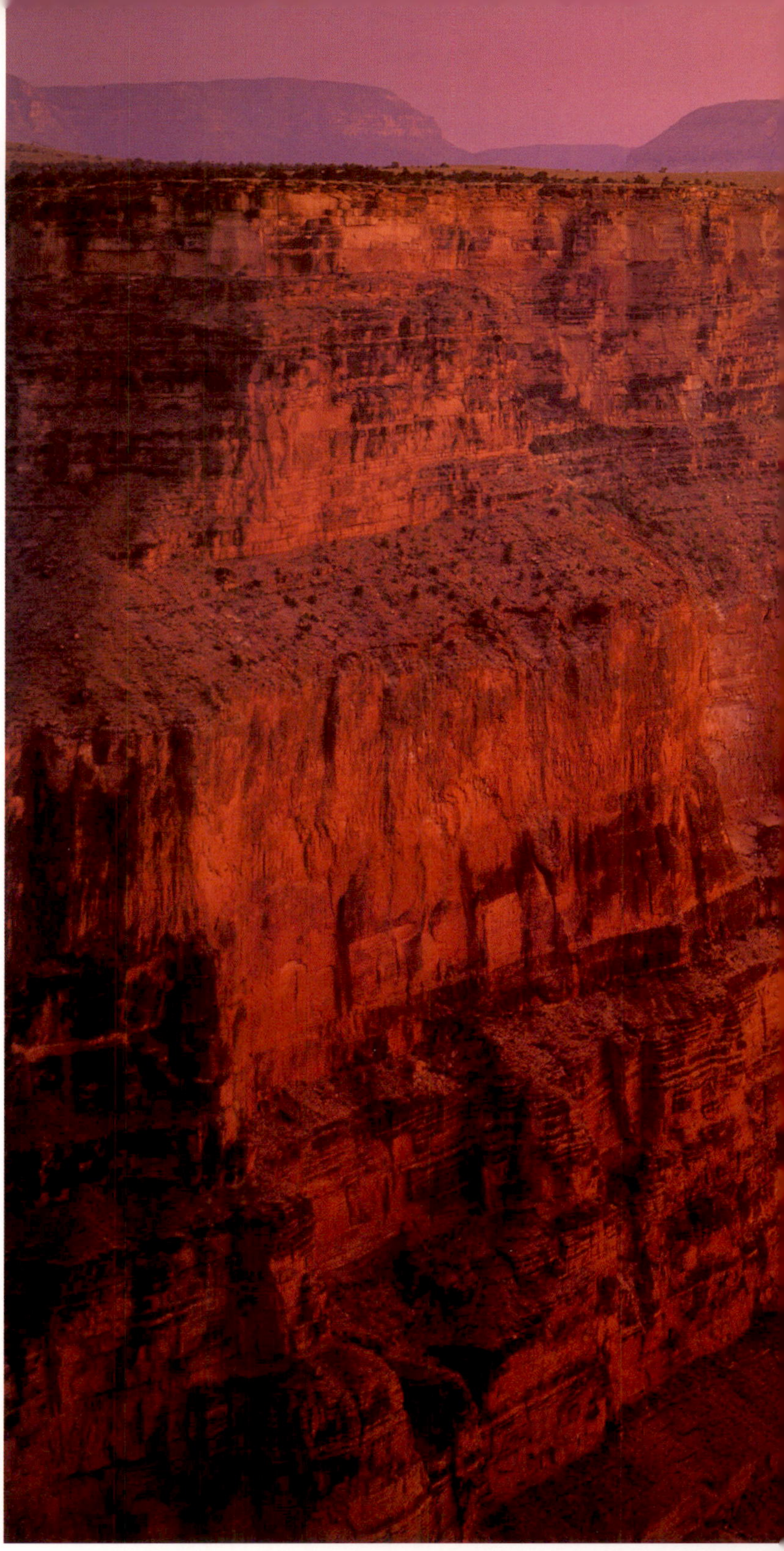

LARRY ULRICH

Again, here is proof that there is more than one Grand Canyon. The Canyon's western portion around Toroweap Point markedly differs from either the South Rim or the North Rim. Here, the high palisades stretch wide. Cliffs step miles and tens of miles back from the inner corridor where the river flows. Red-gray limestone walls line the inner corridor, sometimes called "the refrigerator" for its darkness and its coldness. Up top at Toroweap, the habitat is high desert, a sere, hard country freckled with juniper trees, but not the pines of the South Rim and certainly not the flutter-leafed aspens of the North Rim.

Compared to the nearly urban bustle of the South Rim and the lodge-bespangled North Rim, Toroweap is silent. Only one ranger is stationed here, where the official National Park Service truck rarely lasts more than 3 years patrolling these rough roads. The workshop at the ranger station is decorated with hubcaps sprung free of tourists' cars along the 60-mile dirt road that leads to this point.

On a human scale, this place is an older Grand Canyon. There is no phone here. There is no store, no manicured visitors center. Behind the ranger station rests the grave of ranger John Riffey, who worked at Toroweap for 38 years, until 1980. Next to his headstone is the headstone of his wife, Meribeth Riffey. You won't find such a place at the South Rim, where there are paved roads. There, rangers tend to come and go. Pay phones stand at trailheads. You can buy milk, candy, gasoline, postcards. Some rangers refer to the South Rim as "downtown."

Toroweap is far from town. It presides over an immense wilderness: the long, gaping hallways of Tuckup Canyon and the boulder-clogged floor of National Canyon. Lava Falls, generally considered to be the largest rapid in the Grand Canyon (second to one that is now buried beneath the waters of Lake Mead at the still farther western end of the Canyon), is down there somewhere.

The Hualapai Indian Reservation occupies large portions of the Canyon out here. There are no trails polished by millions of feet and by the hooves of mules.

This is yet another Grand Canyon, one spot of a place that goes on without end.

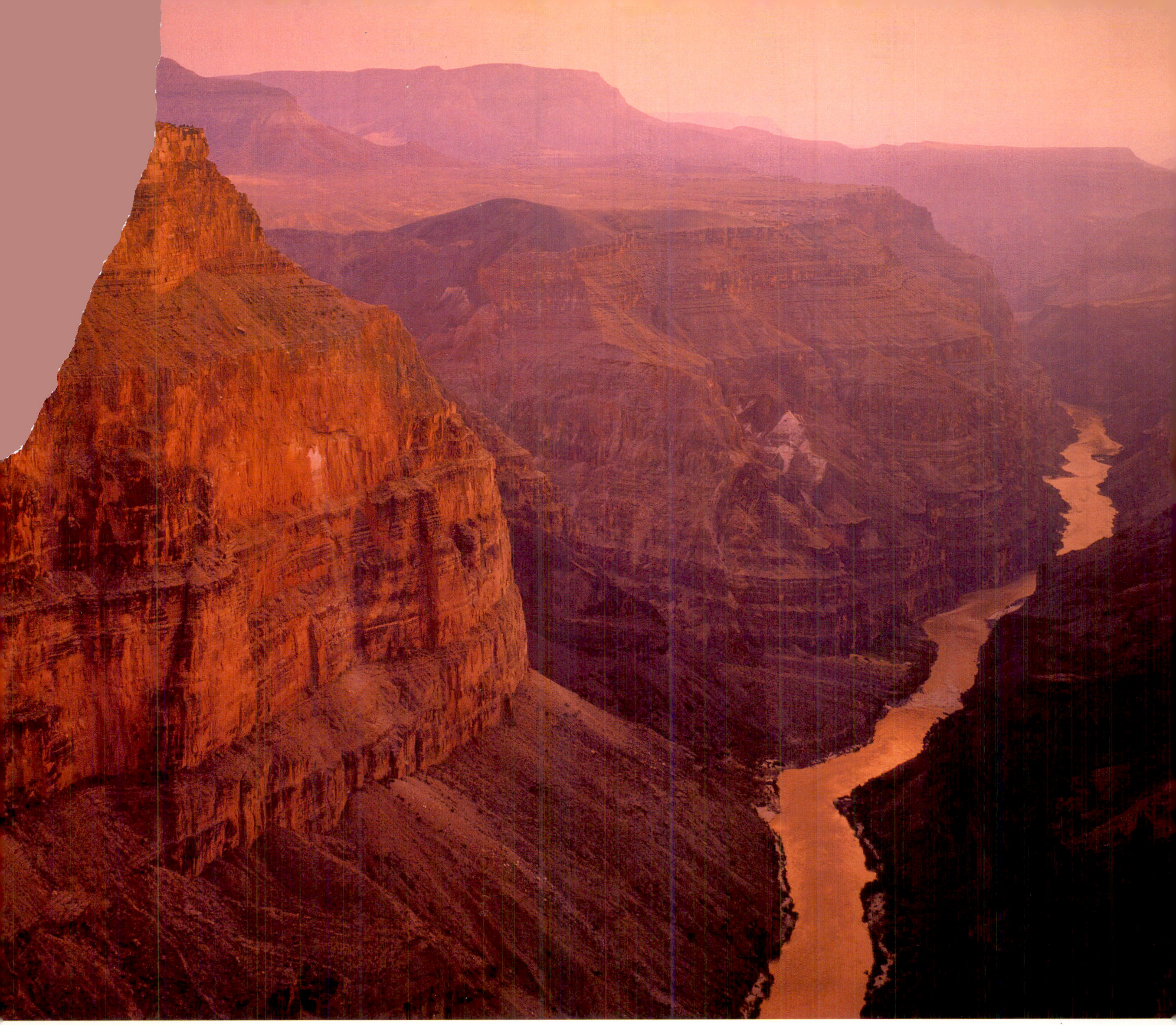

TOROWEAP OVERLOOK AT SUNSET, ABOVE. VIEW, BELOW, FROM SB POINT A FEW MILES EAST OF TOROWEAP OVERLOOK. FAR LEFT, A DORY SURGES THROUGH LAVA FALLS RAPID.

GARY LADD

LARRY LINDAHL

GARY LADD

KERRICK JAMES

ABOVE, LEFT TO RIGHT, FERN GLEN WATERFALL; CONGLOMERATE BRIDGE IN TUCKUP CANYON; TOROWEAP OVERLOOK. BELOW, LAVA FALLS RAPID. NATIONAL CANYON, RIGHT, CURVES EAST OF TOROWEAP.

GARY LADD

INDEX

BOLDFACE INDICATES PHOTOGRAPH